T0206957

Fourier Analysis—A Signal Processing Approach

D. Sundararajan

Fourier Analysis—A Signal Processing Approach

 Springer

D. Sundararajan
Formerly at Department of Electrical
   and Computer Engineering
Concordia University
Montreal, QC, Canada

Additional material to this book can be downloaded from http://extras.springer.com.

ISBN 978-981-13-4665-1   ISBN 978-981-13-1693-7   (eBook)
https://doi.org/10.1007/978-981-13-1693-7

This Springer imprint is published by the registered company Springer Nature Singapore Pte Ltd.
The registered company address is: 152 Beach Road, #21-01/04 Gateway East, Singapore 189721,
Singapore

Preface

Transform methods dominate the study of linear time-invariant systems in all the areas of science and engineering, such as circuit theory, signal/image processing, communications, controls, vibration analysis, remote sensing, biomedical systems, optics, acoustics. The heart of the transform methods is Fourier analysis. Several other often used transforms are generalizations or specific versions of Fourier analysis. It is unique in that it is much used in theoretical studies as well as in practice. The reason for the latter case is the availability of fast algorithms to approximate the Fourier spectrum adequately. For example, the existence and continuing growth of digital signal and image processing are due to the ability to implement the Fourier analysis quickly by digital systems. This book is written for engineering, computer science, and physics students, and engineers and scientists. Therefore, Fourier analysis is presented primarily using physical explanations with waveforms and/or examples, keeping the mathematical form to the extent it is necessary for its practical use. In engineering applications of Fourier analysis, its interpretation and use are relatively more important than rigorous proofs. Plenty of examples, figures, tables, programs, and physical explanations make it easy for the reader to get a good grounding in the basics of Fourier signal representation and its applications.

This book is intended to be a textbook for senior undergraduate- and graduate-level Fourier analysis courses in engineering and science departments and a supplementary textbook for a variety of application courses in science and engineering, such as circuit theory, communications, signal processing, controls, remote sensing, image processing, medical analysis, acoustics, optics, and vibration analysis. For engineering professionals, this book will be useful for self-study. In addition, this book will be a reference for anyone, student or professional, specializing in the practical applications of Fourier analysis. The prerequisite for reading this book is a good knowledge of calculus, linear algebra, signals and systems, and programming at the undergraduate level.

Programming is an important component in learning and practicing Fourier analysis. A set of MATLAB® programs are available at the Web site of the book. While the use of a software package is inevitable in most applications, it is better to use the software in addition to self-developed programs. The effective use of a software package or to develop own programs requires a good grounding in the basic principles of the Fourier analysis. Answers to the selected exercises marked ∗ are given at the end of the book. A Solutions Manual and slides are available for instructors at the Web site of the book.

I assume the responsibility for all the errors in this book and would very much appreciate receiving readers' suggestions and pointing out any errors (email: d_sundararajan@yahoo.com). I am grateful to my Editor and the rest of the team at Springer for their help and encouragement in completing this project. I thank my family for their support during this endeavor.

D. Sundararajan

Contents

About the Author

Dr. D. Sundararajan holds a B.E. in electrical engineering from Madras University and an M.Tech. in electrical engineering from the Indian Institute of Technology Madras (IIT Madras). He obtained his Ph.D. in electrical engineering from Concordia University, Montreal, Canada, in 1988. As the principal inventor of the latest family of discrete Fourier transform (DFT) algorithms, he holds three patents (granted by the USA, Canada, and Britain). Further, he has published several papers in IEEE Transactions and in the Proceedings of the IEEE, and he is the author of five books. He has taught undergraduate and graduate students in digital signal processing, digital image processing, engineering mathematics, programming, operating systems, and digital logic design at Concordia University, Canada; Nanyang Technological University, Singapore; and Adhiyamaan College of Engineering, India. He has also conducted workshops on digital image processing, MATLAB, and LaTeX.

Over the course of his engineering career, he has held positions at the National Aerospace Laboratory, Bangalore, and the National Physical Laboratory, New Delhi, where he worked on the design of digital and analog signal processing systems.

Abbreviations

1-D	One-dimensional
2-D	Two-dimensional
DC	Sinusoid with frequency zero, constant
DFT	Discrete Fourier transform
DIF	Decimation in frequency
DIT	Decimation in time
DTFT	Discrete-time Fourier transform
FIR	Finite impulse response
FS	Fourier series
FT	Fourier transform
IDFT	Inverse discrete Fourier transform
IFT	Inverse Fourier transform
LSB	Least significant bit
LTI	Linear time-invariant
PM	Plus–minus
RDFT	DFT of real-valued data
RIDFT	IDFT of the transform of real-valued data

Chapter 1
Signals

Signals convey some information. Signals are abundant in the applications of science and engineering. Typical signals are audio, video, biomedical, seismic, radar, vibration, communication, and sonar. While the signals are mostly of continuous nature, they are usually converted to digital form and processed by digital systems for efficiency. In signal processing, signals are enhanced to improve their quality with some respect, or some features are extracted, or they are modified in some desired way. A signal, in its mathematical representation, is a function of one or more independent variables. While time is the independent variable most often, it could be anything else, such as distance. The analysis is equally applicable to all types of independent variables. Signal or waveform is used to refer the physical form of a signal. In its mathematical representation, a signal is referred as a function or sequence. This usage is not strictly adhered.

The amplitude profile of most naturally occurring signals is arbitrary, and consequently, it is difficult to analyze, interpret, transmit, and store them in their original form. The idea of a transform is to represent a signal in an alternate, but equivalent, form to gain advantages in its processing. Fourier analysis, the topic of this book, provides a widely used representation of signals. As signal representation is the topic of the book and the most suitable representation of a signal depends on its characteristics, we have to first study the classification of signals. Further, the representation is in terms of some well-defined basis signals, such as the sinusoid, complex exponential, and impulse. Although practical signals are mostly real-valued, it becomes mandatory to use the equivalent complex-valued signals for ease of mathematical manipulation. In addition, operations such as shifting and scaling of signals are often required in signal analysis. All these aspects are presented in this chapter.

© Springer Nature Singapore Pte Ltd. 2018
D. Sundararajan, *Fourier Analysis—A Signal Processing Approach*,
https://doi.org/10.1007/978-981-13-1693-7_1

1

1.1 Basic Signals

The amplitude profile of most naturally occurring signals is arbitrary. These signals
are analyzed using some well-defined basic signals, such as the impulse, step, ramp,
sinusoidal, and exponential signals. In addition, systems, which are hardware or software realizations, modify signals or extract information from them. They are also
characterized by the responses to these signals. The basic signals either have an infinite duration or infinite bandwidth. For practical purposes, they are approximated to
a desired accuracy. Fourier analysis has four versions and each version uses different
type of signals. Therefore, it is necessary to study both the continuous and discrete
type of signals.

1.1.1 Unit-Impulse Signal

The unit-impulse and the sinusoidal signals are the most important signals in the
study of signals and systems. The continuous unit-impulse $\delta(t)$ is a signal with a
shape and amplitude such that its integral at the point $t = 0$ is unity. It is defined, in
terms of an integral, as

$$\int_{-\infty}^{\infty} x(t)\delta(t)\,dt = x(0)$$

It is assumed that $x(t)$ is continuous at $t = 0$ so that the value $x(0)$ is distinct. The
product of $x(t)$ and $\delta(t)$ is

$$x(t)\delta(t) = x(0)\delta(t)$$

since the impulse exists only at $t = 0$. Therefore,

$$\int_{-\infty}^{\infty} x(t)\delta(t)\,dt = x(0)\int_{-\infty}^{\infty} \delta(t)\,dt = x(0)$$

The value of the function $x(t)$, at $t = 0$, is sifted out or sampled by the defining
operation. By using shifted impulses, any value of $x(t)$ can be sifted.

It is obvious that the integral of the unit-impulse is the unit-step. Therefore, the
derivative of the unit-step signal is the unit-impulse signal. The value of the unit-step
is zero for $t < 0$ and 1 for $t > 0$. Therefore, the unit area of the unit-impulse, as
the derivative of the unit-step, must occur at $t = 0$. The unit-impulse and the unit-
step signals enable us to represent and analyze signals with discontinuities as we do
with continuous signals. For example, these signals model the commonly occurring
situations such as opening and closing of switches.

The continuous unit-impulse $\delta(t)$ is difficult to visualize and impossible to realize
in practice. However, the approximation of it by some functions is effective in practice
and can be used to visualize its effect on signals and its properties. While there are

other functions that approach an impulse in the limit, the rectangular function is often used to approximate the impulse. The unit-impulse, for all practical purposes, is essentially a narrow rectangular pulse with unit area. Suppose we compress it by a factor of 2, the area, called its strength, becomes $1/2 = 0.5$. The scaling property of the impulse is given as

$$\delta(at) = \frac{1}{|a|}\delta(t), \ a \neq 0$$

With $a = -1$, $\delta(-t) = \delta(t)$ implying that the impulse is an even-symmetric signal. For example,

$$\delta(3t - 1) = \delta\left(3\left(t - \frac{1}{3}\right)\right) = \frac{1}{3}\delta\left(t - \frac{1}{3}\right)$$

The discrete unit-impulse signal, shown in Fig. 1.1a, is defined as

$$\delta(n) = \begin{cases} 1 \text{ for } n = 0 \\ 0 \text{ for } n \neq 0 \end{cases}$$

The independent variable is n and the dependent variable is $\delta(n)$. The only nonzero value (unity) of the impulse occurs when its argument $n = 0$. The shifted impulse $\delta(n - k)$ has its only nonzero value at $n = k$. Therefore, $\sum_{n=-\infty}^{\infty} x(n)\delta(n - k) = x(k)$ is called the sampling or sifting property of the impulse. For example,

$$\sum_{n=-\infty}^{\infty} 3^n\delta(n) = 1, \ \sum_{n=0}^{2} 9^n\delta(n + 1) = 0, \ \sum_{n=-2}^{0} 4^n\delta(-n - 1) = 0.25,$$

$$\sum_{n=0}^{2} 2^n\delta(n - 1) = 2, \ \sum_{n=-\infty}^{\infty} 2^n\delta(n + 1) = 0.5, \ \sum_{n=-\infty}^{\infty} 3^n\delta(n - 3) = 27$$

The argument $n + 1$ of the impulse, in the second summation, never becomes zero within the limits of the summation.

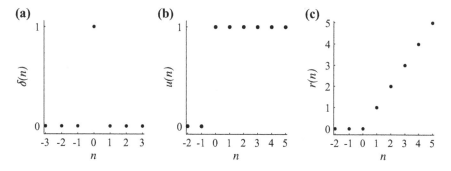

Fig. 1.1 **a** The discrete unit-impulse signal, $\delta(n)$; **b** the discrete unit-step signal, $u(n)$; **c** the discrete unit-ramp signal, $r(n)$

1.1.2 Unit-Step Signal

The discrete unit-step signal, shown in Fig. 1.1b, is defined as

$$u(n) = \begin{cases} 1 \text{ for } n \geq 0 \\ 0 \text{ for } n < 0 \end{cases}$$

For positive values of its argument, the value of the unit-step signal is unity and it is zero otherwise. An arbitrary function can be expressed in terms of appropriately scaled and shifted unit-step or impulse signals. By this way, any signal can be specified, for easier mathematical analysis, by a single expression, valid for all n. For example, a pulse signal, shown in Fig. 1.2a, with its only nonzero values defined as $\{x(1) = 1, x(2) = 1, x(3) = 1\}$ can be expressed as the sum of the two delayed unit-step signals shown in Fig. 1.2b, $x(n) = u(n - 1) - u(n - 4)$. The pulse can also be represented as a sum of delayed impulses.

$$x(n) = u(n - 1) - u(n - 4) = \sum_{k=1}^{3} \delta(n - k) = \delta(n - 1) + \delta(n - 2) + \delta(n - 3)$$

The continuous unit-step signal is defined as

$$u(t) = \begin{cases} 1 & \text{for } t > 0 \\ 0 & \text{for } t < 0 \\ \text{undefined for } t = 0 \end{cases}$$

The value $u(0)$ is undefined and can be assigned a suitable value from 0 to 1 to suit a specific problem. In Fourier analysis, $u(0) = 0.5$. A common application of the unit-step signal is that multiplying a signal with it yields the causal form of the signal. For example, the continuous signal $\sin(t)$ is defined for $-\infty < t < \infty$. The values of $\sin(t)u(t)$ is zero for $t < 0$ and $\sin(t)$ for $t > 0$.

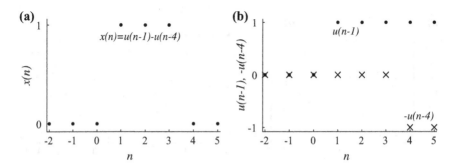

Fig. 1.2 **a** The pulse signal, $x(n) = u(n - 1) - u(n - 4)$; **b** the delayed unit-step signals, $u(n - 1)$ and $-u(n - 4)$

1.1.3 Unit-Ramp Signal

The discrete unit-ramp signal, shown in Fig. 1.1c, is also often used in the analysis of signals and systems. It is defined as

$$r(n) = \begin{cases} n \text{ for } n \geq 0 \\ 0 \text{ for } n < 0 \end{cases}$$

It linearly increases for positive values of its argument and is zero otherwise.

The three signals, the unit-impulse, the unit-step, and the unit-ramp, are related by the operations of sum and difference. The unit-impulse signal $\delta(n)$ is equal to $u(n) - u(n-1)$, the first difference of the unit-step. The unit-step signal $u(n)$ is equal to $\sum_{k=0}^{\infty} \delta(n-k)$, the running sum of the unit-impulse. The shifted unit-step signal $u(n-1)$ is equal to $r(n)$ $r(n-1)$. The unit-ramp signal $r(n)$ is equal to

$$r(n) = nu(n) = \sum_{k=0}^{\infty} k\delta(n-k).$$

Similar relations hold for continuous type of signals.

1.1.4 Sinusoids and Complex Exponentials

Sinusoids

The impulse and the sinusoid are the two most important signals in signal and system analysis. The impulse is the basis for convolution and the sinusoid is the basis for transfer function. The cosine and sine functions are two of the most important functions in trigonometry. As these functions are the basis functions in Fourier analysis, we have study them in detail.

The unit circle, defined by $x^2 + y^2 = 1$ and shown in Fig. 1.3, is a circle with its center located at the origin and radius 1. For each point on the circle defined by the coordinates (x, y), starting at $(1, 0)$ and moving in the counterclockwise direction, with $\theta \geq 0$ (the angle subtended by the x-axis and the line joining the point and the origin), the sine (sin) and cosine (cos) functions are defined in terms of its coordinates (x, y) as

$$\cos(\theta) = x \quad \text{and} \quad \sin(\theta) = y$$

If the point lies on a circle of radius r, then

$$\cos(\theta) = x/r \quad \text{and} \quad \sin(\theta) = y/r, \quad r = \sqrt{x^2 + y^2}$$

Clearly, the sinusoids are of periodic nature. Any function defined on a circle will be a periodic function of an angular variable θ. Therefore, the trigonometric functions are also called circular functions. The argument θ is measured in radians or degrees. The

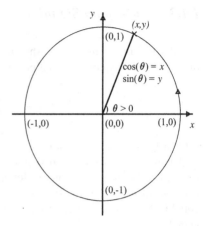

Fig. 1.3 Sine and cosine functions defined on the unit circle

radian is defined as the angle subtended between the x-axis and the line between the point and the origin on the unit circle. One radian is defined as the angle subtended by unit arc length. Since the circumference of the unit circle is 2π, one complete revolution is 2π rad. In degree measure, $2\pi = 360°$ and $\pi = 180°$. One radian is approximately $180/\pi = 57.3°$.

A linear combination of sine and cosine functions is a sinusoid, in rectangular form, given by

$$a\cos(\theta) + b\sin(\theta)$$

where a and b are real numbers with $a \neq 0$ or $b \neq 0$. With $c = \sqrt{a^2 + b^2}$, and $\cos(d) = a/c$ and $\sin(d) = b/c$,

$$a\cos(\theta) + b\sin(\theta) = c\cos(\theta - d)$$

is called the polar form of the sinusoid. For example,

$$\frac{\sqrt{15}}{2}\cos(\theta) + \frac{\sqrt{5}}{2}\sin(\theta) = \sqrt{5}\cos\left(\theta - \frac{\pi}{6}\right)$$

Using the subtraction formula for cosine, we get back the left side from the right side.

Three of the properties of the sinusoids make them the most useful basis functions to represent a signal.

1. The derivative of $\sin(u)$ is defined as

$$\frac{d}{dx}\sin(u) = \cos(u)\frac{du}{dx},$$

where u is a differential function of x. The sinusoids can be repeatedly differentiated or integrated with no changes in their frequencies.

2. The sum of a number of sinusoids of a certain frequency is also a sinusoid of the same frequency.
3. The integral of a sinusoid over an integral number of cycles is zero, as the negative area is equal to the positive area.

Therefore, the steady-state signals in any part of a LTI system, described by a linear differential equation with constant coefficients, are also of the same frequency as that of the input sinusoid. The response of a system for an arbitrary signal, which can be decomposed into a linear combination of sinusoids by Fourier analysis, can be determined from the sinusoidal response of the system. While the same procedure is applicable with the impulse signal as the basis function, the use of sinusoidal basis functions is more advantageous. One advantage is that it reduces the computational complexity in finding the response, as fast algorithms are available to adequately approximate the response. Certain characteristics of the signals are more readily apparent in the Fourier representation. Another advantage is that the interpretation of operations, such as filtering, is often easier. For these reasons, sinusoidal representation plays a dominant role in LTI system analysis.

The general form of a sinusoid is given by

$$x(t) = A \cos(\omega t + \theta), \qquad -\infty < t < \infty$$

where A (half the range of the function) is the amplitude of the sinusoid, ω is its frequency in radians, and θ is the phase. The phase is measured with respect to the reference waveform $A \cos(\omega t)$. For example, the peak value of this waveform occurs at $t = 0$ and its phase is defined to be zero. The first peak of the sine waveform

$$A \sin(\omega t) = A \cos\left(\omega t - \frac{\pi}{2}\right)$$

occurs at $\omega t = \pi/2$ rad and its phase is $-\pi/2$ rad or $-90°$. The period of the waveform is the interval between $\omega t = 0$ to $\omega t = 2\pi$. Therefore, the period is $T = (2\pi)/\omega$ s. The radian frequency is ω radians/second and the cyclic frequency is $f = \omega/(2\pi) = 1/T$ Hz.

Sum of Sinusoids with the Same Frequency

An important property of the sinusoids is that the sum of sinusoids of the same frequency, but with arbitrary amplitudes and phases, is also a sinusoid of the same frequency. In order to find the sum, we have to express the sinusoids in their rectangular form and sum the respective amplitudes of the sine and cosine components. Consider the two sinusoids

$$x(t) = A \cos(\omega t + \theta) \qquad \text{and} \qquad y(t) = B \cos(\omega t + \phi)$$

Then,

$$
\begin{aligned}
z(t) = x(t) + y(t) &= C\cos(\omega t + \psi) = A\cos(\omega t + \theta) + B\cos(\omega t + \phi) \\
&= \cos(\omega t)(A\cos(\theta) + B\cos(\phi)) - \sin(\omega t)(A\sin(\theta) + B\sin(\phi)) \\
&= \cos(\omega t)(C\cos(\psi)) - \sin(\omega t)(C\sin(\psi))
\end{aligned}
$$

Solving for C and ψ, we get

$$
C = \sqrt{A^2 + B^2 + 2AB\cos(\theta - \phi)}
$$
$$
\psi = \tan^{-1}\frac{A\sin(\theta) + B\sin(\phi)}{A\cos(\theta) + B\cos(\phi)}
$$

With $\theta = 0$ and $\phi = -\pi/2$ (one sinusoid being the cosine and the other being sine), the formula reduces to relation between the polar and the rectangular form of a sinusoid.

Example 1.1 Determine the sinusoid that is the sum of two sinusoids

$$
x(t) = 3\cos\left(\omega t + \frac{\pi}{3}\right) \quad \text{and} \quad y(t) = 2\sin\left(\omega t - \frac{\pi}{6}\right)
$$

Solution
The second sinusoid can also be expressed as

$$
y(t) = 2\cos\left(\omega t - \frac{\pi}{6} - \frac{\pi}{2}\right) = 2\cos\left(\omega t - \frac{2\pi}{3}\right)
$$

Now,

$$
A = 3,\, B = 2,\, \theta = \frac{\pi}{3},\, \phi = -\frac{2\pi}{3}
$$

Substituting the numerical values in the equations, we get

$$
C = \sqrt{3^2 + 2^2 + 2(3)(2)\cos\left(\frac{\pi}{3} + \frac{2\pi}{3}\right)} = 1
$$

$$
\psi = \cos^{-1}\frac{3\cos\left(\frac{\pi}{3}\right) + 2\cos\left(-\frac{2\pi}{3}\right)}{1} = \sin^{-1}\frac{3\sin\left(\frac{\pi}{3}\right) + 2\sin\left(-\frac{2\pi}{3}\right)}{1} = 1.0472 \text{ rad} = 60°
$$

$$
z(t) = x(t) + y(t) = \cos(\omega t + 1.0472)
$$

∎

The polar form of a discrete sinusoid is

$$
x(n) = A\cos(\omega n + \theta), \quad n = -\infty, \ldots, -1, 0, 1, \ldots, \infty,
$$

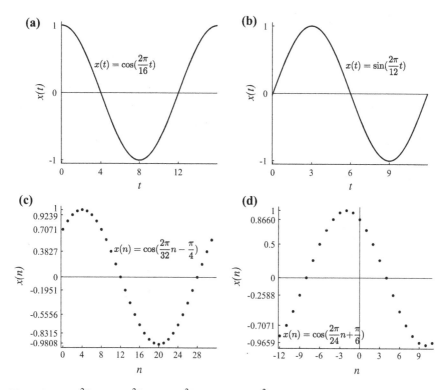

Fig. 1.4 **a** $\cos(\frac{2\pi}{16}t)$; **b** $\sin(\frac{2\pi}{12}t)$; **c** $\cos(\frac{2\pi}{32}n - \frac{\pi}{4})$; **d** $\cos(\frac{2\pi}{24}n + \frac{\pi}{6})$

For discrete sinusoids, the range of frequencies for unique representation is limited due to sampling. Further, the cyclic frequency has to be a rational number (a ratio of two integers, a/b with $b \neq 0$), since the samples of the sinusoid are defined only at the sample points.

Cosine and sine waveforms are two important special cases of the sinusoid. Figure 1.4a shows the continuous cosine waveform $\cos(\frac{2\pi}{16}t)$. The angular frequency is $\omega = \frac{2\pi}{16}$ rad/s. The cyclic frequency is $f = \frac{\omega}{2\pi} = \frac{1}{16}$ Hz. The waveform repeats every 16 s. The period is $T = \frac{1}{f} = 16$ s. The amplitude is one. That is, the maximum distance from the t-axis to either of the peaks of the waveform is one. As the peak occurs at $t = 0$, the phase of the waveform is 0. Figure 1.4b shows the continuous sine waveform $\sin(\frac{2\pi}{12}t)$. The angular frequency is $\omega = \frac{2\pi}{12}$ rad/s. The cyclic frequency is $f = \frac{\omega}{2\pi} = \frac{1}{12}$ Hz. The waveform repeats every 12 s. The period is $T = \frac{1}{f} = 12$ s. The amplitude is one. As the peak occurs at $t = 3$, the phase of the waveform is $-\frac{\pi}{2}$ rad. The occurrence of the peak is delayed by 3 s, compared with the cosine waveform. That is, $\frac{2\pi}{12}(-3) = -\frac{\pi}{2}$. Therefore, the sine waveform can be, equivalently, expressed as $\cos(\frac{2\pi}{12}t - \frac{\pi}{2})$. The sine waveform can be obtained by shifting $\cos(\frac{2\pi}{12}t)$ by $\frac{\pi}{2}$ rad (or 90° or 3 s or a quarter of a cycle) to the right. Alternately, the cosine waveform can be considered as the advanced version of the sine waveform. The cosine waveform leads the sine waveform by 90° or the sine waveform lags the cosine waveform by 90°.

Figure 1.4c shows the discrete sinusoid $\cos(\frac{2\pi}{32}n - \frac{\pi}{4})$. The angular frequency is $\omega = \frac{2\pi}{32}$ rad. The cyclic frequency is $f = \frac{\omega}{2\pi} = \frac{1}{32}$ cycles per sample. The waveform repeats every 32 samples. The period is $T = \frac{1}{f} = 32$ samples. The amplitude is one. As the peak occurs at $n = 4$, the phase of the waveform is $-\frac{\pi}{4}$ rad. Figure 1.4d shows the discrete sinusoid $\cos(\frac{2\pi}{24}n + \frac{\pi}{6})$. The angular frequency is $\omega = \frac{2\pi}{24}$ rad. The cyclic frequency is $f = \frac{\omega}{2\pi} = \frac{1}{24}$ cycles per sample. The waveform repeats every 24 samples. The period is $T = \frac{1}{f} = 24$ samples. The amplitude is one. As the peak occurs at $n = -2$, the phase of the waveform is $\frac{\pi}{6}$ rad.

Exponential Signal

By using sine and cosine functions, signals can be represented. But it involves two basic functions and the two associated constants. It is found that an equivalent representation of signals is obtained using the complex exponential function, in which only one basic function and one associated constant is involved. The compact representation and the ease of manipulating the exponential functions make its use mandatory in the analysis of signals and systems. However, practical devices generate sine and cosine functions. Euler's formula is the bridge between the theory and the practice.

With b any positive real number except 1,

$$x(t) = b^t$$

is called the exponential function with base b. Our primary interest, in this book, is the complex exponential function of the form

$$x(\theta) = Ae^{j\theta}$$

The base is e, which is approximately 2.71828. The exponent is a complex number with its real part zero (pure imaginary number). The coefficient of the exponential A is a complex number.

The exponential $e^{j\theta}$, shown in Fig. 1.5, is a unit rotating vector, rotating in the counterclockwise direction. The exponential carries the same information about a sinusoid in an equivalent form, which is advantageous in the analysis of signals and systems. In combination with the exponential $e^{-j\theta}$, which rotates in the clockwise direction, a real sinusoidal waveform can be obtained. Since

$$e^{j\theta} = \cos(\theta) + j\sin(\theta) \quad \text{and} \quad e^{-j\theta} = \cos(\theta) - j\sin(\theta),$$

solving for $\cos(\theta)$ and $\sin(\theta)$ results in

$$\cos(\theta) = \frac{e^{j\theta} + e^{-j\theta}}{2} \quad \text{and} \quad \sin(\theta) = \frac{e^{j\theta} - e^{-j\theta}}{j2}$$

Fig. 1.5 The complex
exponential $e^{j\theta}$

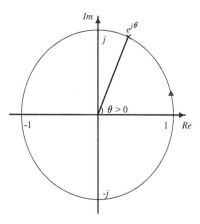

Let $\theta = \omega t + \phi$. Then,

$$R\cos(\omega t + \phi) = \frac{Re^{j(\omega t + \phi)} + Re^{-j(\omega t + \phi)}}{2}$$
$$= 0.5Re^{j\phi}e^{j\omega t} + 0.5Re^{-j\phi}e^{-j\omega t}$$

The exponentials and their coefficients are complex conjugates.

Example 1.2 Express the signal in terms of complex exponentials.

$$x(n) = \cos\left(\frac{2\pi}{32}n - \frac{\pi}{3}\right)$$

Solution
Using Euler's identity, the waveform can be expressed in terms of complex exponentials as

$$x(n) = \frac{1}{2}(e^{j\left(\frac{2\pi}{32}n - \frac{\pi}{3}\right)} + e^{-j\left(\frac{2\pi}{32}n - \frac{\pi}{3}\right)})$$

The complex frequency coefficients of the exponentials are $0.5e^{-j\frac{\pi}{3}}$ and $0.5e^{j\frac{\pi}{3}}$. The real parts of the coefficients are even-symmetric and the imaginary parts are odd-symmetric. This redundancy is expected, since there are only two independent values specifying the sinusoidal waveform (the amplitude and the phase). ∎

Example 1.3 Express the signal in terms of complex exponentials.

$$x(n) = \cos\left(\frac{2\pi}{32}n\right) + j\sqrt{3}\sin\left(\frac{2\pi}{32}n\right)$$

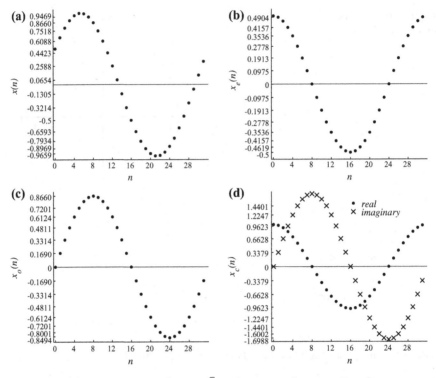

Fig. 1.6 **a** $\cos(\frac{2\pi}{32}n - \frac{\pi}{3})$; **b** $0.5\cos(\frac{2\pi}{32}n)$; **c** $\frac{\sqrt{3}}{2}\sin(\frac{2\pi}{32}n)$; **d** $\cos(\frac{2\pi}{32}n) + j\sqrt{3}\sin(\frac{2\pi}{32}n)$

Solution
Using Euler's identity, the waveform can be expressed in terms of complex exponentials as

$$x(n) = \frac{1}{2}(e^{j(\frac{2\pi}{32}n)} + e^{-j(\frac{2\pi}{32}n)}) + \frac{\sqrt{3}}{2}(e^{j(\frac{2\pi}{32}n)} - e^{-j(\frac{2\pi}{32}n)})$$

$$= \left(\frac{1}{2} + \frac{\sqrt{3}}{2}\right)e^{j(\frac{2\pi}{32}n)} + \left(\frac{1}{2} - \frac{\sqrt{3}}{2}\right)e^{-j(\frac{2\pi}{32}n)}$$

The coefficients of the exponentials are $(\frac{1}{2} + \frac{\sqrt{3}}{2})$ and $(\frac{1}{2} - \frac{\sqrt{3}}{2})$. The waveform is shown in Fig. 1.6d. ∎

1.2 Classification of Signals

Signals are classified into different types based on their characteristics. The classification is an aid in selecting a suitable representation and processing.

Table 1.1 Signal classification

Characteristic	Continuous	Sampled
Unquantized	Continuous	Discrete
Quantized	Quantized continuous	Digital

1.2.1 Continuous, Discrete, and Digital Signals

This type of classification characterizes the type of sampling of the dependent and independent variables. Sampling the amplitude is called quantization. Table 1.1 shows the signal classification based on sampling the amplitude and time. When both the variables of a signal can assume continuum of values, it is called a continuous signal, such as the ambient temperature. Most of the naturally occurring signals are of this type. The temperature measured by a digital thermometer is a quantized continuous signal. This type of signals occurs in the reconstruction of a continuous signal from its sampled version. Sampled continuous-valued signal is a discrete signal. This type of signals, shown in Fig. 1.4c, d, is used in the analysis of discrete signals and systems. A quantized discrete signal is called a digital signal, used in the digital systems.

1.2.2 Periodic and Aperiodic Signals

The sinusoidal signals are defined by the values of the coordinates on a circle in Fig. 1.3. In each rotation of a point on the circle, the same set of values are produced indefinitely. This type of signals, such as the sine and cosine functions, is periodic signals. While only one period of a periodic signal contains new information, periodicity is required to represent signals such as power and communication signals. In communication engineering, the message signal is aperiodic and the carrier signal is periodic. Finite duration signals are represented, by the practically most often used version of the Fourier analysis, assuming periodic extension. The finite signal is considered as the values of one period and concatenation of it indefinitely on either side yields a periodic signal. A signal $x(t)$ is said to be periodic, if $x(t) = x(t + T)$, for all values of t from $-\infty$ to ∞ and $T > 0$ is a positive constant. The minimum value of T that satisfies the constraint is the period. A periodic signal shifted by an integral number of its period remains unchanged. A signal that is not periodic is aperiodic, such as the impulse, step and ramp signals shown in Fig. 1.1 and the real exponential. The period is infinity, so that there is no indefinite repetition. The everlasting definition of a periodic signal is for mathematical convenience. In practice, physical devices are switched on at some time and the response reaches a steady state, after the transient response dies down.

If a discrete signal $x(n)$ satisfies the condition

$$x(n + N) = x(k \bmod N) = x(n) \quad \text{for all} \quad n,$$

then it is said to be periodic, $k = lN + n$ and l is an integer. The smallest $N > 0$ satisfying the constraint is its period. Otherwise, it is aperiodic. The mod function, $r = k \bmod N$, yields the remainder r of dividing k by N.

$$r = k - \lfloor (k/N) \rfloor N, \quad N \neq 0$$

and r has the same sign as N. The floor function rounds the number to the nearest integer less than or equal to its argument. For example, with $N = 3$, $r = k \bmod 3$ yields

k	-8	-7	-6	-5	-4	-3	-2	-1	0	1	2	3	4	5	6	7
r	1	2	0	1	2	0	1	2	0	1	2	0	1	2	0	1

For a negative k, the corresponding positive value in the range 0 to $N - 1$ is obtained. For example, with $k = -1$,

$$\lfloor (-1/3) \rfloor 3 = \lfloor (-0.3333) \rfloor 3 = (-1)3 = -3 \quad \text{and} \quad r = -1 - (-3) = 2$$

With $k = -8$, we get

$$\lfloor (-8/3) \rfloor 3 = \lfloor (-2.6667) \rfloor 3 = (-3)3 = -9 \quad \text{and} \quad r = -8 - (-9) = 1$$

$r = k \bmod N$ is periodic with period N since

$$r = k \bmod N = (k + N) \bmod N$$

Let $\{\check{1}, 2, 3\}$ is a periodic sequence with period 3. The 27th number in the sequence is 1. The index is obtained as 27 mod 3 = 0 and $x(0) = 1$. $x(-4) = x(2) = 3$.

1.2.3 Even- and Odd-Symmetric Signals

Any signal can be decomposed into its even and odd components. Knowing whether a signal is even or odd may reduce computational and storage requirements in its processing. If a signal $x(t)$ satisfies the condition

$$x(-t) = x(t) \quad \text{for all} \quad t,$$

then it is said to be even. The plot of such a signal is symmetrical about the vertical axis at the origin. For example, the cosine waveforms, shown in Figs. 1.4a and 1.6b, are even. For the signal in Fig. 1.6b,

$$0.5 \cos \left(\frac{2\pi}{32}(-n) \right) = 0.5 \cos \left(\frac{2\pi}{32}n \right)$$

If a signal $x(t)$ satisfies the condition

$$x(-t) = -x(t) \quad \text{for all} \quad t,$$

then it is said to be odd. The plot of such a signal is antisymmetrical about the vertical axis at the origin. For example, the sine waveforms, shown in Figs. 1.4b and 1.6c, are odd. For the signal in Fig. 1.6c,

$$\frac{\sqrt{3}}{2} \sin \left(\frac{2\pi}{32}(-n) \right) = -\frac{\sqrt{3}}{2} \sin \left(\frac{2\pi}{32}n \right)$$

Any function can be decomposed into its even and components. Let the even and odd components of $x(t)$ be $x_e(t)$ and $x_o(t)$, respectively. Then,

$$x(t) = x_e(t) + x_o(t) \quad \text{and} \quad x(-t) = x_e(t) - x_o(t)$$

Solving the equations, we get

$$x_e(t) = 0.5(x(t) + x(-t)) \quad \text{and} \quad x_o(t) = 0.5(x(t) - x(-t))$$

The signal, shown in Fig. 1.6a, is neither even nor odd.

$$\cos \left(\frac{2\pi}{32}n - \frac{\pi}{3} \right) = 0.5 \cos \left(\frac{2\pi}{32}n \right) + \frac{\sqrt{3}}{2} \sin \left(\frac{2\pi}{32}n \right)$$

Its even and odd components are shown in Fig. 1.6b, c. The integral of an even signal between symmetric limits $-T$ to T is equal to twice that over the limits 0 to T. The integral of an odd signal between symmetric limits is zero. The product properties are:

$$\text{even signal} \times \text{even signal} = \text{even signal}$$
$$\text{odd signal} \times \text{odd signal} = \text{even signal}$$
$$\text{odd signal} \times \text{even signal} = \text{odd signal}$$

Circular Symmetry

In the DFT, a finite extent signal is extended periodically to make it a periodic signal. As periodic signals are defined on a circle, the even and odd symmetries, defined on a line, can also be defined emphasizing the cyclic nature of the signal. If a N-point signal $x(n)$ satisfies the condition

$$x((-n) \bmod N) = x(n) \quad \text{for all} \quad n,$$

then it is said to be circularly even. For example, the sequences

$$\{x(n), n = 0, 1, \ldots, 7\} = \{8, 2, 6, 4, 6, 4, 6, 2\} \quad \text{and}$$
$$\{x(n), n = 0, 1, \ldots, 6\} = \{8, 2, 6, 4, 4, 6, 2\}$$

are even. The values at the beginning and at the middle can be arbitrary for a signal with even number of elements and the other values satisfy $x(n) = x(-n)$, when placed on a circle. Considering the finite extent alone, the condition is

$$x(N - n) = x(n), \quad 1 \le n \le N - 1$$

The cosine waveform, with period N,

$$x(n) = \cos\left(\frac{2\pi}{N}n\right)$$

is even.

If a N-point signal $x(n)$ satisfies the condition

$$-x((-n) \bmod N) = x(n) \quad \text{for all} \quad n,$$

then it is said to be circularly odd. For example, the sequences

$$\{x(n), n = 0, 1, \ldots, 7\} = \{0, -2, -6, -4, 0, 4, 6, 2\}$$

and

$$\{x(n), n = 0, 1, \ldots, 6\} = \{0, -2, -6, -4, 4, 6, 2\}$$

are odd. The values at the beginning and at the middle must be zero for a signal with even number of elements and the other values satisfy $x(n) = -x(-n)$, when placed on a circle. Considering the finite extent alone, the condition is

$$x(N - n) = -x(n), \quad 1 \le n \le N - 1$$

The sine waveform, with period N,

$$x(n) = \sin\left(\frac{2\pi}{N}n\right)$$

is odd. Even- and odd-symmetric 8-point arbitrary sequences are shown in Fig. 1.7.

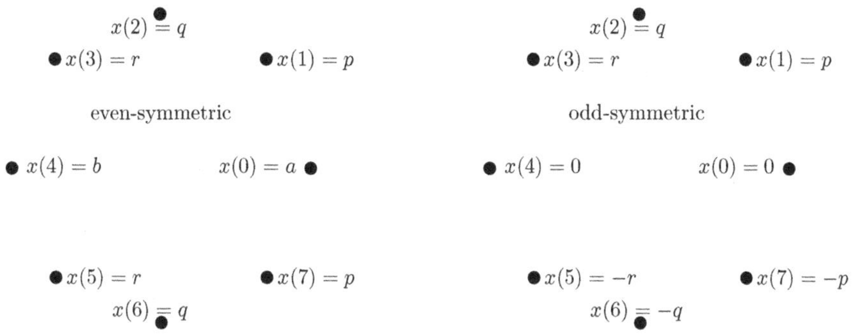

$$x(2) \stackrel{\bullet}{=} q$$
$\bullet\, x(3) = r \qquad\qquad \bullet\, x(1) = p$

even-symmetric

$\bullet\, x(4) = b \qquad\qquad x(0) = a\, \bullet$

$\bullet\, x(5) = r \qquad\qquad \bullet\, x(7) = p$
$$x(6) \underset{\bullet}{=} q$$

$$x(2) \stackrel{\bullet}{=} q$$
$\bullet\, x(3) = r \qquad\qquad \bullet\, x(1) = p$

odd-symmetric

$\bullet\, x(4) = 0 \qquad\qquad x(0) = 0\, \bullet$

$\bullet\, x(5) = -r \qquad\qquad \bullet\, x(7) = -p$
$$x(6) \underset{\bullet}{=} -q$$

Fig. 1.7 Circular sequences, even-symmetric and odd-symmetric

1.2.4 Energy and Power Signals

Power is the rate of doing work, typically measured in watts. It is common to specify devices by their power handling capacity. Devices, such as an electric motor, heater, amplifier and a diesel engine, are characterized by their power rating. The energy of a signal $x(t)$ is measured by the power dissipated in a $1\,\Omega$ resister due to a voltage applied across it or a current passing through. The energy dissipated is

$$E = \int_{-\infty}^{\infty} |x(t)|^2 \, dt$$

assuming that E is finite. Signals satisfying the condition $E < \infty$ are called energy signals. All practical signals are energy signals. The energy of $x(t) = 2e^{-3t},\ t \geq 0$ is

$$E = \int_{0}^{\infty} |2e^{-3t}|^2 dt = \frac{2}{3}$$

For discrete signals,

$$E = \sum_{n=-\infty}^{\infty} |x(n)|^2$$

The energy of $x(n) = 3(0.8)^n u(n)$ is

$$E = \sum_{n=0}^{\infty} |3(0.8)^n|^2 = \frac{9}{1 - 0.64} = \frac{9}{0.36} = 25$$

Sinusoids are not energy signals, since they have infinite energy. Such signals are characterized by their power. The average power of a continuous signal is defined as

$$P = \lim_{T \to \infty} \frac{1}{T} \int_{-\frac{T}{2}}^{\frac{T}{2}} |x(t)|^2 dt$$

For periodic signals, the average power can be computed over one period as

$$P = \frac{1}{T} \int_{-\frac{T}{2}}^{\frac{T}{2}} |x(t)|^2 dt,$$

where T is the period. The average power of the sine wave $2\sin(\frac{2\pi}{8}t)$ is

$$P = \frac{1}{8} \int_{-4}^{4} |2\sin\left(\frac{2\pi}{8}t\right)|^2 dt = \frac{1}{4} \int_{-4}^{4} \left(1 - \cos\left(2\frac{2\pi}{8}t\right)\right) dt = 2$$

The average power of a discrete signal is defined as

$$P = \lim_{N \to \infty} \frac{1}{2N+1} \sum_{n=-N}^{N} |x(n)|^2$$

For a periodic signal with period N, the average power can be determined as

$$P = \frac{1}{N} \sum_{n=0}^{N-1} |x(n)|^2$$

Periodic and aperiodic signals with finite average power are called power signals. The average power of the sine wave $\sin(\frac{2\pi}{8}n)$ is

$$P = \frac{2}{8} \sum_{n=0}^{3} |x(n)|^2 = \frac{1}{4}(0^2 + (1)^2 + 0^2 + (-1)^2) = \frac{1}{2}$$

Signals, such as $x(t) = t$, are neither energy nor power signals.

1.2.5 Deterministic and Random Signals

Signals may be deterministic or random. The values of a deterministic signal are predictable for all time, while the future values of the random signals are not exactly predictable. Their future values can be estimated based on averages of a set of such signals. Typical examples of deterministic signals are the unit-step and sinusoids presented earlier. Most of the naturally occurring signals are of random nature and represented by averages in the time domain and Fourier power spectrum in the frequency domain. Each of the signals produced by persons saying the word Fourier will be different. While the amplitude profile of each of the signals will be different, similarity of the time variation and frequency content are likely to be similar. Deterministic signals are characterized by their amplitude in the time domain and Fourier spectrum in the frequency domain.

1.2.6 Causal and Noncausal Signals

Most signals occur at some finite instant, usually chosen as $t = 0$, and are considered identically zero before this instant. A signal $x(t)$ is said to be causal if

$$x(t) = 0, \quad t < 0$$

If $x(t) \neq 0, \quad t < 0$, then it is a noncausal signal. The unit-step signal is causal. A sinusoid is a noncausal signal. This signal can be made causal by multiplying it with the unit-step signal $u(t)$.

1.3 Signal Operations

In addition to arithmetic operations on the signal samples (dependent variable), operations on the independent variables are also required in signal analysis. Typical operations of this type are shifting, folding, and scaling. While we present these operations assuming that the independent variable is time, these operations can be carried out with any variables such as frequency, distance.

1.3.1 Time Shifting

If the variable t in $x(t)$ is replaced by $(t - t_0)$, then the origin of the signal is shifted to $t = t_0$. If t_0 is positive (negative), the values of the function are retarded (advanced) by t_0. Graphically, it amounts to shifting the plot of the function forward (t_0 positive) or backward by t_0. Examples of time shifting are the waveforms in Figs. 1.2 and 1.4.

Let $x(n) = \{x(0), x(1), x(2), x(3)\} = \{3, 1, 2, 4\}$. By linearly shifting $x(n)$, we get

$$x(n - 1) = \{x(1) = 3, x(2) = 1, x(3) = 2, x(4) = 4\}$$
$$x(n + 1) = \{x(-1) = 3, x(0) = 1, x(1) = 2, x(2) = 4\}$$

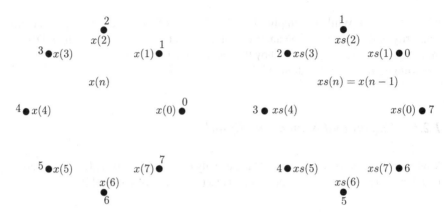

Fig. 1.8 Circular shift of a sequence

Circular Shifting

Circular shifting is simply the shifting of the values of a signal placed on a circle. The right circular shift of a N-point signal $x(n)$ by k sample intervals results in

$$x((n-k) \bmod N)$$

Let $x(n) = \{x(0), x(1), x(2), x(3)\} = \{3, 1, 2, 4\}$. Then,

$$x(n-1) = \{x(3), x(0), x(1), x(2)\} = \{4, 3, 1, 2\}$$
$$x(n+1) = \{x(1), x(2), x(3), x(0)\} = \{1, 2, 4, 3\}$$

Right circular shift of the 8-point sequence $x(n) = \{\check{0}, 1, 2, 3, 4, 5, 6, 7\}$ by one sample interval results in $xs(n) = x(n-1) = \{\check{7}, 0, 1, 2, 3, 4, 5, 6\}$, as shown in Fig. 1.8. The check symbol ˇindicates that the index of that element is 0.

1.3.2 Time Scaling

If the variable t in $x(t)$ is replaced by (at), then the function is scaled. If $a > 1$, the signal is compressed and if $a < 1$, the signal is expanded.

Signal Expansion

A N-point sequence $x(n)$ is expanded by an integer factor a to get its expanded version $x_u(n), n = 0, 1, \ldots, aN - 1$, defined as

$$x_u(n) = \begin{cases} x\left(\frac{n}{a}\right) & \text{for } n = 0, \pm a, \pm 2a, \ldots, \\ 0 & \text{otherwise} \end{cases}$$

The expanded signal $x_u(n)$ is obtained by inserting $a-1$ zeros after each sample in $x(n)$. This operation, also called upsampling, increases the sampling rate. Expansion of the 8-point sequence $x(n) = \{\check{0}, 1, 2, 3, 4, 5, 6, 7\}$ by a factor of 2, $(a = 2)$, results in

$$x_u(n) = \{\check{0}, 0, 1, 0, 2, 0, 3, 0, 4, 0, 5, 0, 6, 0, 7, 0\}$$

Signal Compression

A N-point sequence $x(n)$ is compressed by an integer factor a to get its compressed version $x_d(n), n = 0, 1, \ldots, (N/a) - 1$, defined as

$$x_d(n) = x(na) \text{ for } n = 0, 1, \ldots, \tfrac{N}{a} - 1$$

The compressed signal $x_d(n)$ is obtained by taking every ath sample, starting with $x(0)$. This operation, also called downsampling, decreases the sampling rate. Compression of the 8-point sequence $x(n) = \{\check{0}, 1, 2, 3, 4, 5, 6, 7\}$ by a factor of 2 results in $x_d(n) = x(2n) = \{\check{0}, 2, 4, 6\}$.

Time-Reversal

Time-reversal of a signal $x(t)$ is defined as $xr(t) = x(-t)$ and $xr(-t) = x(t)$. By replacing t by $-t$, we get the time-reversed version of $x(t)$, which is the mirror image of $x(t)$ about the vertical axis at the origin.

Circular time-reversal of a N-point signal $x(n)$ is given by

$$\begin{cases} x(0) & \text{for } n = 0 \\ x(N-n) & 1 \le n \le N - 1 \end{cases}$$

This is just plotting $x(n)$ in the other direction on a circle. Circular time-reversal of the 8-point sequence $x(n) = \{\check{0}, 1, 2, 3, 4, 5, 6, 7\}$ results in $xr(n) = x(N-n) = \{\check{0}, 7, 6, 5, 4, 3, 2, 1\}$, as shown in Fig. 1.9.

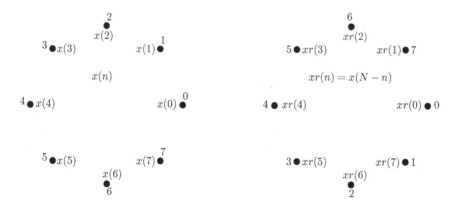

Fig. 1.9 Circular time-reversal of a sequence

Zero Padding

In this operation, a sequence is appended by zero-valued samples. A N-point sequence $x(n)$ is expanded to a M-point, $M > N$, signal $x_z(n)$, $n = 0, 1, \ldots, M - 1$, defined as

$$x_z(n) = \begin{cases} x(n) & \text{for } n = 0, 1, \ldots, N - 1 \\ 0 & \text{for } n = N, N + 1, \ldots, M - 1. \end{cases}$$

1.4 Complex Numbers

Complex number system is an extension of the real number system that suits a large number of applications in science and engineering. The simplest number system is the set of natural numbers $\{1, 2, 3, \ldots\}$, the system we use to count things. By including the negative numbers, we get the integers. Further, the number system is extended by including rational and irrational numbers. With this system, we can only specify the location of a point on a line. There is a necessity in several applications to represent a point on a plane. For example, a sinusoid at a given frequency is characterized by its amplitude and phase. That requires a pair of ordered numbers, a two-element vector. While it is possible to represent a sinusoid with two scalars, the vector representation is so much advantageous that it is used in most of the Fourier analysis, the subject of this book. Another advantage of the complex number system is that it provides solution to all polynomial equations. Further, real integrals, which may or may not be evaluated by real integral calculus, are elegantly evaluated by complex integration. Except that the system is, unfortunately, named as complex number system, there is nothing complex about it and it is just another extension of the number system.

A complex number $z = (x, y)$ is a two-element vector of real numbers, in which the first and second elements are, respectively, called the real and imaginary parts of z. A more commonly used notation for the complex number is $z = x + jy$, where j is the imaginary unit. For example, the real part of $z = (3 - j2)$ is 3 and the imaginary part is -2. This form of a complex number is called its rectangular form. Two complex numbers are equal if and only if their real parts are equal and their imaginary parts are also equal.

The Complex Plane

As complex numbers are 2-element vectors, they can be represented in a plane, called the complex plane. The complex plane, shown in Fig. 1.10, is similar to xy-plane used to plot graphs. In the complex plane, the horizontal axis is called the real axis and the vertical axis is called the imaginary axis. On both axes, the scale is the same. A complex number is plotted as a point in the complex plane with its real and imaginary parts fixing the coordinates of the horizontal and vertical axes, respectively. Several examples are given in the figure. A number lying on the real axis, with its imaginary part zero, is a real number. Therefore, it is obvious that the complex number system is an extension of the real number system. A number lying on the imaginary axis, with

Fig. 1.10 Representation of
complex numbers in the
complex plane

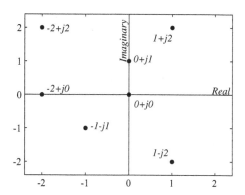

its real part zero, is a pure imaginary number. The number j is called the imaginary unit.

Addition of Complex Numbers
The sum of two complex numbers $p = a + jb$ and $q = c + jd$ is defined as

$$p + q = (a + jb) + (c + jd) = (a + c) + j(b + d) \qquad (1.1)$$

The sum of $3 - j2$ and $4 + j3$ is $7 + j1$.
The difference of p and q is defined as

$$p - q = (a + jb) - (c + jd) = (a - c) + j(b - d) \qquad (1.2)$$

The difference of $7 + j1$ and $3 - j2$ is $4 + j3$.

Multiplication of Complex Numbers
The product of two complex numbers $p = a + jb$ and $q = c + jd$ is defined as

$$pq = (a + jb)(c + jd) = (ac - bd) + j(ad + bc) \qquad (1.3)$$

From this definition, we get

$$(j)(j) = j^2 = -1 \quad \text{and} \quad \sqrt{-1} = j$$

The product of $3 - j2$ and $4 + j2$ is $16 - j2$.

Conjugate of a Complex Number
The conjugate of a complex number $p = a + jb$ is $p^* = a - jb$, obtained by replacing j by $-j$. In the complex plane, it is the mirror image of p about the real axis. The sum of a complex number $a + jb$ and its conjugate $a - jb$ is $p + p^* = 2a$, twice its real part. The result of subtraction is $p - p^* = j2b$, twice its imaginary part multiplied by the imaginary unit j. Therefore, the real and imaginary parts of a complex number $p = a + jb$ are given by

$$\frac{p + p^*}{2} \quad \text{and} \quad \frac{p - p^*}{j2},$$

respectively. Conjugate pair $1 + j2$ and $1 - j2$ is shown in the figure. The product of a complex number p and its conjugate is its magnitude squared, $|p|^2$ (a real number). The product of $1 + j2$ and $1 - j2$ is $1^2 + 2^2 = 5$.

Division of Complex Numbers

The division of two complex numbers $p = (a + jb)$ and $q = (c + jd) \neq 0$ is defined as

$$\frac{p}{q} = \frac{(a + jb)}{(c + jd)} = \frac{(a + jb)(c - jd)}{(c + jd)(c - jd)} = \frac{ac + bd}{c^2 + d^2} + j\frac{bc - ad}{c^2 + d^2} \quad (1.4)$$

Dividing $16 - j2$ by $3 - j2$, we get $4 + j2$.

Polar Form of a Complex Number

Using trigonometric functions, the two parts of a complex number $p = a + jb$ can be expressed as

$$p = a + jb = r\cos(\theta) + jr\sin(\theta) = r(\cos(\theta) + j\sin(\theta)) = re^{j\theta}$$

where $r = \sqrt{a^2 + b^2}$, called the absolute value of p and denoted by $|p|$, is the distance of the point (a, b) in the complex plane from the origin. The angle θ (called the argument) is the angle measured in the counterclockwise direction, between the positive real axis and the vector from the origin and the point, $\theta = \angle(p) = \tan^{-1}(b/a)$. For $p = 0$, the angle θ is undefined. The inverse relations are

$$a = r\cos(\theta) \quad \text{and} \quad b = r\sin(\theta)$$

While r is unique for each complex number, $\theta + 2n\pi$ is an argument for any integer n. Restricting θ between 0 and 2π gives a unique value. For example, let $p = 1 + j\sqrt{3}$. Then,

$$r = \sqrt{1^2 + (\sqrt{3})^2} = 2, \quad \text{and} \quad \tan^{-1}\left(\frac{\sqrt{3}}{1}\right) = 60°$$

Therefore, the polar form is $2e^{j\frac{\pi}{3}}$. Now,

$$2\cos(60°) = 1 \quad \text{and} \quad 2\sin(60°) = \sqrt{3}$$

Multiplication in Polar Form

The product of two complex numbers $re^{j\theta}$ and $se^{j\phi}$ is

$$re^{j\theta}se^{j\phi} = rse^{j(\theta+\phi)}$$

The product of a complex number $re^{j\theta}$ with itself is $r^2e^{j2\theta}$.

$$(2e^{j\frac{\pi}{3}})(2e^{j\frac{\pi}{3}}) = (4e^{j\frac{2\pi}{3}}) = -2 + j2\sqrt{3} = (1 + j\sqrt{3})(1 + j\sqrt{3})$$

The generalization of this result is the De Moivre's theorem. With n a positive integer and $p = re^{j\theta}$, $p^n = r^n e^{jn\theta}$.

Division in Polar Form

The division of two complex numbers $re^{j\theta}$ and $se^{j\phi}$ is

$$\frac{re^{j\theta}}{se^{j\phi}} = \frac{r}{s} e^{j(\theta - \phi)}$$

For example,

$$(4e^{j\frac{2\pi}{3}})/(2e^{j\frac{\pi}{3}}) = (2e^{j\frac{\pi}{3}}) = (-2 + j2\sqrt{3})/(1 + j\sqrt{3}) = (1 + j\sqrt{3})$$

Roots of a Complex Number

With integer $N \geq 1$, the N distinct complex Nth roots of $re^{j\theta}$ are

$$r^{\frac{1}{N}} e^{j\frac{\theta + 2k\pi}{N}}, \quad k = 0, 1, \ldots, N - 1,$$

where $r^{\frac{1}{N}}$ is the positive real Nth root of r. With $r = 1$, we get the N roots of unity, which are the basis functions of the DFT. For example, let us solve the equation $z^4 - 1 = 0$. $z^4 = 1$. Then, $1 = 1e^{j0}$. The four roots are

$$\cos\left(\frac{2k\pi}{4}\right) + j \sin\left(\frac{2k\pi}{4}\right), \quad k = 0, 1, 2, 3$$

$$\cos\left(\frac{2(0)\pi}{4}\right) + j \sin\left(\frac{2(0)\pi}{4}\right) = 1$$

$$\cos\left(\frac{2(1)\pi}{4}\right) + j \sin\left(\frac{2(1)\pi}{4}\right) = j$$

$$\cos\left(\frac{2(2)\pi}{4}\right) + j \sin\left(\frac{2(2)\pi}{4}\right) = -1$$

$$\cos\left(\frac{2(3)\pi}{4}\right) + j \sin\left(\frac{2(3)\pi}{4}\right) = -j$$

$$(1)^4 = (j)^4 = (-1)^4 = (-j)^4 = 1.$$

1.5 Summary

- The amplitude profiles of most naturally occurring signals are arbitrary.
- Signals have to be decomposed in terms of well-defined basis signals for their easier interpretation and manipulation.
- The most often used basis signals are the impulse and the sinusoid.

- The impulse has its strength concentrated at a point, while the sinusoid is an everlasting signal.
- The unit-step signal, which is related to the impulse, is also often used in the representation and manipulation of signals.
- The sinusoidal signal is a linear combination of the well-known trigonometric cosine and sine functions.
- While physical devices generate the sinusoidal signals, a mathematically equivalent form, the complex exponential, is often used in signal and system analysis due to its compact form and ease of manipulation.
- The sinusoidal signal is most often used to represent and manipulate signals, since it provides easier interpretation of the signals and fast processing of operations.
- Signal classification aids the selection of the most suitable transform for its representation and manipulation.
- Signal operations, such as shifting and scaling, are often used in signal manipulation, in addition to arithmetic operations.
- The complex number system is an extension of the real number system and it provides significant advantages in the analysis of signals and systems. A complex number is an ordered pair of real numbers, a two-element vector. More efficient processing is obtained, when closely related quantities are represented in a vector form.

Exercises

1.1 Evaluate the summation.

1.1.1 $\sum_{n=-\infty}^{\infty} 3^n \delta(n)$.

* 1.1.2 $\sum_{n=2}^{\infty} (-0.5)^{(n-1)} \delta(n)$.

1.1.3 $\sum_{n=0}^{\infty} 2^n \delta(n+2)$.

1.1.4 $\sum_{n=-\infty}^{\infty} (0.25)^n \delta(n+1)$.

1.1.5 $\sum_{n=-\infty}^{\infty} (-2)^{(n+1)} \delta(n-3)$.

1.2 Express the signal as a linear combination of scaled and shifted unit-impulses.
1.2.1 $\{x(0) = -2, x(1) = 1, x(2) = -1, x(3) = 3\}$ and $x(n) = 0$ otherwise.
1.2.2 $\{x(0) = 3, x(-1) = 1, x(2) = -2, x(-3) = 4\}$ and $x(n) = 0$ otherwise.
* 1.2.3 $\{x(0) = -2, x(-2) = 3, x(2) = -1, x(-3) = 3\}$ and $x(n) = 0$ otherwise.
1.2.4 $\{x(0) = 3, x(-1) = 1, x(3) = -2, x(4) = 2\}$ and $x(n) = 0$ otherwise.

1.3 Assume that the impulse is approximated by a rectangular pulse, centered at $t = 0$, of width $2a$ and height $\frac{1}{2a}$. Using this quasi-impulse, the signal $x(t)$ is sampled. What are the sample values of $x(t)$ at $t = 0$ with $a = 1$, $a = 0.1$, $a = 0.01$, $a = 0.001$, and $a = 0$.

1.3.1 $x(t) = 2e^{-2t}$.

1.3.2 $x(t) = 3\cos(2\pi t)$.

* 1.3.3 $x(t) = -2\sin\left(t + \frac{\pi}{3}\right)$.

1.3.4 $x(t) = \cos\left(4t + \frac{\pi}{6}\right)$.

1.3.5 $x(t) = |t|$.

1.4 Evaluate the integral.

1.4.1 $\int_{-\infty}^{\infty} \sin(t)\delta(2t - 4)dt$.

1.4.2 $\int_{-\infty}^{\infty} \cos(2t)\delta\left(\frac{1}{2}t - 3\right)dt$.

1.4.3 $\int_{3}^{\infty} e^{t}\delta\left(\frac{1}{3}t + 2\right)dt$.

* 1.4.4 $\int_{-\infty}^{3} e^{-t}\delta\left(-\frac{2}{5}t - 2\right)dt$.

1.4.5 $\int_{-\infty}^{2} \delta(-3t - 2)dt$.

1.5 Define the discrete signal given by a linear combination of scaled and shifted unit-step signals.

* 1.5.1 $x(n) = (2u(-n - 2) - 2u(-n + 1))$.

1.5.2 $x(n) = (-3u(n - 2) + 3u(-n + 3))$.

1.5.3 $x(n) = (4u(n + 2) - 4u(n))$.

1.6 Express the sinusoid in rectangular form. Get back the polar form from the rectangular form and verify that the given sinusoid is obtained. Find the sample values of the sinusoid for $n = 0$ to $n = 7$. What are the amplitude, cyclic frequency, and the phase.

1.6.1 $x(n) = 2\cos\left(\frac{\pi}{4}n - \frac{\pi}{3}\right)$

1.6.2 $x(n) = \cos\left(\frac{\pi}{4}n + \frac{\pi}{4}\right)$

1.6.3 $x(n) = -3\sin\left(\frac{\pi}{4}n - \frac{\pi}{2}\right)$

1.6.4 $x(n) = 4\cos\left(\frac{\pi}{4}n + \frac{\pi}{2}\right)$.

* 1.6.5 $x(n) = 3\cos\left(\frac{\pi}{4}n + \frac{\pi}{6}\right)$

1.7 Determine the sinusoid $c(n)$ that is the sum of the pair of given sinusoids, $a(n)$ and $b(n)$, $c(n) = a(n) + b(n)$. Find the samples of the sinusoid $c(n)$ for $n = 0$ to $n = 7$ and verify that the samples are the same as the sum of the samples of $a(n)$ and $b(n)$, $a(n) + b(n)$.

1.7.1 $a(n) = 2\cos\left(\frac{\pi}{4}n + \frac{\pi}{4}\right)$, $b(n) = -\sin\left(\frac{\pi}{4}n + \frac{\pi}{3}\right)$

1.7.2 $a(n) = 3\cos\left(\frac{\pi}{4}n\right)$, $b(n) = 2\sin\left(\frac{\pi}{4}n\right)$

* 1.7.3 $a(n) = 4\cos\left(\frac{\pi}{4}n + \frac{\pi}{3}\right)$, $b(n) = \cos\left(\frac{\pi}{4}n - \frac{\pi}{4}\right)$

1.7.4 $a(n) = 3\cos\left(\frac{\pi}{4}n\right)$, $b(n) = \cos\left(\frac{\pi}{4}n\right)$

1.7.5 $a(n) = \sin\left(\frac{\pi}{4}n\right)$, $b(n) = 3\sin\left(\frac{\pi}{4}n\right)$

1.8 Is the sinusoid periodic? If so, what is its period?

1.8.1 $x(n) = \cos\left(0.2\pi n - \frac{\pi}{4}\right)$

1.8.2 $x(n) = \cos\left(\sqrt{5}\pi n + \frac{\pi}{3}\right)$

1.8.3 $x(n) = \cos\left(\frac{2\pi}{\sqrt{3}}n + \frac{\pi}{2}\right)$

* 1.8.4 $x(n) = \sin\left(\frac{5\pi}{15}n\right)$

1.8.5 $x(n) = 1 + \cos\left(\frac{\pi}{4}n\right) + \cos(n)$

1.9 Express the signal in terms of complex exponentials. Find the samples of the two forms and verify that they are the same.

1.9.1 $x(n) = \cos(\pi n)$

1.9.2 $x(n) = \cos(2\pi n)$

1.9.3 $x(n) = \cos\left(\frac{2\pi}{8}n + \frac{\pi}{3}\right)$

* 1.9.4 $x(n) = 2\sin\left(\frac{2\pi}{8}n - \frac{\pi}{6}\right)$

1.9.5 $x(n) = 2\cos\left(\frac{2\pi}{8}n\right) + j3\sin\left(\frac{2\pi}{8}n\right)$

1.10 Find the samples of one period of the periodic signal $x(n)$, starting from $n = 0$. Find the kth sample using the mod function.

1.10.1 $x(n) = \cos\left(\frac{2\pi}{8}n + \frac{\pi}{3}\right), \quad k = -11$

1.10.2 $x(n) = \sin\left(\frac{2\pi}{8}n - \frac{\pi}{3}\right), \quad k = 10$

* 1.10.3 $x(n) = \cos\left(\frac{2\pi}{8}n + \frac{\pi}{6}\right), \quad k = -4$

1.10.4 $x(n) = \sin\left(\frac{2\pi}{8}n - \frac{\pi}{6}\right), \quad k = 22$

1.10.5 $x(n) = \cos\left(\frac{2\pi}{8}n - \frac{\pi}{4}\right), \quad k = -27$

1.11 Decompose the signal $x(n)$ into its even and odd components. Find the samples of the signal and its components for $n = -3, -2, -1, 0, 1, 2, 3$. Sum the samples of the components and verify that they are the same as those of the given signal.

1.11.1 $x(0) = -2, x(1) = -2, x(2) = 2, x(3) = 2$, and $x(n) = 0$ otherwise.

1.11.2 $x(n) = 2\sin\left(\frac{\pi}{4}n - \frac{\pi}{3}\right)$

1.11.3 $x(n) = (-0.8)^n u(n)$

* 1.11.4 $x(n) = u(n - 2)$

1.11.5 $x(n) = n u(-n)$

1.12 Is the signal circularly symmetric? If so, is it odd-symmetric or even-symmetric?

1.12.1 $x(0) = -2, x(1) = -1, x(2) = 2, x(3) = 3$.

1.12.2 $x(0) = -2, x(1) = 1, x(2) = 2, x(3) = 1$.

* 1.12.3 $x(0) = 3, x(1) = 2, x(2) = 1, x(3) = 0$.

1.12.4 $x(0) = 7, x(1) = -2, x(2) = 2, x(3) = -2$.

1.12.5 $x(0) = 0, x(1) = 4, x(2) = 0, x(3) = -4$.

1.13 Is $x(n)$ an energy signal, a power signal, or neither? Find the energy of the energy signal. Find the average power of the power signal.

1.13.1 $x(0) = 3, x(-1) = 1, x(-2) = -2, x(-3) = 4$, and $x(n) = 0$ otherwise.

1.13.2 $x(n) = 3^n$.

* 1.13.3 $x(n) = 2\sin\left(\frac{\pi n}{2} - \frac{\pi}{3}\right)$.

1.13.4 $x(n) = 2$.

1.13.5 $x(n) = \frac{1}{n}u(n - 1)$.

1.14 Given the values of the constants a and k, and the sinusoid $x(n)$, find the expression for the sinusoid $x(an + k)$. Find the sample values of the two sinusoids for one cycle, starting from $n = 0$. For expansion, assume that the undefined sample values are zero.

1.14.1 $x(n) = \sin\left(\frac{2\pi}{8}n - \frac{\pi}{6}\right)$, $a = -2$, $k = -1$.

1.14.2 $x(n) = \cos\left(\frac{2\pi}{8}n + \frac{\pi}{3}\right)$, $a = \frac{1}{2}$, $k = 2$.

* 1.14.3 $x(n) = \sin\left(\frac{2\pi}{8}n - \frac{\pi}{3}\right)$, $a = \frac{1}{2}$, $k = -8$.

1.15 Given the values of the constants a and k, and the signal $x(n)$, find the expression for the signal $x(an + k)$. Find the sample values of the two signals for $n = \{-3, -2, -1, 0, 1, 2, 3\}$. For expansion, assume that the undefined sample values are zero.

* 1.15.1 $x(0) = 3$, $x(1) = 1$, $x(2) = -4$, $x(3) = 2$, and $x(n) = 0$ otherwise. $a = 2$, $k = -1$.

1.15.2 $x(n) = (0.8)^n$. $a = -2$, $k = 1$.

1.15.3 $x(n) = (0.6)^n u(n)$. $a = \frac{1}{3}$, $k = -2$.

1.16 Simplify each expression to get it in the form $a + jb$.

1.16.1 $(7 - j3) + (1 + j2)$

* 1.16.2 $(1 + j3) - (3 - j1)$

1.16.3 $(2 - j4) + (3 + 2\sqrt{-4})$

1.17 Simplify each expression to get it in the form $a + jb$.

1.17.1 $(-7 - j3)(1 + j2)$

1.17.2 $(1 + j3) - (-3 - j1)$

* 1.17.3 $\frac{2-j4}{1-j3} - \frac{1-j4}{2-j3}$

1.18 Simplify each expression in polar form to get it in the form $re^{j\theta}$.

* 1.18.1 $(-7 - j3)(1 + j2)$

1.18.2 $\frac{1+j3}{-3-j1}$

1.18.3 $\frac{2-j4}{1-j3}$

1.19 Simplify each expression to get it in the form $a + jb$. Find the conjugate of the expression and simplify that also in the form $a + jb$.

1.19.1 $(1 - j3) + (3 + j2)$

1.19.2 $(2 + j3) - (3 - j1)$

* 1.19.3 $(2 - j4) + (3 + j2)$

1.20

1.20.1 Solve $x^2 - 1.5x + 1 = 0$ for x. Verify the answer by multiplying the resulting factors.

1.20.2 Using De Moivre's theorem, express $(-1 - j\sqrt{3})^5$ in the form $a + jb$. Verify your results by finding the 5th roots of $(-1 - j\sqrt{3})^5$.

* 1.20.3 Find the complex 5th roots of unity. Verify your results by raising each root to its fifth power.

Chapter 2
The Discrete Fourier Transform

Transform means change in form. For example, we use the product rule to change the form of the problem of finding the derivative of the product of two functions, so that its derivative can be found easily. The idea of a transform, in signal analysis, is to approximate practical signals, which usually have arbitrary amplitude profiles and difficult to analyze in their original form, adequately in terms of well-defined basis signals, such as the cosine and sine signals. Then, it is easier to interpret, analyze, transmit, and store them. In the representation of a function in the form $x(t)$, variable t is the independent variable in a certain domain, designated as the time domain. Since the time is the independent variable frequently (but not always), it is named as the time domain. In the representation of a function in the form $X(k)$, variable k, which represents the frequency index of a frequency component, is the independent variable in the frequency domain. Either representation completely specifies the given function. While the frequency-domain representation of signals and systems looks unnatural, it is convenient and efficient in signal and system analysis. For example, a high-quality recording of a music signal requires frequency components in the range 0–20 kHz and the corresponding recording devices, amplifiers, and speakers should have a good frequency response in that frequency range.

Fourier analysis is an indispensable representation of signals and systems in science and engineering. There are many other representations of various entities. Infinite points in a plane are represented by their x-axis and y-axis coordinates. A place on earth is represented by its longitude and latitude. Any color can be specified by its red, green, and blue components. With all the mathematics, Fourier analysis looks complex and difficult. But, it is not so. It is similar to finding the amount of a set of coins. Let us say, we have a box of 1 cent, 10 cent, and 50 cent coins. We can take one by one and add its value to a partial sum. We find the amount after the values of all the coins are added. An alternate way is to decompose the coins into the three denominations and count the number of coins in each. Multiplying the number of different coins by their value and adding results in the amount. In Fourier analysis, we decompose a signal in terms of its sinusoidal components. This decomposition

© Springer Nature Singapore Pte Ltd. 2018
D. Sundararajan, *Fourier Analysis—A Signal Processing Approach*,
https://doi.org/10.1007/978-981-13-1693-7_2

enables the determination of the output of a system faster than other methods, in addition to other advantages. This method is faster due to the orthogonality of the sinusoids and the availability of fast algorithms for its practical realization. While the principle behind Fourier analysis is simple, it is the mathematical details that make it look complex and difficult. With sufficient practice, both paper-and-pencil and programming, one can become proficient in the indispensable Fourier analysis.

Fourier analysis has four different versions to suit the different types of signals. The DFT is the only version in which signals are represented in finite and discrete form in both the domains. Further, fast algorithms are available for its implementation. Therefore, it is most often used in practice. It can approximate the other versions of the Fourier analysis adequately. The DFT uses only a finite number of discrete sinusoids to represent the signals. Therefore, it is easier to understand and the visualization of the reconstruction of the waveforms is much simpler. The other versions involve infinite sums or integrals requiring a detailed study of the convergence properties. As it is the simplest to study the concepts of the Fourier analysis and often used in practice, the DFT is presented in this chapter followed by other versions in later chapters.

2.1 The Exponential Function

Fourier analysis is a representation of arbitrary signals in terms of sinusoidal (or its equivalent complex exponential) basis signals. This representation is similar to that of the logarithmic function. An exponential function is of the form

$$x(n) = b^n$$

where the base $b \neq 1$ is a positive constant and the exponent n is the independent variable. An important property of the exponential function is that

$$b^m b^n = b^{m+n}$$

Therefore, the exponential representation reduces a multiplication operation into a relatively simpler addition operation. Similarly, the Fourier representation reduces a convolution operation into a relatively simpler multiplication operation. Considering the importance of the convolution operation in signal and system analysis, this single advantage alone is sufficient enough to make the Fourier analysis an indispensable tool in science and engineering.

While the detailed description of Fourier analysis is the topic of the book, let us continue with the more familiar exponential function. Let us say that we want to multiply 8 by 16. Assume that a table is available to find the exponent of the exponential function with base 2 of any number. Then, $8 = 2^3$ and $16 = 2^4$ and

$$8 \times 16 = 2^3 2^4 = 2^{3+4} = 2^7 = 128$$

The logarithm n of a positive real number x to the base b is the exponent of the exponential function $x = b^n$. The inverse of the exponential function $x = b^n$ ($b \neq 1$) is the logarithmic function with base b.

$$\log_b x = n \text{ if and only if } x = b^n$$

Some examples are

$$\log_2 8 = 3 \text{ since } 8 = 2^3$$
$$\log_{10} 100 = 2 \text{ since } 100 = 10^2$$
$$\log_e 1 = 0 \text{ since } 1 = e^0, \quad e = \lim_{n \to \infty} \left(1 + \frac{1}{n}\right)^n \approx 2.71828$$

The points to be noted are:

1. Numbers can be represented in their exponential form.
2. In that form of representation, the multiplication operation is reduced to addition operation.

2.2 The Complex Exponential Function

A complex exponential function is an exponential function with a complex exponent. The complex exponential function of the form $e^{j\theta}$ is our primary interest. The exponent is purely imaginary. A complex number is an ordered pair of real numbers and $a + jb$ is one form of its representation. The real part is a, the imaginary part is b and j is the imaginary unit. The points to be noted are:

1. Signals can be represented in their complex exponential form.
2. In this form of representation, the convolution operation is reduced to multiplication operation.

Except for the use of the complex exponentials as the basis function, Fourier analysis is similar, in principle, to the use of real exponentials to reduce the multiplication operation into an addition operation, as presented in the last section. However, a lot of details has to be taken into account in the use of Fourier analysis, as presented in the rest of the book.

2.2.1 Euler's Formula

Oscillators and other physical systems generate waveforms those are a combination of sinusoidal waveforms. However, it is found that the mathematically equivalent complex exponential is found convenient in the analysis of signals. The Euler's formula gives the relation between the sinusoidal functions and the complex exponential

$$e^{j\theta} = \cos\theta + j\sin\theta$$

The Maclaurin series of a function $x(t)$ is given by

$$x(t) = x(0) + \frac{t}{1!}\dot{x}(0) + \frac{t^2}{2!}\ddot{x}(0) + \cdots$$

The dot notation for differentiation places a dot over the dependent variable. That is, if $x(t)$ is a function of t, then the derivative of $x(t)$ with respect to t is $\dot{x}(t)$. $\ddot{x}(t)$ is the second derivative. For $\cos(\theta)$ and $\sin(\theta)$, the series expansions are

$$\cos(\theta) = 1 - \frac{\theta^2}{2!} + \frac{\theta^4}{4!} - \cdots \quad \text{and} \quad \sin(\theta) = \theta - \frac{\theta^3}{3!} + \frac{\theta^5}{5!} - \cdots$$

Since the derivative of $e^{j\theta}$ with respect to $j\theta$ is itself, for any order, the series expansion is

$$e^{j\theta} = 1 + (j\theta) + \frac{(j\theta)^2}{2!} + \frac{(j\theta)^3}{3!} + \frac{(j\theta)^4}{4!} + \cdots$$

$$= \left(1 - \frac{\theta^2}{2!} + \frac{\theta^4}{4!} - \cdots\right) + j\left(\theta - \frac{\theta^3}{3!} + \frac{\theta^5}{5!} - \cdots\right)$$

$$= \cos\theta + j\sin\theta$$

2.2.2 Real Sinusoids in Terms of Complex Exponentials

Solving equations

$$e^{j\theta} = \cos\theta + j\sin\theta \quad \text{and} \quad e^{-j\theta} = \cos\theta - j\sin\theta,$$

sinusoids can be expressed in terms of exponentials as

$$\cos\theta = \frac{e^{j\theta} + e^{-j\theta}}{2} \quad \text{and} \quad \sin\theta = \frac{e^{j\theta} - e^{-j\theta}}{j2}$$

A term such as $7x^2$ is a monomial in a variable x, in which 7 is its coefficient and 2 is its degree. A polynomial is a sum of a number of monomials. The general form of a polynomial in the variable x with degree N is

$$a_N x^N + a_{N-1} x^{N-1} + \cdots + a_2 x^2 + a_1 x + a_0$$

in which a_i are the coefficients and $a_N \neq 0$. The coefficients of a complex polynomial are complex numbers and the variable is also complex. The representation of a signal, using its DFT coefficients, is in this form, with $x = e^{j\theta}$.

2.3 The DFT and the IDFT

Figure 2.1a shows a discrete periodic waveform with period 4 by dots. The corresponding continuous waveform is shown in thin line for clarity only. The independent variable n often represents time, and therefore, the graph n versus $x(n)$, the amplitude of the signal at the instant n, is called the time-domain representation, although n may be other than time such as distance. Signals naturally occur in the time-domain form. Figure 2.1b shows the representation of the same signal in the frequency domain, called the spectrum. The graph k versus $X(k)$ is the DFT representation of the signal. It shows the complex amplitudes, scaled by 4, of the constituent complex exponentials of the signal. For example, $X(0) = 8$ indicates that the amplitude of the DC component is $8/4 = 2$, as shown in Fig. 2.1c by the dash-dot line.

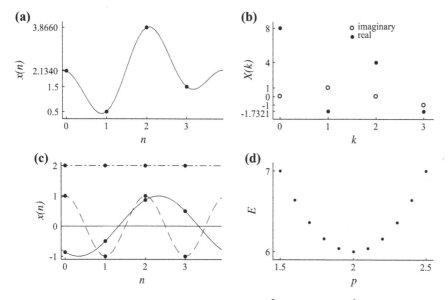

Fig. 2.1 a A discrete periodic waveform, $x(n) = 2 - \cos(\frac{2\pi}{4}n - \frac{\pi}{6}) + \cos(2\frac{2\pi}{4}n)$, with period 4 samples and **b** its frequency-domain representation; **c** the frequency components of the waveform in **a**; **d** the square error in approximating the waveform in **a** using only the DC component with different amplitudes

Similarly, $X(2) = 4$ indicates that the amplitude of the frequency component with frequency index 2 is $4/4 = 1$, as shown in Fig. 2.1c by the dashed line. Coefficients $X(1) = -\sqrt{3} + j$ and $X(3) = -\sqrt{3} - j$ represent the frequency component with index 1, as shown in Fig. 2.1c by the continuous line.

$$\frac{(-\sqrt{3} + j)e^{j\frac{2\pi}{4}n} + (-\sqrt{3} - j)e^{j3\frac{2\pi}{4}n}}{4} = \cos\left(\frac{2\pi}{4}n + \frac{5\pi}{6}\right) = -\cos\left(\frac{2\pi}{4}n - \frac{\pi}{6}\right)$$

Both the time- and frequency-domain representations are equivalent and unique. The DFT finds $X(k)$ from $x(n)$, called the forward transform. The IDFT finds $x(n)$ from $X(k)$, called the inverse transform. The transformation may be regarded as the change of the independent variables n and k.

The discrete periodic function $x(n)$ with period 4, shown in Fig. 2.1a, is given by

$$x(n) = 2 - \cos\left(\frac{2\pi}{4}n - \frac{\pi}{6}\right) + \cos\left(2\frac{2\pi}{4}n\right)$$

in terms of trigonometric functions. Using Euler's formula, we get

$$
\begin{aligned}
x(n) &= 2e^{j0\frac{2\pi}{4}n} - 0.5e^{j\left(\frac{2\pi}{4}n - \frac{\pi}{6}\right)} + e^{j2\frac{2\pi}{4}n} - 0.5e^{-j\left(\frac{2\pi}{4}n - \frac{\pi}{6}\right)} \\
&= 2e^{j0\frac{2\pi}{4}n} + (-0.25\sqrt{3} + j0.25)e^{j\frac{2\pi}{4}n} + e^{j2\frac{2\pi}{4}n} + (-0.25\sqrt{3} - j0.25)e^{-j\frac{2\pi}{4}n} \\
&= 2e^{j0\frac{2\pi}{4}n} + (-0.25\sqrt{3} + j0.25)e^{j\frac{2\pi}{4}n} + e^{j2\frac{2\pi}{4}n} + (-0.25\sqrt{3} - j0.25)e^{j3\frac{2\pi}{4}n}
\end{aligned}
$$

$$(2.1)$$

Due to periodicity,

$$e^{j3\frac{2\pi}{4}n} = e^{j(4-1)\frac{2\pi}{4}n} = e^{j4\frac{2\pi}{4}n}e^{-j1\frac{2\pi}{4}n} = e^{-j\frac{2\pi}{4}n}$$

Equation (2.1) is the representation of $x(n)$ by a complex exponential polynomial of order 3.

For understanding Fourier analysis, we have started with a problem with no unknown variables. Fourier analysis problem is to find the coefficients $X(k)$ in the the complex exponential polynomial representation of a time-domain function $x(n)$, for example with $N = 4$ coefficients,

$$x(n) = X(0)e^{j0\frac{2\pi}{4}n} + X(1)e^{j\frac{2\pi}{4}n} + X(2)e^{j2\frac{2\pi}{4}n} + X(3)e^{j3\frac{2\pi}{4}n}$$

For the given $N = 4$, $x(n)$ and the exponentials are known. The Fourier synthesis problem is to find $x(n)$, given $X(k)$ and the exponentials.

The four samples over one period are obtained from the equations for $n = 0, 1, 2, 3$.

$$\{x(0) = (3 - 0.5\sqrt{3}), x(1) = 0.5, x(2) = (3 + 0.5\sqrt{3}), x(3) = 1.5\}$$

For example, letting $n = 1$, we get

$$x(1) = 2 - \cos\left(\frac{2\pi}{4} - \frac{\pi}{6}\right) + \cos\left(2\frac{2\pi}{4}1\right) = 2 - 0.5 - 1 = 0.5$$

Figure 2.1b shows the frequency-domain representation of the waveform in (a), scaled by 4.

$$\{X(0) = 8, X(1) = (-\sqrt{3} + j), X(2) = 4, X(3) = (-\sqrt{3} - j)\}$$

It shows the complex amplitudes of its constituent complex exponentials (Eq. (2.1)) scaled by 4 (the period). While the form

$$x(n) = 2 - \cos\left(\frac{2\pi}{4}n - \frac{\pi}{6}\right) + \cos\left(2\frac{2\pi}{4}n\right)$$

itself is a frequency-domain representation of $x(n)$, expanding the middle term we get

$$x(n) = 2 - 0.5\sqrt{3}\cos\left(\frac{2\pi}{4}n\right) + \cos\left(2\frac{2\pi}{4}n\right) - 0.5\sin\left(\frac{2\pi}{4}n\right)$$

This form uses the rectangular form of the sinusoid, while the earlier form uses the polar form. Fourier representation expresses a function in terms of sinusoidal functions. The last two equations are trigonometric polynomials. The frequency indices of the terms are all integer multiple of the smallest one, called the fundamental. The other terms are called the harmonics.

Given the samples of a waveform, the DFT is a tool to find the coefficients of its constituent frequency components. The decomposition of a waveform is obtained using the orthogonality property of the complex exponentials and sinusoids. Orthogonality of two complex sequences is that the sum of pointwise products of a sequence and the conjugate of the other sequence is zero or a constant over a specified interval. For the two complex exponential signals $e^{j\frac{2\pi}{N}kn}$ and $e^{j\frac{2\pi}{N}ln}$ over a period of N samples, the orthogonality condition is given by

$$\sum_{n=0}^{N-1} e^{j\frac{2\pi}{N}(k-l)n} = \begin{cases} N & \text{for } k = l \\ 0 & \text{for } k \neq l \end{cases}$$

where $k, l = 0, 1, \ldots, N - 1$. For $k = l$, the summation evaluates to N. Using the closed-form expression for the sum of a geometric progression, we get

$$\sum_{n=0}^{N-1} e^{j\frac{2\pi}{N}(k-l)n} = \frac{1 - e^{j2\pi(k-l)}}{1 - e^{j\frac{2\pi(k-l)}{N}}} = 0, \text{ for } k \neq l$$

This is also obvious from the fact that the sine and cosine waveforms are symmetrical about the x-axis. The sum of the equidistant samples of a cosine or sine waveform,

Table 2.1 Samples of complex exponentials $e^{j\frac{2\pi}{4}kn}$, $k, n = 0, 1, 2, 3$ and their conjugates $e^{-j\frac{2\pi}{4}kn}$

Frequency index k	Sample index n				Sample index n			
	0	1	2	3	0	1	2	3
0	1	1	1	1	1	1	1	1
1	1	j	-1	$-j$	1	$-j$	-1	j
2	1	-1	1	-1	1	-1	1	-1
3	1	$-j$	-1	j	1	j	-1	$-j$

over an integral number of periods, with a nonzero frequency index is always zero. The orthogonality property can be verified using the samples of $e^{j\frac{2\pi}{4}kn}$ and $e^{-j\frac{2\pi}{4}kn}$ given in Table 2.1. The sum of the pointwise product of the values of each row in the left table with the corresponding values of the rows in the right table yields zero, except for the same row. For example, the sum of pointwise product of the last row $\{1, -j, -1, j\}$ in the left table with the corresponding row in the right table yields 4. In the three other cases, the sum is zero.

If we multiply both sides of Eq. (2.1) by $e^{-jk\frac{2\pi}{4}n}$, $k = 0, 1, 2, 3$ and sum the products from $n = 0$ to $n = 3$, we get

$$\sum_{n=0}^{3} x(n)e^{-jk\frac{2\pi}{4}n} =$$

$$\sum_{n=0}^{3}(2e^{j0\frac{2\pi}{4}n} + (-0.25\sqrt{3} + j0.25)e^{j\frac{2\pi}{4}n} + e^{j2\frac{2\pi}{4}n} + (-0.25\sqrt{3} - j0.25)e^{j3\frac{2\pi}{4}n})e^{-jk\frac{2\pi}{4}n}$$

Letting $k = 0, 1, 2, 3$ in the last equation in turn, we get the DFT coefficients.

$$X(0) = 2\sum_{n=0}^{3} e^{j0\frac{2\pi}{4}n}e^{-j0\frac{2\pi}{4}n} + 0 + 0 + 0 = (2)(4) = 8, \ k = 0$$

$$X(1) = 0 + (-0.25\sqrt{3} + j0.25)\sum_{n=0}^{3} e^{j\frac{2\pi}{4}n}e^{-j\frac{2\pi}{4}n} + 0 + 0$$

$$= (-0.25\sqrt{3} + j0.25)(4) = -\sqrt{3} + j, \ k = 1$$

$$X(2) = \sum_{n=0}^{3} 0 + 0 + e^{j2\frac{2\pi}{4}n}e^{-j2\frac{2\pi}{4}n} + 0 = 4, \ k = 2$$

$$X(3) = 0 + 0 + 0 + (-0.25\sqrt{3} - j0.25)\sum_{n=0}^{3} e^{j3\frac{2\pi}{4}n}e^{-j3\frac{2\pi}{4}n}$$

$$= (-0.25\sqrt{3} - j0.25)(4) = -\sqrt{3} - j, \ k = 3$$

We get the coefficients of the frequency components of $x(n)$, scaled by a factor of 4, the length of the sequence.

2.3.1   The DFT and the IDFT

The DFT of a sequence $x(n)$ of length N is defined as

$$X(k) = \sum_{n=0}^{N-1} x(n)e^{-jk\frac{2\pi}{N}n}, \quad k = 0, 1, 2, \ldots, N-1 \tag{2.2}$$

The complex exponential is also written in its abbreviated form.

$$e^{-j\frac{2\pi}{N}kn} = W_N^{kn}, \quad W = e^{-j\frac{2\pi}{N}}$$

The sum of the complex exponentials $e^{jk\frac{2\pi}{N}n}$ multiplied by their respective DFT coefficients gets back the time-domain waveform. Therefore, the inverse DFT (IDFT) equation is defined as

$$x(n) = \frac{1}{N} \sum_{k=0}^{N-1} X(k)e^{jk\frac{2\pi}{N}n}, \quad n = 0, 1, 2, \ldots, N-1 \tag{2.3}$$

As the DFT coefficients have been scaled by N, the factor $\frac{1}{N}$ appears in the IDFT definition. There are other ways to take care of this constant.

Let us prove that the DFT and IDFT definitions form a transform pair. Substituting the definition for $X(k)$ in the definition of the IDFT, we get

$$x(n) = \frac{1}{N} \sum_{k=0}^{N-1} e^{jk\frac{2\pi}{N}n} \sum_{l=0}^{N-1} x(l)e^{-jk\frac{2\pi}{N}l}$$

Changing the order of summation, we get

$$x(n) = \frac{1}{N} \sum_{l=0}^{N-1} x(l) \sum_{k=0}^{N-1} e^{jk\frac{2\pi}{N}(n-l)} = x(n)$$

since

$$\sum_{k=0}^{N-1} e^{jk\frac{2\pi}{N}(n-l)} \begin{cases} N & \text{for } n = l \\ 0 & \text{for } n \neq l \end{cases}$$

Center-Zero Format of the DFT and IDFT

The center-zero format places $x(0)$ and $X(0)$ values approximately in the middle, while these values are placed in the beginning in the usual format, as can be seen in Eqs. (2.2) and (2.3). For N even, the DFT and IDFT definitions, in the center-zero format, are

$$X(k) = \sum_{n=(-\frac{N}{2})}^{\frac{N}{2}-1} x(n)e^{-jk\frac{2\pi}{N}n}, \quad k = -\left(\frac{N}{2}\right), -\left(\frac{N}{2}-1\right), \ldots, \frac{N}{2}-1 \quad (2.4)$$

$$x(n) = \frac{1}{N} \sum_{k=-(\frac{N}{2})}^{\frac{N}{2}-1} X(k)e^{jk\frac{2\pi}{N}n}, \quad n = -\left(\frac{N}{2}\right), -\left(\frac{N}{2}-1\right), \ldots, \frac{N}{2}-1 \quad (2.5)$$

This form is convenient in some derivations. When the spectrum is displayed in this format, the interpretation of the spectral features is easier. The conversion of the data or spectrum from one format to another involves circular shift by half the number of samples. The spectrum

$$\{X(0) = 2, X(1) = 2 - j\sqrt{2}, X(2) = 3, X(3) = 2 + j\sqrt{2}\}$$

in center-zero format is

$$\{X(-2) = 3, X(-1) = 2 + j\sqrt{2}, X(0) = 2, X(1) = 2 - j\sqrt{2}\}$$

In the trigonometric form, a signal is represented, at a given frequency, by two functions (cosine and sine) and the two associated coefficients. The exponential form has one function and one coefficient. Because of the compactness of the exponential form of the Fourier representation of signals and the ease of manipulation of expressions involving exponential functions, the exponential form is mostly used in the analysis. However, physical systems, such as oscillators, generate sinusoidal output. Further, visualization of the reconstruction of the waveforms and other operations is easier in the trigonometric form of the Fourier representation of signals. Therefore, both the forms are important and they are related by the Euler's formula. The exponential form is redundant by a factor of 2 for real signals. But, this redundancy can be almost eliminated in the implementation of the algorithms.

2.3.2   The Criterion of Approximation

In Fourier analysis, the signal is reconstructed with respect to the least squares error criterion. The DFT $X(k)$ of the sequence $x(n)$ of period N is also periodic of period N. Then, with N even,

$$x(n) = \frac{1}{N} \sum_{k=-\left(\frac{N}{2}\right)}^{\frac{N}{2}-1} X(k)e^{jk\frac{2\pi}{N}n}, \quad n = -\left(\frac{N}{2}\right), -\left(\frac{N}{2}-1\right), \ldots, \frac{N}{2}-1$$

Let $x_a(n)$ be an approximation of $x(n)$ using $M < N$ frequency coefficients $X_a(k)$, where M is odd. Then,

$$x_a(n) = \frac{1}{N} \sum_{k=-\left(\frac{M-1}{2}\right)}^{\frac{M-1}{2}} X_a(k)e^{jk\frac{2\pi}{N}n}, \quad n = -\left(\frac{N}{2}\right), -\left(\frac{N}{2}-1\right), \ldots, \frac{N}{2}-1$$

Now, the least squares error is

$$E = \sum_{n=-\left(\frac{N}{2}\right)}^{\frac{N}{2}-1} \left(x(n) - \frac{1}{N} \sum_{k=-\left(\frac{M-1}{2}\right)}^{\frac{M-1}{2}} X_a(k)e^{jk\frac{2\pi}{N}n}\right)^2$$

Differentiating this expression with respect to $X_a(k_0)$ and setting the derivative equal to zero, we get

$$\sum_{n=-\left(\frac{N}{2}\right)}^{\frac{N}{2}-1} \left(x(n) - \frac{1}{N} \sum_{k=-\left(\frac{M-1}{2}\right)}^{\frac{M-1}{2}} X_a(k)e^{jk\frac{2\pi}{N}n}\right) e^{jk_0\frac{2\pi}{N}n} = 0$$

Interchanging the summation order, we get

$$\frac{1}{N} \sum_{k=-\left(\frac{M-1}{2}\right)}^{\frac{M-1}{2}} X_a(k) \sum_{n=-\left(\frac{N}{2}\right)}^{\frac{N}{2}-1} e^{j(k+k_0)\frac{2\pi}{N}n} = \sum_{n=-\left(\frac{N}{2}\right)}^{\frac{N}{2}-1} x(n)e^{jk_0\frac{2\pi}{N}n}$$

Since $(k + k_0)$ is an integer, the only nonzero term in the inner summation occurs when $k = -k_0$. Therefore, we get

$$X_a(-k_0) = \sum_{n=-\left(\frac{N}{2}\right)}^{\frac{N}{2}-1} x(n)e^{jk_0\frac{2\pi}{N}n}$$

Replacing k_0 by $-k$, we get

$$X_a(k) = \sum_{n=-\left(\frac{N}{2}\right)}^{\frac{N}{2}-1} x(n)e^{-jk\frac{2\pi}{N}n} = X(k)$$

This property also holds for complex signals.

2.3.3 The Matrix Form of the DFT and IDFT

For each k, the sum of pointwise product of $x(n)$ and $e^{-jk\frac{2\pi}{N}n}$, with n varying from 0 to $N - 1$, is computed in Eq. (2.2). The values of $e^{-jk\frac{2\pi}{N}n}$ are the conjugate samples of the basis function for that k. The sum of products can be expressed as a matrix formulation of the DFT definition. With $N = 2$, the DFT definition is

$$\begin{bmatrix} X(0) \\ X(1) \end{bmatrix} = \begin{bmatrix} 1 & 1 \\ 1 & -1 \end{bmatrix} \begin{bmatrix} x(0) \\ x(1) \end{bmatrix}$$

The IDFT is defined as

$$\begin{bmatrix} x(0) \\ x(1) \end{bmatrix} = \frac{1}{2} \begin{bmatrix} 1 & 1 \\ 1 & -1 \end{bmatrix} \begin{bmatrix} X(0) \\ X(1) \end{bmatrix}$$

The DFT of $\{x(0) = 2, x(1) = 3\}$ is $\{X(0) = 5, X(1) = -1\}$. It can be verified that IDFT gets back the time-domain samples.

With $N = 4$, the DFT definition is

$$\begin{bmatrix} X(0) \\ X(1) \\ X(2) \\ X(3) \end{bmatrix} = \begin{bmatrix} 1 & 1 & 1 & 1 \\ 1 & -j & -1 & j \\ 1 & -1 & 1 & -1 \\ 1 & j & -1 & -j \end{bmatrix} \begin{bmatrix} x(0) \\ x(1) \\ x(2) \\ x(3) \end{bmatrix}$$

Using vector and matrix quantities, the DFT definition is given by

$$X = Wx$$

where X is the coefficient vector, W is the transform matrix, and x is the input vector. The transform matrix W is

$$\begin{bmatrix} e^{-j0\frac{2\pi}{4}0} & e^{-j0\frac{2\pi}{4}1} & e^{-j0\frac{2\pi}{4}2} & e^{-j0\frac{2\pi}{4}3} \\ e^{-j1\frac{2\pi}{4}0} & e^{-j1\frac{2\pi}{4}1} & e^{-j1\frac{2\pi}{4}2} & e^{-j1\frac{2\pi}{4}3} \\ e^{-j2\frac{2\pi}{4}0} & e^{-j2\frac{2\pi}{4}1} & e^{-j2\frac{2\pi}{4}2} & e^{-j2\frac{2\pi}{4}3} \\ e^{-j3\frac{2\pi}{4}0} & e^{-j3\frac{2\pi}{4}1} & e^{-j3\frac{2\pi}{4}2} & e^{-j3\frac{2\pi}{4}3} \end{bmatrix} = \begin{bmatrix} e^{-j0} & e^{-j0} & e^{-j0} & e^{-j0} \\ e^{-j0} & e^{-j\frac{\pi}{2}} & e^{-j\pi} & e^{-j\frac{3\pi}{2}} \\ e^{-j0} & e^{-j\pi} & e^{-j2\pi} & e^{-j3\pi} \\ e^{-j0} & e^{-j\frac{3\pi}{2}} & e^{-j3\pi} & e^{-j\frac{9\pi}{2}} \end{bmatrix} = \begin{bmatrix} 1 & 1 & 1 & 1 \\ 1 & -j & -1 & j \\ 1 & -1 & 1 & -1 \\ 1 & j & -1 & -j \end{bmatrix}$$

The IDFT definition is

$$\begin{bmatrix} x(0) \\ x(1) \\ x(2) \\ x(3) \end{bmatrix} = \frac{1}{4} \begin{bmatrix} 1 & 1 & 1 & 1 \\ 1 & j & -1 & -j \\ 1 & -1 & 1 & -1 \\ 1 & -j & -1 & j \end{bmatrix} \begin{bmatrix} X(0) \\ X(1) \\ X(2) \\ X(3) \end{bmatrix}$$

Concisely,

$$x = \frac{1}{4}W^{-1}X = \frac{1}{4}(W^*)X$$

The inverse and forward transform matrices are orthogonal. That is,

$$
\frac{1}{4}
\begin{bmatrix}
1 & 1 & 1 & 1 \\
1 & j & -1 & -j \\
1 & -1 & 1 & -1 \\
1 & -j & -1 & j
\end{bmatrix}
\begin{bmatrix}
1 & 1 & 1 & 1 \\
1 & -j & -1 & j \\
1 & -1 & 1 & -1 \\
1 & j & -1 & -j
\end{bmatrix}
=
\begin{bmatrix}
1 & 0 & 0 & 0 \\
0 & 1 & 0 & 0 \\
0 & 0 & 1 & 0 \\
0 & 0 & 0 & 1
\end{bmatrix}
$$

With $N = 8$, the transform matrix W is

$$
W =
\begin{bmatrix}
1 & 1 & 1 & 1 & 1 & 1 & 1 & 1 \\
1 & 0.71 - j0.71 & -j & -0.71 - j0.71 & -1 & -0.71 + j0.71 & j & 0.71 + j0.71 \\
1 & -j & -1 & j & 1 & -j & -1 & j \\
1 & -0.71 - j0.71 & j & 0.71 - j0.71 & -1 & 0.71 + j0.71 & -j & -0.71 + j0.71 \\
1 & -1 & 1 & -1 & 1 & -1 & 1 & -1 \\
1 & -0.71 + j0.71 & -j & 0.71 + j0.71 & -1 & 0.71 - j0.71 & j & -0.71 - j0.71 \\
1 & j & -1 & -j & 1 & j & -1 & -j \\
1 & 0.71 + j0.71 & j & -0.71 + j0.71 & -1 & -0.71 - j0.71 & -j & 0.71 - j0.71
\end{bmatrix}
$$

$$
=
\begin{bmatrix}
1 & 1 & 1 & 1 & 1 & 1 & 1 & 1 \\
1 & \frac{\sqrt{2}}{2} - j\frac{\sqrt{2}}{2} & -j & -\frac{\sqrt{2}}{2} - j\frac{\sqrt{2}}{2} & -1 & -\frac{\sqrt{2}}{2} + j\frac{\sqrt{2}}{2} & j & \frac{\sqrt{2}}{2} + j\frac{\sqrt{2}}{2} \\
1 & -j & -1 & j & 1 & -j & -1 & j \\
1 & -\frac{\sqrt{2}}{2} - j\frac{\sqrt{2}}{2} & j & \frac{\sqrt{2}}{2} - j\frac{\sqrt{2}}{2} & -1 & \frac{\sqrt{2}}{2} + j\frac{\sqrt{2}}{2} & -j & -\frac{\sqrt{2}}{2} + j\frac{\sqrt{2}}{2} \\
1 & -1 & 1 & -1 & 1 & -1 & 1 & -1 \\
1 & -\frac{\sqrt{2}}{2} + j\frac{\sqrt{2}}{2} & -j & \frac{\sqrt{2}}{2} + j\frac{\sqrt{2}}{2} & -1 & \frac{\sqrt{2}}{2} - j\frac{\sqrt{2}}{2} & j & -\frac{\sqrt{2}}{2} - j\frac{\sqrt{2}}{2} \\
1 & j & -1 & -j & 1 & j & -1 & -j \\
1 & \frac{\sqrt{2}}{2} + j\frac{\sqrt{2}}{2} & j & -\frac{\sqrt{2}}{2} + j\frac{\sqrt{2}}{2} & -1 & -\frac{\sqrt{2}}{2} - j\frac{\sqrt{2}}{2} & -j & \frac{\sqrt{2}}{2} - j\frac{\sqrt{2}}{2}
\end{bmatrix}
$$

The samples of the complex exponential $e^{-j\frac{2\pi}{N}kn}$ are called the twiddle factors. They are the Nth roots of unity. The twiddle factors are periodic of periodic N. They are called twiddle factors because multiplying a complex number with them is changing the phase of the number only. The periodicity of the twiddle factors with $N = 8$ and the discrete frequencies are shown in Fig. 2.2. At these discrete frequencies, called the bins, the DFT coefficients are computed. The frequency increment of the DFT spectrum is $1/8$ cycles per sample, which is also the fundamental frequency. The frequency index 0 corresponds to the DC, the average value of the time-domain waveform.

Example 2.1 The DFT of 4-point impulse signal

$$
\delta(n) = \{x(0) = 1, x(1) = 0, x(2) = 0, x(3) = 0\}
$$

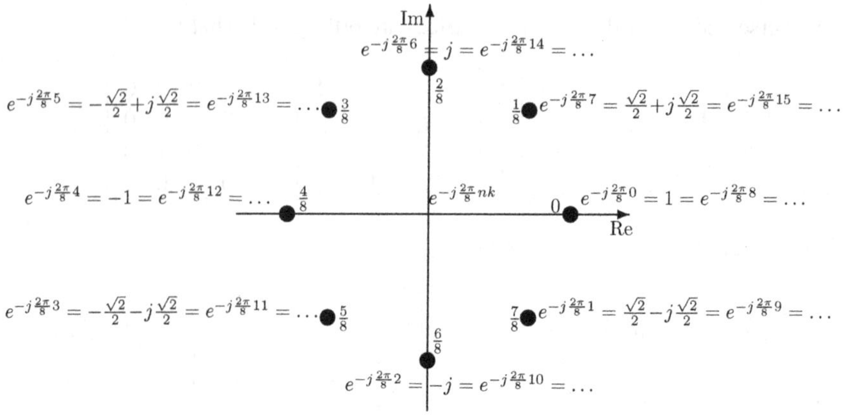

Fig. 2.2 Periodicity of the twiddle factors with $N = 8$ and the discrete frequencies

is computed as

$$
\begin{bmatrix} X(0) \\ X(1) \\ X(2) \\ X(3) \end{bmatrix} = \begin{bmatrix} 1 & 1 & 1 & 1 \\ 1 & -j & -1 & j \\ 1 & -1 & 1 & -1 \\ 1 & j & -1 & -j \end{bmatrix} \begin{bmatrix} 1 \\ 0 \\ 0 \\ 0 \end{bmatrix} = \begin{bmatrix} 1 \\ 1 \\ 1 \\ 1 \end{bmatrix}
$$

The DFT spectrum of the impulse $\{X(0) = 1, X(1) = 1, X(2) = 1, X(3) = 1\}$ is uniform. The magnitude of all the frequency components is 1 and the phase is zero. The reconstructed waveform from the spectrum is given by

$$
\begin{aligned}
x(n) &= \frac{1}{4}(1 + e^{j\frac{2\pi}{4}n} + e^{j2\frac{2\pi}{4}n} + e^{j3\frac{2\pi}{4}n}) \\
&= \frac{1}{4}\left(1 + 2\cos\left(\frac{\pi}{2}n\right) + \cos(\pi n)\right) = \begin{cases} 1 \text{ for } n = 0 \\ 0 \text{ for } n = 1, 2, 3 \end{cases}
\end{aligned}
$$

The IDFT of the spectrum formally gets back the input $\delta(n)$.

$$
\begin{bmatrix} x(0) \\ x(1) \\ x(2) \\ x(3) \end{bmatrix} = \frac{1}{4} \begin{bmatrix} 1 & 1 & 1 & 1 \\ 1 & j & -1 & -j \\ 1 & -1 & 1 & -1 \\ 1 & -j & -1 & j \end{bmatrix} \begin{bmatrix} 1 \\ 1 \\ 1 \\ 1 \end{bmatrix} = \begin{bmatrix} 1 \\ 0 \\ 0 \\ 0 \end{bmatrix}
$$

The DFT of the delayed impulses are

$$
\begin{aligned}
\delta(n - 1) &= \{0, 1, 0, 0\} \leftrightarrow X(k) = \{1, -j, -1, j\} \\
\delta(n - 2) &= \{0, 0, 1, 0\} \leftrightarrow X(k) = \{1, -1, 1, -1\} \\
\delta(n - 3) &= \{0, 0, 0, 1\} \leftrightarrow X(k) = \{1, j, -1, -j\}
\end{aligned}
$$

Therefore, the DFT of the rows or columns of the identity matrix

$$
\begin{bmatrix}
1 & 0 & 0 & 0 \\
0 & 1 & 0 & 0 \\
0 & 0 & 1 & 0 \\
0 & 0 & 0 & 1
\end{bmatrix}
$$

is the DFT transform matrix.

$$
\begin{bmatrix}
1 & 1 & 1 & 1 \\
1 & -j & -1 & j \\
1 & -1 & 1 & -1 \\
1 & j & -1 & -j
\end{bmatrix}
$$

The transform matrix can be generated using this procedure, for any N, in software implementation. That is, form the identity matrix of the required order and compute the DFT of its rows or columns to get the DFT transform matrix.

Example 2.2 The DFT of the DC signal, $x(n) = e^{j0\frac{2\pi}{4}n}$,

$$
\{x(0) = 1, x(1) = 1, x(2) = 1, x(3) = 1\}
$$

is computed as

$$
\begin{bmatrix}
X(0) \\
X(1) \\
X(2) \\
X(3)
\end{bmatrix}
=
\begin{bmatrix}
1 & 1 & 1 & 1 \\
1 & -j & -1 & j \\
1 & -1 & 1 & -1 \\
1 & j & -1 & -j
\end{bmatrix}
\begin{bmatrix}
1 \\
1 \\
1 \\
1
\end{bmatrix}
=
\begin{bmatrix}
4 \\
0 \\
0 \\
0
\end{bmatrix}
$$

The DFT spectrum is $\{X(0) = 4, X(1) = 0, X(2) = 0, X(3) = 0\}$. The spectrum of the DC signal is nonzero only at $k = 0$, since its frequency index is zero. Using the IDFT, we get back the input $x(n)$.

$$
\begin{bmatrix}
x(0) \\
x(1) \\
x(2) \\
x(3)
\end{bmatrix}
=
\frac{1}{4}
\begin{bmatrix}
1 & 1 & 1 & 1 \\
1 & j & -1 & -j \\
1 & -1 & 1 & -1 \\
1 & -j & -1 & j
\end{bmatrix}
\begin{bmatrix}
4 \\
0 \\
0 \\
0
\end{bmatrix}
=
\begin{bmatrix}
1 \\
1 \\
1 \\
1
\end{bmatrix}
$$

Example 2.3 The DFT of the alternating signal, $x(n) = e^{j2\frac{2\pi}{4}n}$,

$$
\{x(0) = 1, x(1) = -1, x(2) = 1, x(3) = -1\}
$$

is computed as

$$
\begin{bmatrix} X(0) \\ X(1) \\ X(2) \\ X(3) \end{bmatrix} = \begin{bmatrix} 1 & 1 & 1 & 1 \\ 1 & -j & -1 & j \\ 1 & -1 & 1 & -1 \\ 1 & j & -1 & -j \end{bmatrix} \begin{bmatrix} 1 \\ -1 \\ 1 \\ -1 \end{bmatrix} = \begin{bmatrix} 0 \\ 0 \\ 4 \\ 0 \end{bmatrix}
$$

The DFT spectrum is

$$
X(k) = 4\delta(k-2) = \{X(0) = 0, X(1) = 0, X(2) = 4, X(3) = 0\}
$$

The spectrum of the alternating signal is nonzero only at $k = 2$, since its frequency index is 2. Using the IDFT, we get back the input $x(n)$.

$$
\begin{bmatrix} x(0) \\ x(1) \\ x(2) \\ x(3) \end{bmatrix} = \frac{1}{4} \begin{bmatrix} 1 & 1 & 1 & 1 \\ 1 & j & -1 & -j \\ 1 & -1 & 1 & -1 \\ 1 & -j & -1 & j \end{bmatrix} \begin{bmatrix} 0 \\ 0 \\ 4 \\ 0 \end{bmatrix} = \begin{bmatrix} 1 \\ -1 \\ 1 \\ -1 \end{bmatrix}
$$

Example 2.4 The DFT of

$$
\{x(0) = (3 - 0.5\sqrt{3}), x(1) = 0.5, x(2) = (3 + 0.5\sqrt{3}), x(3) = 1.5\}
$$

is computed as

$$
\begin{bmatrix} X(0) \\ X(1) \\ X(2) \\ X(3) \end{bmatrix} = \begin{bmatrix} 1 & 1 & 1 & 1 \\ 1 & -j & -1 & j \\ 1 & -1 & 1 & -1 \\ 1 & j & -1 & -j \end{bmatrix} \begin{bmatrix} (3 - 0.5\sqrt{3}) \\ 0.5 \\ (3 + 0.5\sqrt{3}) \\ 1.5 \end{bmatrix} = \begin{bmatrix} 8 \\ -\sqrt{3} + j \\ 4 \\ -\sqrt{3} - j \end{bmatrix}
$$

The DFT spectrum, as shown in Fig. 2.1b, is

$$
\{X(0) = 8, X(1) = (-\sqrt{3} + j), X(2) = 4, X(3) = (-\sqrt{3} - j)\}
$$

Using the IDFT, we get back the input $x(n)$.

$$
\begin{bmatrix} x(0) \\ x(1) \\ x(2) \\ x(3) \end{bmatrix} = \frac{1}{4} \begin{bmatrix} 1 & 1 & 1 & 1 \\ 1 & j & -1 & -j \\ 1 & -1 & 1 & -1 \\ 1 & -j & -1 & j \end{bmatrix} \begin{bmatrix} 8 \\ -\sqrt{3} + j \\ 4 \\ -\sqrt{3} - j \end{bmatrix} = \begin{bmatrix} (3 - 0.5\sqrt{3}) \\ 0.5 \\ (3 + 0.5\sqrt{3}) \\ 1.5 \end{bmatrix}
$$

From the preceding examples, we find that

$$
e^{jk_0 \frac{2\pi}{N} n} \leftrightarrow N\delta(k - k_0)
$$

Example 2.5 The samples of one period of the complex exponential $x(n) = e^{j3\frac{2\pi}{4}n}$
are

$$\{x(0) = 1, x(1) = -j, x(2) = -1, x(3) = j\}$$

and

$$\begin{bmatrix} X(0) \\ X(1) \\ X(2) \\ X(3) \end{bmatrix} = \begin{bmatrix} 1 & 1 & 1 & 1 \\ 1 & -j & -1 & j \\ 1 & -1 & 1 & -1 \\ 1 & j & -1 & -j \end{bmatrix} \begin{bmatrix} 1 \\ -j \\ -1 \\ j \end{bmatrix} = \begin{bmatrix} 0 \\ 0 \\ 0 \\ 4 \end{bmatrix}$$

$$\begin{bmatrix} x(0) \\ x(1) \\ x(2) \\ x(3) \end{bmatrix} = \frac{1}{4}\begin{bmatrix} 1 & 1 & 1 & 1 \\ 1 & j & -1 & -j \\ 1 & -1 & 1 & -1 \\ 1 & -j & -1 & j \end{bmatrix} \begin{bmatrix} 0 \\ 0 \\ 0 \\ 4 \end{bmatrix} = \begin{bmatrix} 1 \\ -j \\ -1 \\ j \end{bmatrix}$$

Example 2.6 The samples of one period of the complex exponential $x(n) = e^{j(\frac{2\pi}{4}n+\frac{\pi}{4})} = e^{j\frac{\pi}{4}}e^{j\frac{2\pi}{4}n}$ are

$$\left\{x(0) = \frac{1}{\sqrt{2}} + j\frac{1}{\sqrt{2}}, x(1) = -\frac{1}{\sqrt{2}} + j\frac{1}{\sqrt{2}}, x(2) = -\frac{1}{\sqrt{2}} - j\frac{1}{\sqrt{2}}, x(3) = \frac{1}{\sqrt{2}} - j\frac{1}{\sqrt{2}}\right\}$$

and

$$\begin{bmatrix} X(0) \\ X(1) \\ X(2) \\ X(3) \end{bmatrix} = \begin{bmatrix} 1 & 1 & 1 & 1 \\ 1 & -j & -1 & j \\ 1 & -1 & 1 & -1 \\ 1 & j & -1 & -j \end{bmatrix} \begin{bmatrix} \frac{1}{\sqrt{2}} + j\frac{1}{\sqrt{2}} \\ -\frac{1}{\sqrt{2}} + j\frac{1}{\sqrt{2}} \\ -\frac{1}{\sqrt{2}} - j\frac{1}{\sqrt{2}} \\ \frac{1}{\sqrt{2}} - j\frac{1}{\sqrt{2}} \end{bmatrix} = \begin{bmatrix} 0 \\ 2\sqrt{2} + j2\sqrt{2} \\ 0 \\ 0 \end{bmatrix}$$

$$\begin{bmatrix} x(0) \\ x(1) \\ x(2) \\ x(3) \end{bmatrix} = \frac{1}{4}\begin{bmatrix} 1 & 1 & 1 & 1 \\ 1 & j & -1 & -j \\ 1 & -1 & 1 & -1 \\ 1 & -j & -1 & j \end{bmatrix} \begin{bmatrix} 0 \\ 2\sqrt{2} + j2\sqrt{2} \\ 0 \\ 0 \end{bmatrix} = \begin{bmatrix} \frac{1}{\sqrt{2}} + j\frac{1}{\sqrt{2}} \\ -\frac{1}{\sqrt{2}} + j\frac{1}{\sqrt{2}} \\ -\frac{1}{\sqrt{2}} - j\frac{1}{\sqrt{2}} \\ \frac{1}{\sqrt{2}} - j\frac{1}{\sqrt{2}} \end{bmatrix}$$

The complex exponential is the standard unit in Fourier analysis. The DFT of a real sinusoid can be obtained using the Euler's formula.

$$x(n) = \cos\left(\frac{2\pi}{N}k_0 n + \theta\right) = \frac{1}{2}(e^{j\theta}e^{j\frac{2\pi}{N}k_0 n} + e^{-j\theta}e^{-j\frac{2\pi}{N}k_0 n})$$

Since the DFT of complex exponentials with unit amplitude is N, we get

$$x(n) = \cos\left(\frac{2\pi}{N}k_0 n + \theta\right) \leftrightarrow X(k) = \frac{N}{2}(e^{j\theta}\delta(k - k_0) + e^{-j\theta}\delta(k - (N - k_0)))$$

As the signals are periodic, $e^{-j\frac{2\pi}{N}k_0 n} = e^{j\frac{2\pi}{N}n(N-k_0)}$. With $\theta = 0$ and $\theta = -\frac{\pi}{2}$, we get

$$\cos\left(\frac{2\pi}{N}k_0 n\right) \leftrightarrow \frac{N}{2}\left(\delta(k-k_0) + \delta(k-(N-k_0))\right)$$

$$\sin\left(\frac{2\pi}{N}k_0 n\right) \leftrightarrow \frac{N}{2}\left(-j\delta(k-k_0) + j\delta(k-(N-k_0))\right)$$

Example 2.7 The samples of one period of the cosine waveform $x(n) = \cos(\frac{2\pi}{4}n)$ are $\{x(0) = 1, x(1) = 0, x(2) = -1, x(3) = 0\}$ and

$$\begin{bmatrix} X(0) \\ X(1) \\ X(2) \\ X(3) \end{bmatrix} = \begin{bmatrix} 1 & 1 & 1 & 1 \\ 1 & -j & -1 & j \\ 1 & -1 & 1 & -1 \\ 1 & j & -1 & -j \end{bmatrix} \begin{bmatrix} 1 \\ 0 \\ -1 \\ 0 \end{bmatrix} = \begin{bmatrix} 0 \\ 2 \\ 0 \\ 2 \end{bmatrix}$$

$$\begin{bmatrix} x(0) \\ x(1) \\ x(2) \\ x(3) \end{bmatrix} = \frac{1}{4}\begin{bmatrix} 1 & 1 & 1 & 1 \\ 1 & j & -1 & -j \\ 1 & -1 & 1 & -1 \\ 1 & -j & -1 & j \end{bmatrix} \begin{bmatrix} 0 \\ 2 \\ 0 \\ 2 \end{bmatrix} = \begin{bmatrix} 1 \\ 0 \\ -1 \\ 0 \end{bmatrix}$$

Example 2.8 The samples of one period of the sine waveform $x(n) = \sin(\frac{2\pi}{4}n)$ are $\{x(0) = 0, x(1) = 1, x(2) = 0, x(3) = -1\}$ and

$$\begin{bmatrix} X(0) \\ X(1) \\ X(2) \\ X(3) \end{bmatrix} = \begin{bmatrix} 1 & 1 & 1 & 1 \\ 1 & -j & -1 & j \\ 1 & -1 & 1 & -1 \\ 1 & j & -1 & -j \end{bmatrix} \begin{bmatrix} 0 \\ 1 \\ 0 \\ -1 \end{bmatrix} = \begin{bmatrix} 0 \\ -j2 \\ 0 \\ j2 \end{bmatrix}$$

$$\begin{bmatrix} x(0) \\ x(1) \\ x(2) \\ x(3) \end{bmatrix} = \frac{1}{4}\begin{bmatrix} 1 & 1 & 1 & 1 \\ 1 & j & -1 & -j \\ 1 & -1 & 1 & -1 \\ 1 & -j & -1 & j \end{bmatrix} \begin{bmatrix} 0 \\ -j2 \\ 0 \\ j2 \end{bmatrix} = \begin{bmatrix} 0 \\ 1 \\ 0 \\ -1 \end{bmatrix}$$

Example 2.9 The samples of one period of the sinusoid $x(n) = \cos(\frac{2\pi}{4}n - \frac{\pi}{3})$ are $\{x(0) = 0.5, x(1) = 0.5\sqrt{3}, x(2) = -0.5, x(3) = -0.5\sqrt{3}\}$ and

$$\begin{bmatrix} X(0) \\ X(1) \\ X(2) \\ X(3) \end{bmatrix} = \begin{bmatrix} 1 & 1 & 1 & 1 \\ 1 & -j & -1 & j \\ 1 & -1 & 1 & -1 \\ 1 & j & -1 & -j \end{bmatrix} \begin{bmatrix} 0.5 \\ 0.5\sqrt{3} \\ -0.5 \\ -0.5\sqrt{3} \end{bmatrix} = \begin{bmatrix} 0 \\ 1 - j\sqrt{3} \\ 0 \\ 1 + j\sqrt{3} \end{bmatrix}$$

$$
\begin{bmatrix} x(0) \\ x(1) \\ x(2) \\ x(3) \end{bmatrix} = \frac{1}{4} \begin{bmatrix} 1 & 1 & 1 & 1 \\ 1 & j & -1 & -j \\ 1 & -1 & 1 & -1 \\ 1 & -j & -1 & j \end{bmatrix} \begin{bmatrix} 0 \\ 1 - j\sqrt{3} \\ 0 \\ 1 + j\sqrt{3} \end{bmatrix} = \begin{bmatrix} 0.5 \\ 0.5\sqrt{3} \\ -0.5 \\ -0.5\sqrt{3} \end{bmatrix}
$$

A finite or infinite sequence

$$
az^n = \{a, az, az^2, \dots, \}
$$

is a geometric sequence, where a and z are some fixed numbers. The first value is a constant. The rest of the values are the product of the preceding value by the common ratio. A geometric series is a series composed of the terms of a geometric sequence. For example,

$$
S_N = \sum_{n=0}^{N-1} = 1 + z + z^2 + \cdots + z^{N-1}, \quad z = e^{-j\frac{2\pi}{N}}
$$

is a geometric series. To find the sum in a closed form, we multiply S_N by z to get $z S_N$. Now,

$$
S_N - z S_N = 1 - z^N \quad \text{and} \quad S_N = \frac{1 - z^N}{1 - z}
$$

The DFT of the rectangular waveform is derived as follows.

$$
x(n) = \begin{cases} 1 & \text{for } n = 0, 1, \dots, L - 1 \\ 0 & \text{for } n = L, L+1, \dots, N - 1 \end{cases}
$$

$$
X(k) = \sum_{n=0}^{L-1} e^{-j\frac{2\pi}{N}nk} = \frac{1 - e^{-j\frac{2\pi}{N}Lk}}{1 - e^{-j\frac{2\pi}{N}k}} = \frac{e^{j\frac{2\pi}{N}\frac{L}{2}k} - e^{-j\frac{2\pi}{N}\frac{L}{2}k}}{e^{j\frac{2\pi}{N}\frac{k}{2}} - e^{-j\frac{2\pi}{N}\frac{k}{2}}}
$$

$$
= e^{-j\frac{2\pi}{N}\frac{(L-1)}{2}k} \frac{\sin\left(\frac{2\pi}{N}\frac{L}{2}k\right)}{\sin\left(\frac{2\pi}{N}\frac{k}{2}\right)} = e^{-j\frac{\pi}{N}(L-1)k} \frac{\sin\left(\frac{\pi}{N}Lk\right)}{\sin\left(\frac{\pi}{N}k\right)}
$$

Verify the DFT of $\delta(n)$ and DC signals, obtained in Examples 2.1 and 2.2, using this formula.

Example 2.10 The samples of a signal are $\{x(0) = 1, x(1) = 1, x(2) = 0, x(3) = 0\}$ and

$$
\begin{bmatrix} X(0) \\ X(1) \\ X(2) \\ X(3) \end{bmatrix} = \begin{bmatrix} 1 & 1 & 1 & 1 \\ 1 & -j & -1 & j \\ 1 & -1 & 1 & -1 \\ 1 & j & -1 & -j \end{bmatrix} \begin{bmatrix} 1 \\ 1 \\ 0 \\ 0 \end{bmatrix} = \begin{bmatrix} 2 \\ 1 - j \\ 0 \\ 1 + j \end{bmatrix}
$$

$$
\begin{bmatrix} x(0) \\ x(1) \\ x(2) \\ x(3) \end{bmatrix} = \frac{1}{4} \begin{bmatrix} 1 & 1 & 1 & 1 \\ 1 & j & -1 & -j \\ 1 & -1 & 1 & -1 \\ 1 & -j & -1 & j \end{bmatrix} \begin{bmatrix} 2 \\ 1-j \\ 0 \\ 1+j \end{bmatrix} = \begin{bmatrix} 1 \\ 1 \\ 0 \\ 0 \end{bmatrix}
$$

Example 2.11 The samples of a signal are $\{x(0) = 1 + j1, x(1) = 2 - j1, x(2) = 1 + j2, x(3) = 2 + j2\}$ and

$$
\begin{bmatrix} X(0) \\ X(1) \\ X(2) \\ X(3) \end{bmatrix} = \begin{bmatrix} 1 & 1 & 1 & 1 \\ 1 & -j & -1 & j \\ 1 & -1 & 1 & -1 \\ 1 & j & -1 & -j \end{bmatrix} \begin{bmatrix} 1+j1 \\ 2-j1 \\ 1+j2 \\ 2+j2 \end{bmatrix} = \begin{bmatrix} 6+j4 \\ -3-j \\ -2+j2 \\ 3-j \end{bmatrix}
$$

$$
\begin{bmatrix} x(0) \\ x(1) \\ x(2) \\ x(3) \end{bmatrix} = \frac{1}{4} \begin{bmatrix} 1 & 1 & 1 & 1 \\ 1 & j & -1 & -j \\ 1 & -1 & 1 & -1 \\ 1 & -j & -1 & j \end{bmatrix} \begin{bmatrix} 6+j4 \\ -3-j \\ -2+j2 \\ 3-j \end{bmatrix} = \begin{bmatrix} 1+j1 \\ 2-j1 \\ 1+j2 \\ 2+j2 \end{bmatrix}
$$

2.4   Applications of the DFT and the IDFT

Fourier analysis has innumerable applications in science and engineering. All the applications are based on the easier interpretation and characterization of the signal characteristics through the Fourier amplitude and power spectrum, its ability to provide signal compression with adequate accuracy and reduce the computational complexity of important operations such as convolution and correlation. One has to adapt these abilities to suit the specific application. In this and later chapters, we present some examples of the applications of Fourier analysis.

2.4.1   Fourier Boundary Descriptor

In image processing, objects in an image have to be identified. For that purpose, the image is segmented using the properties of the objects. The segmented objects have to be compactly represented. One way of characterizing an object is by its boundary representation. A boundary can be described by its coordinates. The objective is to minimize the storage requirements in representing it. Fourier boundary descriptor is one of the effective methods to represent the boundary of an object.

A closed boundary is represented by a set of its coordinates in the spatial domain. At each point on the boundary, a complex number is formed with its real part being the x-coordinate and the imaginary part being the y-coordinate. The set of the complex numbers is a periodic complex data, the period being the number of points on the boundary. Let the N boundary coordinates be

$$\{(x(0), y(0)), (x(1), y(1)), \ldots, (x(N-1), y(N-1))\}$$

The complex number representation of the boundary becomes

$$b(n) = \{(x(0) + jy(0)), (x(1) + jy(1)), \ldots, (x(N-1) + jy(N-1))\}$$

The 1-D DFT of this set, $B(k)$, is the Fourier descriptor of the boundary with significant advantages. The 2-D data is represented by a 1-D data.

It is desirable that the descriptor is as much insensitive as possible for scaling, translation, and rotation of the boundary. The properties of the DFT make it to relate the Fourier descriptors of a boundary and its modified versions. Consider the 4×4 binary image $x(m, n)$ and its shifted version $x(m-1, n-1)$

$$x(m, n) = \begin{bmatrix} 1 & 1 & 1 & 0 \\ 1 & 0 & 1 & 0 \\ 1 & 1 & 1 & 0 \\ 0 & 0 & 0 & 0 \end{bmatrix} \quad x(m-1, n-1) = \begin{bmatrix} 0 & 0 & 0 & 0 \\ 0 & 1 & 1 & 1 \\ 0 & 1 & 0 & 1 \\ 0 & 1 & 1 & 1 \end{bmatrix}$$

The complex data formed from the boundary coordinates of $x(m, n)$ is

$$b(n) = \{0 + j0, 1 + j0, 2 + j0, 2 + j1, 2 + j2, 1 + j2, 0 + j2, 0 + j1\}$$

It is assumed that the top left corner is the origin. The DFT of $b(n)$ is

$$B(k) = \{8 + j8, -6.8284 - j6.8284, 0, 0, 0, -1.1716 - j1.1716, 0, 0\}$$

The complex data formed from the boundary coordinates of $x(m-1, n-1)$ is

$$bs(n) = \{1 + j1, 2 + j1, 3 + j1, 3 + j2, 3 + j3, 2 + j3, 1 + j3, 1 + j2\}$$

The DFT of $bs(n)$ is

$$BS(k) = \{16 + j16, -6.8284 - j6.8284, 0, 0, 0, -1.1716 - j1.1716, 0, 0\}$$

The translation of the boundary can be found by the change in the DC coefficients of the two DFTs, the remaining being the same.

Consider the 4×4 binary image $x(m, n)$ and its $90°$ anticlockwise rotated version $xr(m, n)$

$$x(m, n) = \begin{bmatrix} 0 & 0 & 0 & 0 \\ 1 & 1 & 1 & 1 \\ 1 & 1 & 1 & 1 \\ 0 & 0 & 0 & 0 \end{bmatrix} \quad xr(m, n) = \begin{bmatrix} 0 & 1 & 1 & 0 \\ 0 & 1 & 1 & 0 \\ 0 & 1 & 1 & 0 \\ 0 & 1 & 1 & 0 \end{bmatrix}$$

The complex data formed from the boundary coordinates of $x(m, n)$ is

$$b(n) = \{1 + j0, 1 + j1, 1 + j2, 1 + j3, 2 + j0, 2 + j1, 2 + j2, 2 + j3\}$$

The DFT of $b(n)$ is

$$B(k) = \{12 + j12, -1 + j2.4142, -4 - j4, -1 + j0.4142,$$
$$- j4, -1 - j0.4142, 4 - j4, -1 - j2.4142\}$$

The complex data formed from the boundary coordinates of $br(n)$ is

$$br(n) = \{0 + j1, -1 + j1, -2 + j1, -3 + j1, 0 + j2,$$
$$- 1 + j2, -2 + j2, -3 + j2\} = e^{j\frac{\pi}{2}} b(n)$$

Note that the coordinates are shifted due to rotation. The DFT of $br(n)$ is

$$BR(k) = \{-12 + j12, -2.4142 - j1, 4 - j4, -0.4142 - j1,$$
$$4, 0.4142 - j1, 4 + j4, 2.4142 - j1\} = e^{j\frac{\pi}{2}} B(k)$$

The DFT of the starting-point shifted complex data formed from the coordinates of the boundary of $b(n)$ can be obtained using the time-domain shift theorem. The DFT of scaled complex data formed from the coordinates of the boundary of $b(n)$ is also the scaled version of $B(k)$, due to the linearity of the DFT.

Figure 2.3a, b show, respectively, a triangular boundary and the normalized magnitude of part of its Fourier descriptor. The first and the last 16 DFT coefficients are shown by dots and crosses, respectively. A 464-point complex data sequence is formed from the pair of 464 boundary coordinates and its DFT is computed. As the data is arbitrary and complex, the DFT coefficients do not exhibit any symmetry. Figure 2.3c, d show the reconstructed boundary using only 32 and 16, respectively, of the largest of the 464 DFT coefficients. Despite the use of much fewer coefficients, the reconstructed boundary is quite close to the original.

In Fourier analysis, a real signal is decomposed in terms of real sinusoids or complex sinusoids with conjugate frequencies. The most famous example is the reconstruction of a square wave from its Fourier spectrum. As the number of frequency components is increased in the reconstruction, the synthesized waveform becomes closer to the original waveform by constructive and destructive interferences.

In the Fourier boundary descriptor, the input signal is arbitrary and complex. Therefore, the components to reconstruct a waveform are complex exponentials with pure imaginary exponents, whose shape is a circle. The diameter of the circle is proportional to the magnitude of the corresponding Fourier coefficient. As in the case of real signals, the magnitude of the coefficients become negligible over a considerable range of the spectrum. Therefore, a boundary can be represented with fewer coefficients with a required accuracy.

Figure 2.4a shows the circles corresponding to three of the largest DFT coefficients, in addition to the DC. The reconstructed boundary using these DFT

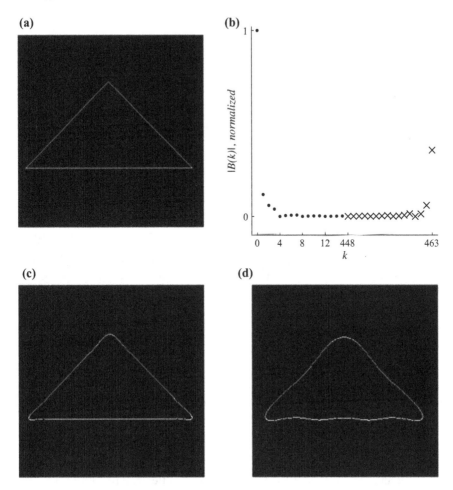

Fig. 2.3 **a** A triangular boundary; **b** the magnitude of the largest 32 of the 464 DFT coefficients (normalized) of the boundary points; **c** and **d** the reconstructed boundaries using only the largest 32 and 16 DFT coefficients, respectively

coefficients is also shown, which is an approximation of the actual boundary. The smallest circle corresponds to a higher frequency and it suitably adds negative and positive values to the larger circles, thereby making it closer to the actual boundary. Now, the addition of the DC component of the DFT, which represents the average of the two coordinates, fixes the center of the boundary. With just the DC and five other DFT coefficients, the approximation of the boundary is quite good, as shown in Fig. 2.4b. Therefore, a major advantage, as in other applications of the DFT, is that the shape can be adequately described by much fewer than the N coefficients. As noise is generally present at all frequencies, this leads to reduce the noise affecting the boundary points. A further advantage is that the 2-D shape is described by 1-D data.

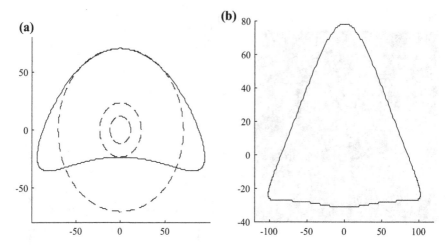

Fig. 2.4 **a** Three circles corresponding to three DFT coefficients and the boundary corresponding to their combined effect; **b** the reconstructed boundary using the DC and five other DFT coefficients only

2.5   Summary

- Transform methods change a problem into another equivalent form so that it is relatively easier to interpret and solve the problem.
- Fourier analysis represents an arbitrary signal in terms of sinusoidal (or its mathematically equivalent complex exponential) signals to gain significant advantages in its processing.
- While physical devices generate sinusoidal signals, the complex exponential is advantageous for analysis purposes.
- Fourier analysis has four different versions to suit different types of signals.
- The DFT is the only version in which the signal is finite and discrete in both the time and frequency domains. Therefore, it is the only version suitable for implementation on digital systems and, consequently, it is most often used in practice.
- The DFT is the simplest version of the Fourier analysis to understand and it is easier to visualize the decomposition of the waveforms in that form.
- The decomposition of the waveforms, in Fourier analysis, is based on the orthogonality property of the sinusoids and complex exponentials.
- The Fourier representation of a signal is with respect to the least squares error criterion.
- The complex exponential basis signals of the DFT are equidistant samples on the unit circle.
- The DFT has both algebraic and matrix forms.
- While the theory of Fourier analysis requires frequency components of infinite order to represent an arbitrary waveform exactly, it is found that adequate

approximation of the practical signals is possible with a finite number of frequency components. This is due to the rate of fall of the magnitude of the frequency components of practical signals with increasing frequency.

- The DFT assumes periodic extension of a finite length signal and represents it in terms of a finite number of discrete sinusoids.
- The major advantages of Fourier analysis are the importance of the magnitude and power spectra of signals in large number of applications and reducing indispensable operations, such as convolution and correlation, into relatively simpler operations.
- Fourier analysis is dominating in both the theoretical and practical study of signals and LTI systems. The reasons are the insight it provides about the characteristics of signals and systems and the availability of fast algorithms for its practical implementation.

Exercises

2.1 Find the product of the two numbers after representing them in exponential form with base 2.

2.1.1 256 and 128.

* 2.1.2 64 and 512.

2.1.3 32 and 128.

2.2 Using the Euler's formula, find the complex exponential representation of the signal in terms of: (i) the real and imaginary parts and (ii) amplitude and phase. Find the samples of all the three forms and verify that they are the same.

2.2.1 $x(n) = 4 \cos \left(\frac{\pi}{4} n - \frac{\pi}{3} \right)$

* 2.2.2 $x(n) = 2 \cos \left(\frac{\pi}{4} n + \frac{\pi}{6} \right)$

2.2.3 $x(n) = \sqrt{3} \cos \left(\frac{\pi}{4} n \right) - \sin \left(\frac{\pi}{4} n \right)$

2.3 Compute the DFT, $X(k)$, of $x(n)$ using the matrix form of the DFT. Compute the IDFT of $X(k)$ to get back $x(n)$.

2.3.1 $x(n) = \{1, 2, 3, 4\}$

2.3.2 $x(n) = \{-2, -1, 3, 4\}$

2.3.3 $x(n) = \{3, -1, 2, 4\}$

2.3.4 $x(n) = \{1 + j3, 2 + j2, 1 - j1, 3 + j2\}$

* 2.3.5 $x(n) = \{1 + j3, 2 + j2, 3 + j1, 4 - j4\}$

2.3.6 $x(n) = \{3 + j3, 2 + j2, -1 + j1, 2 - j4\}$

2.4 Find the samples of $x(n)$ over one period and use the matrix form of the DFT to compute the spectrum $X(k)$ of the set of samples. Using $X(k)$, find the exponential form of $x(n)$ and reduce the expression to real form. Verify that the input $x(n)$ is obtained.

2.4.1 $x(n) = 1 + 3 \cos \left(\frac{2\pi}{4} n + \frac{\pi}{6} \right) + 2 \cos(\pi n)$.

* 2.4.2 $x(n) = 1 - 2 \sin \left(\frac{2\pi}{4} n - \frac{\pi}{3} \right) - 3 \cos(\pi n)$.

2.4.3 $x(n) = 1 - \cos \left(\frac{2\pi}{4} n - \frac{\pi}{4} \right) + \cos(\pi n)$.

2.4.4 $x(n) = 3 + \sin \left(\frac{2\pi}{4} n - \frac{\pi}{4} \right) - \cos(\pi n)$.

2.4.5 $x(n) = -1 - 2 \cos \left(\frac{2\pi}{4} n - \frac{\pi}{3} \right) - 3 \cos(\pi n)$.

Chapter 3
Properties of the DFT

In signal and system analysis, we frequently carry out operations on signals, such as shifting, scaling, multiplication, differentiation, integration. The properties relate the effect of an operation in one domain in the other. It is of interest to know the time–frequency-domain correlations. It helps to select the appropriate domain for interpretation and operation of signals. Further, the transforms of a set of related signals can be determined more easily from the knowledge of the transform of some simple signals.

3.1 Linearity

Let $x(n) \leftrightarrow X(k)$ and $y(n) \leftrightarrow Y(k)$, both the sequences with period N. Then,

$$ax(n) + by(n) \leftrightarrow aX(k) + bY(k)$$

where a and b are arbitrary constants. The DFT of a linear combination of a set of signals is the same linear combination of the DFT of the individual signals. If

$$x(n) = \{\check{2}, 1, 3, 4\} \leftrightarrow X(k) = \{\check{10}, -1 + j3, 0, -1 - j3\} \text{ and}$$

$$y(n) = \{\check{1}, -2, 3, -1\} \leftrightarrow Y(k) = \{\check{1}, -2 + j1, 7, -2 - j1\}$$

then

$$2x(n) - 3y(n) = \{\check{1}, 8, -3, 11\} \leftrightarrow 2X(k) - 3Y(k) = \{\check{17}, 4 + j3, -21, 4 - j3\}$$

The value of the signal with index zero is indicated by a check mark on it. The length of the two signals must be the same. If necessary, zero padding can be employed.

© Springer Nature Singapore Pte Ltd. 2018

D. Sundararajan, *Fourier Analysis—A Signal Processing Approach*,

https://doi.org/10.1007/978-981-13-1693-7_3

3.2 Periodicity

Let $x(n) \leftrightarrow X(k)$ with period N. Then,

$$x(n + aN) = x(n), \text{ for all } n \quad \text{and} \quad X(k + aN) = X(k), \text{ for all } k$$

where a is an arbitrary integer. If a signal $x(n)$ is periodic with period N samples, then its values over any successive N samples are the same. Let

$$x(n) = \{\breve{3}, 1, 2, 4\} \leftrightarrow X(k) = \{\breve{1}0, 1 + j3, 0, 1 - j3\}$$

As $-4 \bmod 4 = 0$, $x(-4) = x(0) = 3$. Similarly, $X(5) = X(1) = 1 + j3$.

With the exponential $e^{j\frac{2\pi}{N}nk_0}$ also periodic with period N, $x(n)e^{j\frac{2\pi}{N}nk_0}$ is also periodic with period N. Therefore, the DFT coefficients are periodic and can be determined using the expression

$$X(k) = \sum_{n=a}^{a+N-1} x(n)e^{-jk\frac{2\pi}{N}n}, \quad k = 0, 1, 2, \ldots, N - 1 \tag{3.1}$$

where a is an arbitrary integer. Therefore, given a set of samples starting with index other than zero, Eq. (3.1) can be used. Alternately, the given N-point sequence can be periodically extended and the samples starting with index zero can be obtained. Then, Eq. (3.1), with $a = 0$, can be used. For example,

$$x(n) = \{\breve{2}, 1, 2, 4\} \leftrightarrow X(k) = \{\breve{9}, j3, -1, -j3\}$$

The periodic extension of $x(n)$ is

$$\{\ldots, 2, 1, 2, 4, 2, 1, 2, 4, \ \breve{2}, 1, 2, 4, \ 2, 1, 2, 4, 2, 1, 2, 4, \ldots\}$$

Then, for example,

$$x(n) = \{2, 4, \breve{2}, 1\} \leftrightarrow X(k) = \{\breve{9}, j3, -1, -j3\}$$

using Eq. (3.1) with $a = -2$. Similarly, the IDFT can be computed using any successive N samples of the periodic DFT spectrum.

Another implication of the periodicity is that the convolution and other operations implemented using the DFT are periodic or circular or cyclic. However, the linear convolution is often required. The DFT can still be used with sufficient zero padding. Further, the periodic extension of a N-point sequence may create discontinuities at the boundaries. The result is that the convergence of its spectral coefficients may be slow. It requires more number of frequency components for a fairly good representation of the signal. This is a disadvantage in applications such as signal compression. With suitable extensions, such as making an even-symmetric extension, this problem can be alleviated.

3.3 Circular Time Shifting

Let $x(n) \leftrightarrow X(k)$ with period N. Then,

$$x(n \pm n_0) \leftrightarrow e^{\pm j \frac{2\pi}{N} k n_0} X(k)$$

where n_0 is an arbitrary number of sampling intervals. The shift of a sinusoidal waveform affects only its phase. There is no change in the magnitude. Therefore, the shift of a waveform in the time domain results in adding increments to the phase of the frequency components, which are linearly proportional to the respective frequency indices. The increment in phase is $e^{\pm j \frac{2\pi}{N} k n_0}$ for a frequency component with frequency index k due to a shift of $\pm n_0$ sampling intervals in the time domain. For example,

$$x(n) = \{\overset{\vee}{1}, 2, 1, 4\} \leftrightarrow X(k) = \{\overset{\vee}{8}, j2, -4, -j2\}$$

$$x(n-1) = \{\overset{\vee}{4}, 1, 2, 1\} \leftrightarrow X(k) = \{\overset{\vee}{8}, j2(-j), -4(-1), -j2(j)\} = \{\overset{\vee}{8}, 2, 4, 2\}$$

After shifting, the signal becomes even-symmetric and, therefore, its DFT is also even-symmetric. The computational complexity of computing the DFT of $x(n)$ can be reduced by shifting it to get $xs(n)$, computing its DFT $Xs(k)$ and, then, deducing the DFT of $x(n)$ from $Xs(k)$ using the shift theorem.

With $x(n)$ advanced by 2 sampling intervals, we get

$$x(n+2) = \{\overset{\vee}{1}, 4, 1, 2\} \leftrightarrow X(k) = \{\overset{\vee}{8}, j2(-1), -4(1), -j2(-1)\} = \{\overset{\vee}{8}, -j2, -4, j2\}$$

3.4 Circular Frequency Shifting

Let $x(n) \leftrightarrow X(k)$ with period N. Then,

$$e^{\pm j \frac{2\pi}{N} k_0 n} x(n) \leftrightarrow X(k \mp k_0)$$

where k_0 is an arbitrary number of sampling intervals. Since

$$e^{-j \frac{2\pi}{N} kn} e^{j \frac{2\pi}{N} k_0 n}$$

becomes

$$e^{-j \frac{2\pi}{N} (k - k_0) n}$$

in the DFT definition, the spectral values get delayed by k_0 sampling intervals and occurs at frequency index $k + k_0$ in the shifted spectrum. For example,

$$x(n) = \{\check{3}, 2, 1, 4\} \leftrightarrow X(k) = \{\check{10}, 2 + j2, -2, 2 - j2\}$$

$$e^{j\frac{2\pi}{4}n}x(n) = \{\check{3}, j2, -1, -j4\} \leftrightarrow X(k) = \{2 - j2, \check{10}, 2 + j2, -2\}$$

$$e^{j\frac{2\pi}{4}2n}x(n) = (-1)^n x(n) = \{\check{3}, -2, 1, -4\} \leftrightarrow X(k) = \{-2, 2 - j2, \check{10}, 2 + j2\}$$

By multiplying $x(n)$ by $(-1)^n$ and taking the DFT, we get the spectrum in the center-zero format.

3.5 Circular Time-Reversal

Let $x(n) \leftrightarrow X(k)$ with period N. Then,

$$x(N - n) \leftrightarrow X(N - k)$$

For a DFT with $N = 8$, the function $\mathrm{mod}(nk, 8)$ for each nk in $e^{-j\frac{2\pi}{8}nk}$ in the DFT definition, with $k = \{0, 1, 2, 3, 4, 5, 6, 7\}$, yields

$$\mathrm{mod}(nk, 8) = \begin{bmatrix} 0 & 0 & 0 & 0 & 0 & 0 & 0 & 0 \\ 0 & 1 & 2 & 3 & 4 & 5 & 6 & 7 \\ 0 & 2 & 4 & 6 & 0 & 2 & 4 & 6 \\ 0 & 3 & 6 & 1 & 4 & 7 & 2 & 5 \\ 0 & 4 & 0 & 4 & 0 & 4 & 0 & 4 \\ 0 & 5 & 2 & 7 & 4 & 1 & 6 & 3 \\ 0 & 6 & 4 & 2 & 0 & 6 & 4 & 2 \\ 0 & 7 & 6 & 5 & 4 & 3 & 2 & 1 \end{bmatrix}$$

The mod function returns the remainder of nk divided by 8. Each row of values, for a specific k, is the time-reversal of that of $(N - k)$. Therefore, $x(N - n) \leftrightarrow X(N - k)$. For example,

$$x(n) = \{\check{1}, 2, 3, 4\} \leftrightarrow X(k) = \{\check{10}, -2 + j2, -2, -2 - j2\}$$

$$x(4 - n) = \{\check{1}, 4, 3, 2\} \leftrightarrow X(k) = \{\check{10}, -2 - j2, -2, -2 + j2\}$$

3.6 Duality

Duality is the interchangeability of the time-domain and frequency-domain variables in the DFT definition. Let $x(n) \leftrightarrow X(k)$ with period N. Then,

$$X(n) \leftrightarrow Nx(N - k)$$

Computing the DFT twice in succession of a signal $x(n)$ yields N times the time-reversal of $x(n)$. The product of the transform matrix with itself yields a matrix whose elements below the reverse diagonal are N (except the first entry) and, therefore, when the input vector is multiplied by this matrix, we get a scaled and time-reversed version of the input. With $N = 8$,

$$
\begin{bmatrix}
1 & 1 & 1 & 1 & 1 & 1 & 1 & 1 \\
1 & \frac{\sqrt{2}}{2}-j\frac{\sqrt{2}}{2} & -j & -\frac{\sqrt{2}}{2}-j\frac{\sqrt{2}}{2} & -1 & -\frac{\sqrt{2}}{2}+j\frac{\sqrt{2}}{2} & j & \frac{\sqrt{2}}{2}+j\frac{\sqrt{2}}{2} \\
1 & -j & -1 & j & 1 & -j & -1 & j \\
1 & -\frac{\sqrt{2}}{2}-j\frac{\sqrt{2}}{2} & j & \frac{\sqrt{2}}{2}-j\frac{\sqrt{2}}{2} & -1 & \frac{\sqrt{2}}{2}+j\frac{\sqrt{2}}{2} & -j & -\frac{\sqrt{2}}{2}+j\frac{\sqrt{2}}{2} \\
1 & -1 & 1 & -1 & 1 & -1 & 1 & -1 \\
1 & -\frac{\sqrt{2}}{2}+j\frac{\sqrt{2}}{2} & -j & \frac{\sqrt{2}}{2}+j\frac{\sqrt{2}}{2} & -1 & \frac{\sqrt{2}}{2}-j\frac{\sqrt{2}}{2} & j & -\frac{\sqrt{2}}{2}-j\frac{\sqrt{2}}{2} \\
1 & j & -1 & -j & 1 & j & -1 & -j \\
1 & \frac{\sqrt{2}}{2}+j\frac{\sqrt{2}}{2} & j & -\frac{\sqrt{2}}{2}+j\frac{\sqrt{2}}{2} & -1 & -\frac{\sqrt{2}}{2}-j\frac{\sqrt{2}}{2} & -j & \frac{\sqrt{2}}{2}-j\frac{\sqrt{2}}{2}
\end{bmatrix} \times
$$

$$
\begin{bmatrix}
1 & 1 & 1 & 1 & 1 & 1 & 1 & 1 \\
1 & \frac{\sqrt{2}}{2}-j\frac{\sqrt{2}}{2} & -j & -\frac{\sqrt{2}}{2}-j\frac{\sqrt{2}}{2} & -1 & -\frac{\sqrt{2}}{2}+j\frac{\sqrt{2}}{2} & j & \frac{\sqrt{2}}{2}+j\frac{\sqrt{2}}{2} \\
1 & -j & -1 & j & 1 & -j & -1 & j \\
1 & -\frac{\sqrt{2}}{2}-j\frac{\sqrt{2}}{2} & j & \frac{\sqrt{2}}{2}-j\frac{\sqrt{2}}{2} & -1 & \frac{\sqrt{2}}{2}+j\frac{\sqrt{2}}{2} & -j & -\frac{\sqrt{2}}{2}+j\frac{\sqrt{2}}{2} \\
1 & -1 & 1 & -1 & 1 & -1 & 1 & -1 \\
1 & -\frac{\sqrt{2}}{2}+j\frac{\sqrt{2}}{2} & -j & \frac{\sqrt{2}}{2}+j\frac{\sqrt{2}}{2} & -1 & \frac{\sqrt{2}}{2}-j\frac{\sqrt{2}}{2} & j & -\frac{\sqrt{2}}{2}-j\frac{\sqrt{2}}{2} \\
1 & j & -1 & -j & 1 & j & -1 & -j \\
1 & \frac{\sqrt{2}}{2}+j\frac{\sqrt{2}}{2} & j & -\frac{\sqrt{2}}{2}+j\frac{\sqrt{2}}{2} & -1 & -\frac{\sqrt{2}}{2}-j\frac{\sqrt{2}}{2} & -j & \frac{\sqrt{2}}{2}-j\frac{\sqrt{2}}{2}
\end{bmatrix} \times
\begin{bmatrix} x(0) \\ x(1) \\ x(2) \\ x(3) \\ x(4) \\ x(5) \\ x(6) \\ x(7) \end{bmatrix}
$$

$$
=
\begin{bmatrix}
8 & 0 & 0 & 0 & 0 & 0 & 0 & 0 \\
0 & 0 & 0 & 0 & 0 & 0 & 0 & 8 \\
0 & 0 & 0 & 0 & 0 & 0 & 8 & 0 \\
0 & 0 & 0 & 0 & 0 & 8 & 0 & 0 \\
0 & 0 & 0 & 0 & 8 & 0 & 0 & 0 \\
0 & 0 & 0 & 8 & 0 & 0 & 0 & 0 \\
0 & 0 & 8 & 0 & 0 & 0 & 0 & 0 \\
0 & 8 & 0 & 0 & 0 & 0 & 0 & 0
\end{bmatrix}
\begin{bmatrix} x(0) \\ x(1) \\ x(2) \\ x(3) \\ x(4) \\ x(5) \\ x(6) \\ x(7) \end{bmatrix}
= 8
\begin{bmatrix} x(0) \\ x(7) \\ x(6) \\ x(5) \\ x(4) \\ x(3) \\ x(2) \\ x(1) \end{bmatrix}
= 8x(8-n)
$$

The first row of the first matrix, with $k = 0$, multiplied by the first column of the second matrix and summed yields 8. For all the other rows, the sum of the product is 8 only with the $(N - k)$th column. In all other cases, rows and columns are orthogonal and sum of the products is zero. For example, the DFT of $\{\breve{0}, 1, 2, 3\}$ is $\{\breve{6}, -2 + j2, -2, -2 - j2\}$. The DFT of this is $4\{\breve{0}, 3, 2, 1\}$.

3.7 Transform of Complex Conjugates

Let $x(n) \leftrightarrow X(k)$ with period N. Then,

$$x^*(n) \leftrightarrow X^*(N - k) \quad \text{and} \quad x^*(N - n) \leftrightarrow X^*(k)$$

Conjugating both sides of Eq. (2.2), we get

$$X^*(k) = \sum_{n=0}^{N-1} x^*(n)e^{j\frac{2\pi}{N}nk} = \sum_{n=0}^{N-1} x^*(N-n)e^{-j\frac{2\pi}{N}nk}$$

Conjugating both sides of Eq. (2.2) and replacing k by $N - k$, we get

$$X^*(N-k) = \sum_{n=0}^{N-1} x^*(n)e^{-j\frac{2\pi}{N}nk}$$

For example,

$$x(n) = \{\overset{\vee}{1}, 4, j, 3\} \leftrightarrow X(k) = \{8 + j, 1 - j2, -6 + j, 1\}$$

$$x^*(n) = \{\overset{\vee}{1}, 4, -j, 3\} \leftrightarrow X^*(4-k) = \{8 - j, 1, -6 - j, 1 + j2\}$$

$$x^*(4-n) = \{\overset{\vee}{1}, 3, -j, 4\} \leftrightarrow X^*(k) = \{8 - j, 1 + j2, -6 - j, 1\}$$

3.8   Circular Convolution and Correlation

3.8.1   Circular Convolution of Time-Domain Sequences

Let $x(n) \leftrightarrow X(k)$ and $h(n) \leftrightarrow H(k)$, both with period N. Then, the circular convolution of the sequences is defined as

$$y(n) = \sum_{m=0}^{N-1} x(m)h(n-m) = \sum_{m=0}^{N-1} h(m)x(n-m), \quad n = 0, 1, \ldots, N-1$$

The major difference between linear and circular convolutions is that, as the sequences are periodic, the limits are changed from $\pm\infty$ to one period. Otherwise, the convolution may become undefined as the sum may not remain finite. Summation over additional periods is unnecessary, as it yields integer multiples of that of over one period. The circular convolution of two 8-point periodic sequences $x(n)$ and $h(n)$ is computed as follows. The sequences are placed on a circle, as shown in Fig. 3.1, with one of the sequences time-reversed, $h(n)$ as shown in (a). The sum of the products of the corresponding values is the convolution output $y(0)$. The time-reversed sequence is shifted by one sample, as shown in (b). The sum of the products of the corresponding values is the convolution output $y(1)$. The procedure is repeated to find the rest of the 6 outputs. The output will repeat after 8 shifts.

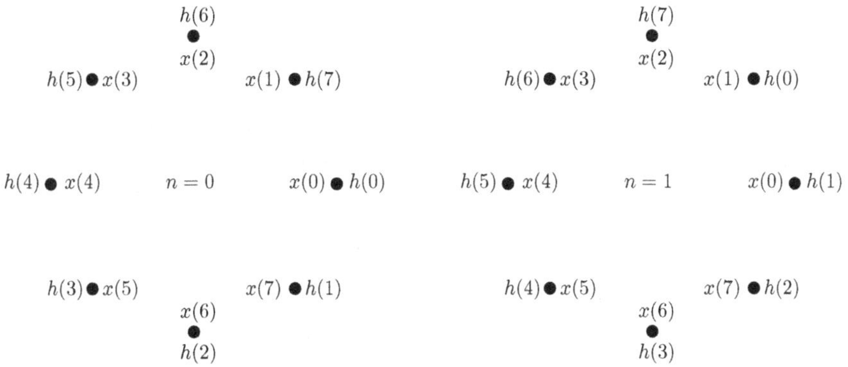

Fig. 3.1 Circular convolution

The convolution of two complex exponentials, with the same frequency $k_0\omega_0 = \frac{2\pi}{N}k_0$ but with different complex coefficients $X(k_0)$ and $H(k_0)$, $X(k_0)e^{jk_0\omega_0 n}$ and $H(k_0)e^{jk_0\omega_0 n}$

$$\sum_{m=0}^{N-1} X(k_0)e^{jk_0\omega_0 m} H(k_0)e^{jk_0\omega_0(n-m)} = X(k_0)H(k_0)e^{jk_0\omega_0 n} \sum_{m=0}^{N-1} e^{jk_0\omega_0(m-m)}$$

$$= NX(k_0)H(k_0)e^{jk_0\omega_0 n}$$

is the same complex exponential with the coefficient $NX(k_0)H(k_0)$. The coefficient is the scaled product of the two coefficients of the given exponentials.

Let $x(n) = e^{j\frac{2\pi}{4}n}$. The samples of one period are $\{1, j, -1, -j\}$. The DFT $X(k) = \{0, 4, 0, 0\}$. The pointwise product of $X(k)$ with itself is $\{0, 16, 0, 0\}$, the IDFT of which is $4\{1, j, -1, -j\}$. We have used the convolution theorem to find the convolution. We could have also used the defining equation of convolution in the time domain, given above, to find the convolution. If we convolve $x(n)$ with $h(n) = e^{j3\frac{2\pi}{4}n}$, the result is zero.

The convolution output of two complex exponentials, with frequencies $k_0\omega_0 = \frac{2\pi}{N}k_0$ and $k_1\omega_0 = \frac{2\pi}{N}k_1$ with complex coefficients $X(k_0)$ and $H(k_1)$, $X(k_0)e^{jk_0\omega_0 n}$ and $H(k_1)e^{jk_1\omega_0 n}$

$$\sum_{m=0}^{N-1} X(k_0)e^{jk_0\omega_0 m} H(k_1)e^{jk_1\omega_0(n-m)} = X(k_0)H(k_1)e^{jk_1\omega_0 n} \sum_{m=0}^{N-1} e^{jm\omega_0(k_0-k_1)}$$

$$= NX(k_0)H(k_1)e^{jk_1\omega_0 n}0 = 0$$

is zero.

The DFT representations of the sequences $x(n)$ and $h(n)$, in terms of complex exponentials, are

$$x(n) = \frac{1}{N} \sum_{k=0}^{N-1} X(k)e^{j\frac{2\pi}{N}nk} \quad \text{and} \quad h(n) = \frac{1}{N} \sum_{k=0}^{N-1} H(k)e^{j\frac{2\pi}{N}nk}$$

Since the coefficients of the convolution of the two signals in the frequency domain are $NX(k)H(k)$, the convolution of the sequences is given by

$$y(n) = \frac{1}{N} \sum_{k=0}^{N-1} X(k)H(k)e^{j\frac{2\pi}{N}nk}$$

That is, the IDFT of $X(k)H(k)$ is the convolution of $x(n)$ and $h(n)$.

$$x(n) \circledast h(n) \leftrightarrow X(k)H(k)$$

Therefore, convolution in the time domain corresponds to multiplication in the frequency domain. This fact, along with the characterization of signals by the amplitude and power spectrum, the compression of signals using the spectrum and the fast computation of the DFT are the major reasons for the dominance of the Fourier analysis in the study of signals and systems.

For example,

$$x(n) = \{\breve{1}, 4, 1, 3\} \leftrightarrow X(k) = \{\breve{9}, -j, -5, j\}$$

$$h(n) = \{\breve{3}, 2, 1, 4\} \leftrightarrow X(k) = \{\breve{10}, 2+j2, -2, 2-j2\}$$

$$x(n) \circledast h(n) = \{\breve{26}, 21, 24, 19\} \leftrightarrow X(k)H(k) = \{\breve{90}, 2-j2, 10, 2+j2\}$$

3.8.2 Circular Convolution of Frequency-Domain Sequences

The DFT representations of two 4-point sequences, $x(n)$ and $h(n)$, are

$$x(n) = \frac{1}{4}(X(0) + X(1)e^{j\frac{2\pi}{4}n} + X(2)e^{j2\frac{2\pi}{4}n} + X(3)e^{j3\frac{2\pi}{4}n})$$

$$h(n) = \frac{1}{4}(H(0) + H(1)e^{j\frac{2\pi}{4}n} + H(2)e^{j2\frac{2\pi}{4}n} + H(3)e^{j3\frac{2\pi}{4}n})$$

Then,

$$x(n)h(n) = c(n) =$$
$$\frac{1}{16}(C(0) + C(1)e^{j\frac{2\pi}{4}n} + C(2)e^{j2\frac{2\pi}{4}n} + C(3)e^{j3\frac{2\pi}{4}n}$$
$$+ C(4)e^{j4\frac{2\pi}{4}n} + C(5)e^{j5\frac{2\pi}{4}n} + C(6)e^{j6\frac{2\pi}{4}n})$$

where

$$C(0) = X(0)H(0)$$
$$C(1) = X(0)H(1) + X(1)H(0)$$
$$C(2) = X(0)H(2) + X(1)H(1) + X(2)H(0)$$
$$C(3) = X(0)H(3) + X(1)H(2) + X(2)H(1) + X(3)H(0)$$
$$C(4) = X(1)H(3) + X(2)H(2) + X(3)H(1)$$
$$C(5) = X(2)H(3) + X(3)H(2)$$
$$C(6) = X(3)H(3)$$

This is the linear convolution of the two coefficient sequences. Since the signals are 4-point periodic, only the first four frequency components have unique representation. Frequency components with indices 4, 5, and 6 are added with those with indices 0, 1, and 2, respectively. Therefore, we get

$$C(0) = X(0)H(0) + X(1)H(3) + X(2)H(2) + X(3)H(1)$$
$$C(1) = X(0)H(1) + X(1)H(0) + X(2)H(3) + X(3)H(2)$$
$$C(2) = X(0)H(2) + X(1)H(1) + X(2)H(0) + X(3)H(3)$$
$$C(3) = X(0)H(3) + X(1)H(2) + X(2)H(1) + X(3)H(0)$$

This is the circular convolution of the two coefficient sequences. Therefore, the scaled circular convolution of the two frequency-domain sequences $X(k)$ and $H(K)$ corresponds to multiplication of the corresponding sequences in the time domain.

$$x(n)h(n) \leftrightarrow \frac{1}{N}(X(k) \circledast H(k))$$

For example,

$$x(n) = \{\check{1}, 4, -1, 3\} \leftrightarrow X(k) = \{\check{7}, 2 - j, -7, 2 + j\}$$

$$h(n) = \{\check{3}, -2, 1, 4\} \leftrightarrow X(k) = \{\check{6}, 2 + j6, 2, 2 - j6\}$$

$$x(n)h(n) = \{\check{3}, -8, -1, 12\} \leftrightarrow \frac{1}{4}X(k) \circledast H(k) = \frac{1}{4}\{\check{24}, 16 + j80, -8, 16 - j80\}$$

3.8.3 Circular Correlation of Time-Domain Sequences

Let $x(n) \leftrightarrow X(k)$ and $h(n) \leftrightarrow H(k)$, both with period N. The circular cross-correlation of $x(n)$ and $h(n)$ is given by

$$r_{xh}(n) = \sum_{p=0}^{N-1} x(p)h^*(p-n) , n = 0, 1, \ldots, N-1 \leftrightarrow X(k)H^*(k)$$

Since $h^*(N-n) \leftrightarrow H^*(k)$, correlation operation is the same as convolution of $x(n)$ and $h^*(N-n)$. Unlike convolution, correlation operation is not commutative, in general.

$$r_{hx}(n) = r_{xh}(N-n) = \text{IDFT}(X^*(k)H(k))$$

For example,

$$x(n) = \{\check{1}, 4, 1, -3\} \leftrightarrow X(k) = \{\check{3}, -j7, 1, j7\}$$

$$h(n) = \{\check{3}, 2, 1, -4\} \leftrightarrow H(k) = \{\check{2}, 2 - j6, 6, 2 + j6\}$$

The cross-correlation output of $x(n)$ and $h(n)$ and its DFT are

$$\{\check{24}, 7, -18, -7\} \leftrightarrow X(k)H^*(k) = \{\check{6}, 42 - j14, 6, 42 + j14\}$$

The cross-correlation output of $h(n)$ and $x(n)$ and its DFT are

$$\{\check{24}, -7, -18, 7\} \leftrightarrow H(k)X^*(k) = \{\check{6}, 42, +j14, 6, 42 - j14\}$$

Correlation of a signal $x(n)$ with itself is the autocorrelation operation.

$$r_{xx}(n) = \text{IDFT}(|X(k)|^2)$$

The autocorrelation of $x(n)$ is

$$\{\check{27}, 2, -22, 2\} \leftrightarrow |X(k)|^2 = \{\check{9}, 49, 1, 49\}$$

3.9 Sum and Difference of Sequences

Since, with $k = 0$, the value of all the transform matrix coefficients is unity, $X(0)$ is sum of the input sequence values, $x(n)$. With N even and $k = N/2$, the transform matrix coefficients form the alternating sequence, $\{1, -1, 1, -1, \ldots, -1\}$. Therefore, $X(N/2)$ is the difference between the sum of the even- and odd-indexed values of $x(n)$.

$$X(0) = \sum_{n=0}^{N-1} x(n) \quad \text{and} \quad X\left(\frac{N}{2}\right) = \sum_{n=0,2}^{N-2} x(n) - \sum_{n=1,3}^{N-1} x(n)$$

Coefficient $X(0)$ is the sum of $x(n)$. Coefficient $X(\frac{N}{2})$ is the alternating sum of $x(n)$. Similarly, in the frequency domain,

$$x(0) = \frac{1}{N} \sum_{k=0}^{N-1} X(k) \quad \text{and} \quad x\left(\frac{N}{2}\right) = \frac{1}{N}\left(\sum_{k=0,2}^{N-2} X(k) - \sum_{k=1,3}^{N-1} X(k)\right)$$

Sample $x(0)$ is the average of $X(k)$. Sample $x(\frac{N}{2})$ is the alternating average of $X(k)$. For example,

$$x(n) = \{\check{1}, -4, 1, 3\} \leftrightarrow X(k) = \{\check{1}, j7, 3, -j7\}$$

$$X(0) = (\check{1} - 4 + 1 + 3) = 1 \qquad X(2) = (\check{1} + 4 + 1 - 3) = 3$$

$$x(0) = (\check{1} + j7 + 3 - j7)/4 = 1 \qquad x(2) = (\check{1} - j7 + 3 + j7)/4 = 1$$

3.10 Upsampling of a Sequence

Let $x(n) \leftrightarrow X(k)$, $n, k = 0, 1, \ldots, N - 1$. If we upsample $x(n)$ with zeros to get $x_u(n)$, $n = 0, 1, \ldots, LN - 1$ defined as

$$x_u(n) = \begin{cases} x(\frac{n}{L}) \text{ for } n = 0, L, 2L, \ldots, L(N-1) \\ 0 \qquad \text{otherwise} \end{cases}$$

where L is any positive integer, then

$$X_u(k) = X(k \bmod N), \quad k = 0, 1, \ldots, LN - 1$$

The DFT of the sequence $x_u(n)$ is given by

$$X_u(k) = \sum_{n=0}^{LN-1} x_u(n)e^{-j\frac{2\pi}{LN}nk}, \quad k = 0, 1, \ldots, LN - 1$$

Since we have nonzero input values only at intervals of L, we can substitute $n = mL$. Then, we get

$$X_u(k) = \sum_{m=0}^{N-1} x_u(mL)e^{-j\frac{2\pi}{LN}mLk}$$

$$= \sum_{m=0}^{N-1} x(m)e^{-j\frac{2\pi}{N}mk} = X(k \bmod N), \quad k = 0, 1, \ldots, LN - 1$$

since N-point DFT is periodic of period N. $X_u(k)$ is the L-fold repetition of $X(k)$.

Example 3.1 Let $L = 3$ and $x(n) = \{\check{-2}, 1, 3, 4\}$. Then, $X(k) = \{\check{6}, -5, +j3, -4,$
$-5 - j3\}$.

$$x_u(n) = \{\check{-2}, 0, 0, 1, 0, 0, 3, 0, 0, 4, 0, 0\}$$

$$\begin{aligned} X_u(k) = \{&\check{6}, -5, +j3, -4, -5 - j3, \quad 6, -5, +j3, -4, -5 - j3, \\ &6, -5, +j3, -4, -5 - j3\} \end{aligned}$$ ∎

Figure 3.2a shows the upsampled version of one cycle of $x(n) = \sin(\frac{2\pi}{8}n)$ by a factor $L = 2$. Its replicated spectrum is shown in (b). The spectrum of $x(n) = \cos(\frac{2\pi}{8}n)$ upsampled by a factor $L = 3$ is shown in (c). The replicated waveform,

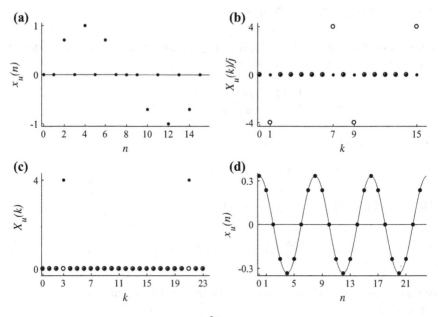

Fig. 3.2 a Upsampled version of $x(n) = \sin(\frac{2\pi}{8}n)$ by a factor $L = 2$; **b** its replicated spectrum; **c** the spectrum of $x(n) = \cos(\frac{2\pi}{8}n)$ upsampled by a factor $L = 3$; **d** the corresponding scaled and replicated waveform

with its amplitude reduced by a factor of 3, is shown in (d). Since the power of the spectrum is reduced by a factor of 3 after upsampling, the amplitude of the waveform is also reduced by a factor of 3.

3.11 Zero Padding the Data

Let $x(n) \leftrightarrow X(k)$, $n, k = 0, 1, \ldots, N - 1$. If we append $x(n)$ with zeros to get $x_z(n)$, $n = 0, 1, \ldots, LN - 1$ defined as

$$x_z(n) = \begin{cases} x(n) \text{ for } n = 0, 1, \ldots, N - 1 \\ 0 \quad \text{otherwise} \end{cases}$$

where L is any positive integer, then

$$X_z(Lk) = X(k), \ k = 0, 1, \ldots, N - 1$$

The DFT of the signal $x_z(n)$ is given by

$$X_z(k) = \sum_{n=0}^{LN-1} x_z(n) e^{-j\frac{2\pi}{LN}nk}, \ k = 0, 1, \ldots, LN - 1$$

Since $x_z(n)$ is zero for $n > N - 1$, we get

$$X_z(k) = \sum_{n=0}^{N-1} x_z(n) e^{-j\frac{2\pi}{LN}nk}, \ k = 0, 1, \ldots, LN - 1$$

Replacing k by Lk, we get

$$X_z(Lk) = \sum_{n=0}^{N-1} x_z(n) e^{-j\frac{2\pi}{N}nk} = X(k), \ k = 0, 1, \ldots, N - 1$$

Example 3.2 Let $L = 2$ and $x(n) = \{\check{1}, 2, 3, -4\}$. $X(k) = \{\check{2}, -2 - j6, 6, -2 + j6\}$. Then,

$$x_z(n) = \{\check{1}, 2, 3, 4, 0, 0, 0, 0\} \leftrightarrow X_z(k) = \{\check{2}, *, -2 - j6, *, 6, *, -2 + j6, *\}$$

The frequency increment of the spectrum is halved due to zero padding. Therefore, the spectral values in $X(k)$ become the even-indexed spectral values in $X_z(k)$. ∎

Figure 3.3a, b shows, respectively, a signal with eight samples and its spectrum. Figure 3.3c, d shows, respectively, the same signal padded up with eight zeros at the end and the corresponding spectrum. The even-indexed spectral values are the same

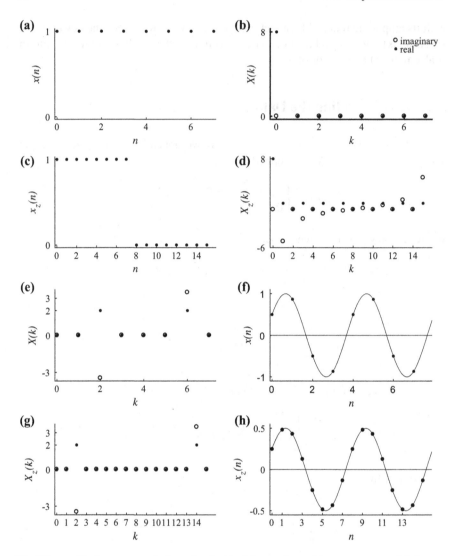

Fig. 3.3 **a** and **b** shows, respectively, a signal with eight samples and its spectrum; **c** and **d** shows, respectively, the same signal padded up with eight zeros at the end and the corresponding spectrum; **e** and **f** shows, respectively, a spectrum with eight samples and the corresponding time-domain signal; **g** and **h** shows, respectively, the same spectrum padded up with eight zeros in the middle of the spectrum and the corresponding time-domain signal

as those shown in Fig. 3.3b. The odd-indexed spectral values are not specified by this theorem. By zero padding at the end, we get interpolation of the spectral values.

A similar effect is observed in zero padding a spectrum. Fig. 3.3e, f shows, respectively, a spectrum with eight samples and the corresponding time-domain signal. Figure 3.3g, h shows, respectively, the same spectrum padded up with eight zeros in the middle of the spectrum (at the end in the center-zero format) and the corresponding time-domain signal. The even-indexed signal values are one-half of those shown in Fig. 3.3f. By zero padding in the frequency domain, we get interpolation of the time-domain samples.

3.12 Symmetry Properties

Let the N-point sequence and its DFT be complex-valued. Writing the DFT definition in the more explicit form using the rectangular form of the complex numbers with

$$x(n) = x_r(n) + jx_i(n), \quad X(k) = X_r(k) + jX_i(k),$$

$$e^{-j\frac{2\pi}{N}nk} = \cos\left(\frac{2\pi}{N}nk\right) - j\sin\left(\frac{2\pi}{N}nk\right),$$

we get

$$X_r(k) + jX_i(k) = \sum_{n=0}^{N-1} (x_r(n) + jx_i(n))\left(\cos\left(\frac{2\pi}{N}nk\right) - j\sin\left(\frac{2\pi}{N}nk\right)\right) \quad (3.2)$$

With $x(n)$ real-valued and even, the definition reduces to

$$X(k) = \sum_{n=0}^{N-1} x_r(n)\cos\left(\frac{2\pi}{N}nk\right)$$

and $X(k)$ is real and even, as $x_r(n)\cos\left(\frac{2\pi}{N}nk\right)$ is even and $x_r(n)\sin\left(\frac{2\pi}{N}nk\right)$ is odd.

$$x(n) \text{ real and even } \leftrightarrow X(k) \text{ real and even}$$

Figure 3.4c, d shows a real and even-symmetric signal and its real and even-symmetric spectrum. For example,

$$x(n) = \{1, 2, 3, 2\} \leftrightarrow X(k) = \{8, -2, 0, -2\}$$

With $x(n)$ real-valued and odd, the definition reduces to

$$X(k) = -j\sum_{n=0}^{N-1} x_r(n)\sin\left(\frac{2\pi}{N}nk\right)$$

and $X(k)$ is purely imaginary and odd, as $x_r(n)\cos\left(\frac{2\pi}{N}nk\right)$ is odd and $x_r(n)\sin\left(\frac{2\pi}{N}nk\right)$ is even.

$$x(n) \text{ real and odd } \leftrightarrow X(k) \text{ imaginary and odd}$$

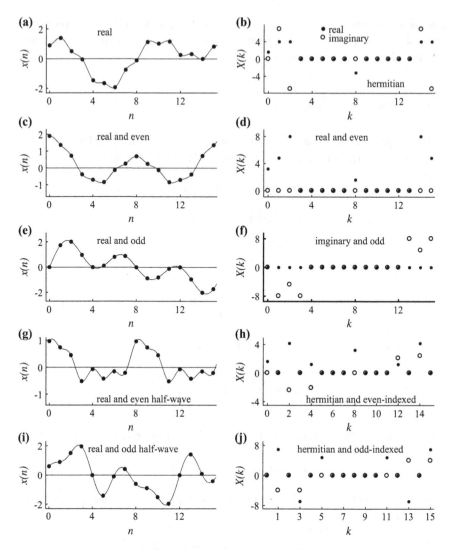

Fig. 3.4 a A real signal and **b** its hermitian-symmetric spectrum; **c** an even-symmetric real signal and **d** its real and even-symmetric spectrum; **e** an odd-symmetric real signal and **f** its imaginary and odd-symmetric spectrum; **g** a real signal with even half-wave symmetry and **h** its hermitian-symmetric spectrum with zero-valued odd-indexed harmonics; **i** a real signal with odd half-wave symmetry and **j** its hermitian-symmetric spectrum with zero-valued even-indexed harmonics

Figure 3.4e, f shows a real and odd-symmetric signal and its imaginary and odd-symmetric spectrum. For example,

$$x(n) = \{0, 2, 0, -2\} \leftrightarrow X(k) = \{0, -j4, 0, j4\}$$

In all these cases, the summation range can be reduced by a factor of 2 and the result multiplied by 2. Note that the real parts of the twiddle factors are the same for frequency index k and $N - k$, while they are the negatives of each other for the imaginary part.

The DFT coefficients for a real and even signal are real and even and those for a real and odd signal are imaginary and odd. Therefore, the real part of the DFT coefficients, $\text{Re}(X(k))$, of an arbitrary real signal $x(n)$ are the DFT coefficients for its even component $x_e(n)$ and $j\,\text{Im}(X(k))$ are those for its odd component $x_o(n)$. It follows that the DFT of a complex signal $x(n) + jy(n)$, where both the real and imaginary parts are even, is also even.

$$x(n) \text{ complex and even } \leftrightarrow X(k) \text{ complex and even}$$

The DFT of a complex signal $x(n) + jy(n)$, where both the real and imaginary parts are odd, is also odd.

$$x(n) \text{ complex and odd } \leftrightarrow X(k) \text{ complex and odd}$$

Since the real sinusoids are related to conjugate complex exponentials by the Euler's formula

$$A\cos(\omega n + \theta) = 0.5A\angle\theta\, e^{j\omega n} + 0.5A\angle-\theta\, e^{-j\omega n},$$

the transform of real-valued signals exhibits conjugate (or hermitian)- symmetry. Therefore, $X_r(k) = X_r(N - k)$ and $X_i(k) = -X_i(N - k)$. Replacing k by $\frac{N}{2} - k$, we get $X_r(\frac{N}{2} - k) = X_r(\frac{N}{2} + k)$ and $X_i(\frac{N}{2} - k) = -X_i(\frac{N}{2} + k)$.

$$x(n) \text{ real } \leftrightarrow X(k) \text{ hermitian}$$

Figure 3.4a, b shows a real signal and its conjugate-symmetric spectrum. For example,

$$x(n) = \{1, 2, 3, 4\} \leftrightarrow X(k) = \{10, -2 + j2, -2, -2 - j2\}$$

If a given periodic function $x(n)$ with period N satisfies the condition

$$x\left(n \pm \frac{N}{2}\right) = x(n),$$

then it is said to be even half-wave symmetric. Figure 3.4g, h shows an even half-wave symmetric signal and its even-indexed spectrum. For example,

$$x(n) = \{2, 0, 2, 0\} \leftrightarrow X(k) = \{4, 0, 4, 0\}$$

If a given periodic function $x(n)$ with period N satisfies the condition

$$x\left(n \pm \frac{N}{2}\right) = -x(n),$$

then it is said to be odd half-wave symmetric. Figure 3.4i, j shows an odd half-wave symmetric signal and its odd-indexed spectrum. For example,

$$x(n) = \left\{\frac{\sqrt{3}}{2}, 0.5, -\frac{\sqrt{3}}{2}, -0.5\right\} \leftrightarrow X(k) = \{0, \sqrt{3} - j, 0, \sqrt{3} + j\}$$

3.13 Parseval's Theorem

The representation of $x(n)$ by the DFT coefficients $X(k)$, which is obtained by an orthogonal transform, is the specification of the same signal $x(n)$ in the frequency domain. Therefore, $X(k)$ is, in every respect, just as complete and specific representation of $x(n)$. It is just a representation of $x(n)$ in another domain or just change of the independent variable. Consequently, the power of the signal over one period can be uniquely determined in either representation.

Let $x(n) \leftrightarrow X(k)$ with sequence length N. In the DFT, a signal is represented by complex exponentials with harmonic frequencies. Since the samples of a complex exponential lies on the unit circle, the power of an exponential over one period is N. The DFT coefficients $X(k)/N$ are the actual amplitudes of the exponentials. Therefore, the power of a complex exponential is $\frac{|X(k)|^2}{N^2} N = \frac{|X(k)|^2}{N}$. Summing the powers of all the constituent complex exponentials of the signal, we get the power over one period. Therefore,

$$\sum_{n=0}^{N-1} |x(n)|^2 = \frac{1}{N} \sum_{k=0}^{N-1} |X(k)|^2$$

Orthogonal transforms, such as the Fourier analysis, have the power preservation property.

Example 3.3 Consider the DFT pair

$$x(n) = \{3 + j2, -1 + j2, 2 - j1, -4 + j2\} \leftrightarrow X(k) = \{j5, 1, 10 - j3, 1 + j6\}$$

The sums of the squared magnitude of the data sequence and that of the DFT coefficients divided by 4 are equal and it is 43.   ∎

The generalized form of this theorem holds for two different signals as given by

$$\sum_{n=0}^{N-1} x(n)y^*(n) = \frac{1}{N} \sum_{k=0}^{N-1} X(k)Y^*(k)$$

This result follows from the circular correlation evaluated at output index zero.

3.14 Summary

- Properties relate the effect of an operation in one domain in the other. It helps to select the appropriate domain for interpretation and operation of signals. Further, the transforms of a set of related signals can be determined more easily.
- The DFT of a linear combination of a set of signals is the same linear combination of the DFT of the individual signals.
- The DFT of a N-point signal is periodic of period N in the frequency domain. The IDFT of the DFT is also periodic of period N in the time domain.
- The shift of a waveform in the time domain results in adding increments to the phase of its frequency components, which are linearly proportional to the respective frequency indices. There is no change in the amplitude.
- Multiplying a signal by a complex exponential with frequency index k_0 results in the shift of its spectrum by k_0 sample intervals.
- The DFT of the time-reversal $x(N - n)$ of a N-point signal is the time-reversed version of the DFT $X(k)$ of $x(n)$, $X(N - n)$.
- Computing the DFT twice in succession of a signal $x(n)$ yields N times the time-reversal of $x(n)$.
- Convolution in the time domain corresponds to multiplication in the frequency domain.
- The scaled circular convolution of the two frequency-domain sequences $X(k)$ and $H(K)$ corresponds to multiplication of the corresponding sequences in the time domain.
- The DFT of the conjugate of $x(n)$ of length N, $x^*(n)$, is the frequency-reversed and conjugated version of that of $x(n)$, $X^*(N - k)$.
- The DFT of the correlation of two sequences is the product of the DFT of the first sequence multiplied by the conjugate of the DFT of the second sequence.
- Transform values $X(0)$ and $X(N/2)$ of a N-point sequence are, respectively, the sum and alternating sum of $x(n)$. Values $x(0)$ and $x(N/2)$ of a N-point sequence are, respectively, the sum and alternating sum of the DFT of $x(n)$, $X(k)$, divided by N.
- Upsampling a signal by a factor of L results in the L-fold repetition of its DFT.
- Zero padding a signal by a factor of L and computing its DFT results in a spectrum, every Lth sample of which is the same as those of the DFT of the signal.
- The DFT of a real-valued signal is conjugate-symmetric.

- The DFT of a real-valued and even-symmetric signal is real-valued and even-symmetric.
- The DFT of a real-valued and odd-symmetric signal is imaginary-valued and odd-symmetric.
- The average power of a signal can be obtained either from its time-domain or frequency-domain representation.

Exercises

3.1 Given two 4-point sequences $x(n)$ and $y(n)$, find the sequence $z(n) = 3x(n) - 2y(n)$. Find the DFT $Z(k)$ of $z(n)$ using the linearity property with $X(k)$ and $Y(k)$ and also directly from the definition. Verify that they are the same. Find the IDFT of $(Z(k) + 2Y(k))/3$ to get back $x(n)$.

3.1.1

$$x(n) = \{\check{2}, 1, 3, 4\}, y(n) = \{\check{2}, 1, -3, 4\}$$

* 3.1.2

$$x(n) = \{\check{1}, 2, 3, 4\}, y(n) = \{\check{-2}, 1, -3, 4\}$$

3.1.3

$$x(n) = \{\check{2}, 2, 1, 4\}, y(n) = \{\check{2}, 1, 3, -2\}$$

3.2 Given a 4-point sequence $x(n)$, find its DFT $X(k)$. Find the values of $x(23)$ and $X(-7)$ using the periodicity property. Compute the IDFT of $X(k)$ with index $n = 23$. Compute the DFT of $x(n)$ with index $k = -7$. Verify that these values are the same as those found using the periodicity property.

3.2.1

$$x(n) = \{\check{2}, 1, 3, 3\}$$

3.2.2

$$x(n) = \{\check{-1}, 1, 4, 3\}$$

* 3.2.3

$$x(n) = \{\check{4}, 2, 3, 2\}$$

3.3 Given a 4-point sequence $x(n)$, find its DFT $X(k)$. Using the time-shift property, find the DFT of $x(n - k)$. Form the sequence $x(n - k)$ and find its DFT. Verify that both the DFTs are the same.

* 3.3.1

$$x(n) = \{\check{3}, 1, 4, 3\}, \quad k = -1$$

3.3.2

$$x(n) = \{\check{-1}, 1, 1, 3\}, \quad k = 3$$

3.3.3
$$x(n) = \{\check{3}, 2, 1, 2\}, \quad k = 2$$

3.4 Given a 4-point sequence $x(n)$, find its DFT $X(k)$. Using the frequency-shift property, find the DFT of $xm(n) = e^{j\frac{2\pi}{4}kn}x(n)$. Form the sequence $xm(n)$ and find its DFT. Verify that both the DFTs are the same.

3.4.1
$$x(n) = \{\check{1}, 2, 3, 1\}, \quad k = 1$$

* 3.4.2
$$x(n) = \{\check{-2}, 2, 1, 3\}, \quad k = -3$$

3.4.3
$$x(n) = \{\check{1}, 4, 1, 2\}, \quad k = 2$$

3.5 Given a 4-point sequence $x(n)$, find its DFT $X(k)$. Using the time-reversal property, find the DFT of $xr(n) = x(4 - n)$. Form the sequence $xr(n)$ and find its DFT. Verify that both the DFTs are the same.

3.5.1
$$x(n) = \{\check{3}, -2, 3, 1\}$$

* 3.5.2
$$x(n) = \{\check{-3}, 1, 1, 3\}$$

3.5.3
$$x(n) = \{\check{2}, -4, 1, 2\}$$

3.6 Given a 4-point sequence $x(n)$, find its DFT $X(k)$. Find the DFT of $X(k)$. Verify that it is equal to $4x(4 - n)$.

3.6.1
$$x(n) = \{\check{3}, -2, 3, 1\}$$

3.6.2
$$x(n) = \{\check{2}, -2, 1, 1\}$$

* 3.6.3
$$x(n) = \{\check{4}, -2, 2, 1\}$$

3.7 Given a 4-point sequence $x(n)$, find its DFT $X(k)$. Verify that $x^*(n) \leftrightarrow X^*(4 - k)$.

3.7.1
$$x(n) = \{\check{1}, j, 3, 2\}$$

3.7.2

$$x(n) = \{\check{2}, -3, j, 2\}$$

* 3.7.3

$$x(n) = \{\check{1}, -4, -2, j\}$$

3.8 Given two 4-point sequences $x(n)$ and $h(n)$, find the circular convolution output $y(n) = x(n) \circledast h(n)$, using the DFT and IDFT. Verify the output by using the time-domain convolution expression.

* 3.8.1

$$x(n) = \{\check{1}, 4, 3, 2\}, \quad h(n) = \{\check{1}, -1, 3, 2\}$$

3.8.2

$$x(n) = \{\check{3}, 4, 3, 2\}, \quad h(n) = \{\check{1}, -1, 4, 2\}$$

3.8.3

$$x(n) = \{\check{1}, 4, -3, -2\}, \quad h(n) = \{\check{2}, -2, 3, 2\}$$

3.9 Given two 4-point sequences $x(n)$ and $h(n)$, find the DFT of their product $y(n) = x(n)h(n)$, using the frequency-domain convolution property. Verify the computed DFT with the DFT of $y(n)$.

3.9.1

$$x(n) = \{\check{3}, 1, 4, 2\}, \quad h(n) = \{\check{3}, -2, 1, 2\}$$

* 3.9.2

$$x(n) = \{\check{1}, 2, 1, 2\}, \quad h(n) = \{\check{3}, -3, 4, 2\}$$

3.9.3

$$x(n) = \{\check{2}, 1, 3, -2\}, \quad h(n) = \{\check{2}, 1, 4, 3\}$$

3.10 Given two 4-point sequences $x(n)$ and $h(n)$, find the circular cross-correlation output $r_{xh}(n)$, using the DFT and IDFT. Verify the output by using the time-domain correlation expression.

3.10.1

$$x(n) = \{\check{2}, 3, 3, 2\}, \quad h(n) = \{\check{1}, -3, 3, 2\}$$

* 3.10.2

$$x(n) = \{\check{3}, -1, 3, 2\}, \quad h(n) = \{\check{1}, -1, 2, 2\}$$

3.10.3

$$x(n) = \{\check{1}, 3, -3, -2\}, \quad h(n) = \{\check{2}, -2, 1, 2\}$$

3.11 Given a 4-point sequence $x(n)$, find its circular autocorrelation output $r_{xx}(n)$, using the DFT and IDFT. Verify the output by using the time-domain correlation expression.

3.11.1
$$x(n) = \{\check{4}, 3, 2, 1\}$$

3.11.2
$$x(n) = \{\check{1}, -2, 3, 4\}$$

* 3.11.3
$$x(n) = \{\check{1}, -3, -2, -2\}$$

3.12 Given a 4-point sequence $x(n)$, find its DFT $X(k)$. Verify that $x(0)$ and $x(2)$ are, respectively, the sum and alternating sum of the its DFT $X(k)$ divided by 4. Verify that $X(0)$ and $X(2)$ are, respectively, the sum and alternating sum of $x(n)$.

* 3.12.1
$$x(n) = \{\check{1}, 3, 2, 1\}$$

3.12.2
$$x(n) = \{\check{3}, -2, 3, 4\}$$

3.12.3
$$x(n) = \{\check{4}, -3, -2, -2\}$$

3.13 Given a 2-point sequence $x(n)$, find its DFT $X(k)$. Find the DFT of the upsampled $x(n)$ by a factor of 2 using the theorem. Find the IDFT of this DFT and verify that we get upsampled $x(n)$.

3.13.1
$$x(n) = \{\check{1}, 3\}$$

3.13.2
$$x(n) = \{\check{3}, -2\}$$

* 3.13.3
$$x(n) = \{\check{4}, -3\}$$

3.14 Given a 2-point sequence $x(n)$, find its DFT $X(k)$. Find the IDFT, $x\_rep(n)$, of the upsampled $X(k)$ by a factor of 2. Verify that $2x\_rep(n)$ is the replicated $x(n)$.

3.14.1
$$x(n) = \{\check{4}, 2\}$$

3.14.2
$$x(n) = \{\check{1}, 3\}$$

* 3.14.3
$$x(n) = \{\check{-2}, 3\}$$

3.15 Given a 2-point sequence $x(n)$, find its DFT $X(k)$. Zero pad $x(n)$ by 2 zeros to get $xz(n)$. Find the DFT, $XZ(k)$, of $xz(n)$ and verify that the even-indexed samples of $XZ(k)$ are the same as $X(k)$.

* 3.15.1
$$x(n) = \{\check{1}, 2\}$$

3.15.2
$$x(n) = \{\check{5}, 3\}$$

3.15.3
$$x(n) = \{\check{-4}, 3\}$$

3.16 Given a 4-point sequence $x(n)$, find its DFT $X(k)$. What is the type of $x(n)$.

3.16.1
$$x(n) = \{\check{1}, 2, 1, 2\}$$

3.16.2
$$x(n) = \{\check{3}, 1, -3, -1\}$$

3.16.3
$$x(n) = \{\check{-4}, 3, 1, 3\}$$

3.16.4
$$x(n) = \{\check{5}, 3, 2, 4\}$$

* 3.16.5
$$x(n) = \{\check{0}, -3, 0, 3\}$$

3.17 Given a 4-point sequence $x(n)$, find its DFT $X(k)$. Verify Parseval's theorem.

3.17.1
$$x(n) = \{\check{1}, 2, 3, 1\}$$

* 3.17.2
$$x(n) = \{\check{5}, 3, 2, 4\}$$

3.17.3
$$x(n) = \{\check{-4}, 3, 1, 3\}$$

Chapter 4
Two-Dimensional DFT

In 1-D Fourier analysis, an arbitrary waveform, which is a curve, is represented in terms of sinusoidal waveforms of harmonic frequencies. For example, the DFT coefficient $X(k)$ is the coefficient of the complex exponential with frequency index k. Fourier analysis can be extended to higher dimensions. The L-dimensional DFT of

$$x(n_1, n_2, \ldots, n_L), \; n_1 = 0, 1, \ldots, N_1 - 1,$$
$$n_2 = 0, 1, \ldots, N_2 - 1, \ldots, n_L = 0, 1, \ldots, N_L - 1$$

is defined as

$$X(k_1, k_2, \ldots, k_L) = \sum_{n_1=0}^{N_1-1} \sum_{n_2=0}^{N_2-1} \cdots \sum_{n_L=0}^{N_L-1} x(n_1, n_2, \ldots, n_L) e^{-j2\pi\left(\frac{k_1}{N_1}n_1 + \frac{k_2}{N_2}n_2 + \cdots + \frac{k_L}{N_L}n_L\right)}$$

$$(4.1)$$

In 2-D Fourier analysis, an arbitrary waveform, which is a surface, is represented in terms of sinusoidal surfaces of harmonic frequencies in two orthogonal directions. For example, the 2-D DFT coefficient $X(k, l)$ is the coefficient of the 2-D complex exponential with frequency indices (k, l). In determining the coefficients, the orthogonality principle is used, as in the case of 1-D signals. However, it is easier to interpret and compute the 2-D DFT as two sets of 1-D DFTs in two orthogonal directions. The 1-D DFT can be generalized to three or more dimensions in a way similar to that of the 2-DFT. The task in 2-D DFT is to represent images, such as that shown in Fig. 4.1, in terms of sinusoidal surfaces.

© Springer Nature Singapore Pte Ltd. 2018
D. Sundararajan, *Fourier Analysis—A Signal Processing Approach*,
https://doi.org/10.1007/978-981-13-1693-7_4

Fig. 4.1 A 256 × 256 8-bit image

4.1   Two-Dimensional DFT as Two Sets of 1-D DFTs

The samples of a N-point 1-D sinusoid

$$x(m) = \cos\left(\frac{2\pi}{N}(mk)\right), \ m = 0, 1, \ldots, N - 1$$

constitute a 1-D sequence of numbers. The samples of a $N \times N$ 2-D sinusoid

$$x(m, n) = \cos\left(\frac{2\pi}{N}(mk + nl)\right), \ m, n = 0, 1, \ldots, N - 1$$

constitute a 2-D sequence of numbers. A 2-D sinusoidal sequence is a stack of a number of phase-shifted 1-D sinusoidal sequences.

$$\left\{x_0(m) = \cos\left(\frac{2\pi}{N}(mk + 0l)\right), \ x_1(m) = \cos\left(\frac{2\pi}{N}(mk + 1l)\right),\right.$$
$$\left. x_2(m) = \cos\left(\frac{2\pi}{N}(mk + 2l)\right), \ \ldots\right\}$$

The complex exponential

$$x(m, n) = e^{j\left(\frac{2\pi}{N}(mk+nl)\right)}$$

constitute a 2-D sequence of complex numbers. A 2-D complex exponential sequence is a stack of a number of phase-shifted 1-D complex exponential sequences.

$$\left\{ x_0(m) = e^{j\left(\frac{2\pi}{N}(mk+0l)\right)}, \quad x_1(m) = e^{j\left(\frac{2\pi}{N}(mk+1l)\right)}, \quad x_2(m) = e^{j\left(\frac{2\pi}{N}(mk+2l)\right)}, \quad \ldots \right\}$$

The 8×8 complex exponential, with $k, l = 1$,

$$x(m, n) = e^{j\left(\frac{2\pi}{8}(m+n)\right)}, \quad m, n = 0, 1, \ldots, 7$$

can be expressed as a stack of eight phase-shifted 1-D complex exponential sequences.

$$\left\{ x_0(m) = e^{j\left(\frac{2\pi}{8}(m+0)\right)} = e^{j\frac{2\pi}{8}0} e^{j\frac{2\pi}{8}m}, \quad x_1(m) = e^{j\left(\frac{2\pi}{8}(m+1)\right)} - e^{j\frac{2\pi}{8}1} e^{j\frac{2\pi}{8}m}, \right.$$

$$x_2(m) = e^{j\left(\frac{2\pi}{8}(m+2)\right)} = e^{j\frac{2\pi}{8}2} e^{j\frac{2\pi}{8}m}, \quad x_3(m) = e^{j\left(\frac{2\pi}{8}(m+3)\right)} = e^{j\frac{2\pi}{8}3} e^{j\frac{2\pi}{8}m},$$

$$x_4(m) = e^{j\left(\frac{2\pi}{8}(m+4)\right)} = e^{j\frac{2\pi}{8}4} e^{j\frac{2\pi}{8}m}, \quad x_5(m) = e^{j\left(\frac{2\pi}{8}(m+5)\right)} = e^{j\frac{2\pi}{8}5} e^{j\frac{2\pi}{8}m},$$

$$\left. x_6(m) = e^{j\left(\frac{2\pi}{8}(m+6)\right)} = e^{j\frac{2\pi}{8}6} e^{j\frac{2\pi}{8}m}, \quad x_7(m) = e^{j\left(\frac{2\pi}{8}(m+7)\right)} = e^{j\frac{2\pi}{8}7} e^{j\frac{2\pi}{8}m} \right\}$$

The real and imaginary values of $x(m, n)$, respectively, are

$$\text{Re}(x(m, n)) = \frac{1}{\sqrt{2}} \begin{bmatrix} \sqrt{2} & 1 & 0 & -1 & -\sqrt{2} & -1 & 0 & 1 \\ 1 & 0 & -1 & -\sqrt{2} & -1 & 0 & 1 & \sqrt{2} \\ 0 & -1 & -\sqrt{2} & -1 & 0 & 1 & \sqrt{2} & 1 \\ -1 & -\sqrt{2} & -1 & 0 & 1 & \sqrt{2} & 1 & 0 \\ -\sqrt{2} & -1 & 0 & 1 & \sqrt{2} & 1 & 0 & -1 \\ -1 & 0 & 1 & \sqrt{2} & 1 & 0 & -1 & -\sqrt{2} \\ 0 & 1 & \sqrt{2} & 1 & 0 & -1 & -\sqrt{2} & -1 \\ 1 & \sqrt{2} & 1 & 0 & -1 & -\sqrt{2} & -1 & 0 \end{bmatrix}$$

$$\text{Im}(x(m, n)) = \frac{1}{\sqrt{2}} \begin{bmatrix} 0 & 1 & \sqrt{2} & 1 & 0 & -1 & -\sqrt{2} & -1 \\ 1 & \sqrt{2} & 1 & 0 & -1 & -\sqrt{2} & -1 & 0 \\ \sqrt{2} & 1 & 0 & -1 & -\sqrt{2} & -1 & 0 & 1 \\ 1 & 0 & -1 & -\sqrt{2} & -1 & 0 & 1 & \sqrt{2} \\ 0 & -1 & -\sqrt{2} & -1 & 0 & 1 & \sqrt{2} & 1 \\ -1 & -\sqrt{2} & -1 & 0 & 1 & \sqrt{2} & 1 & 0 \\ -\sqrt{2} & -1 & 0 & 1 & \sqrt{2} & 1 & 0 & -1 \\ -1 & 0 & 1 & \sqrt{2} & 1 & 0 & -1 & -\sqrt{2} \end{bmatrix}$$

The 8×8 image representations of the real and imaginary parts of $x(m, n) = e^{j\left(\frac{2\pi}{8}(m+n)\right)}$ are shown, respectively, in Fig. 4.2a, b. The white and black pixels correspond, respectively, to 1 and -1. The frequency in both the directions is 1/8

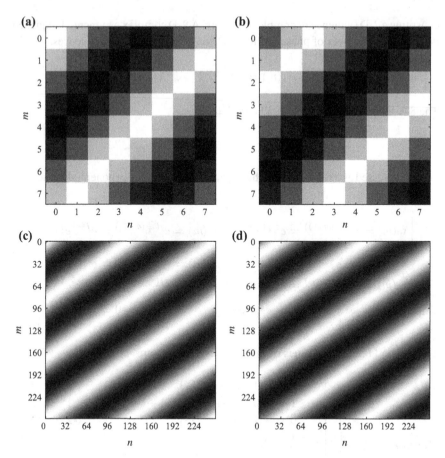

Fig. 4.2 **a** The real part of the 8×8 image representation of $x(m, n) = e^{j(\frac{2\pi}{8}(m+n))} = \cos(\frac{2\pi}{8}(m + n)) + j \sin(\frac{2\pi}{8}(m + n))$; **b** the imaginary part; **c** the real part of the 256×256 image representation of $x(m, n) = e^{j(\frac{2\pi}{256}(3m+2n))} = \cos(\frac{2\pi}{256}(3m + 2n)) + j \sin(\frac{2\pi}{256}(3m + 2n))$; **d** the imaginary part

cycles/sample. The 256×256 image representations of the real and imaginary parts of $x(m, n) = e^{j(\frac{2\pi}{256}(3m+2n))}$ are shown, respectively, in Fig. 4.2c, d. The frequency along the m and n directions are 3/256 cycles/sample and 2/256 cycles/sample, respectively.

4.1.1 Computation of the 2-D DFT

Taking the 1-D DFT of the exponentials along the columns of $x(m, n) = e^{j(\frac{2\pi}{8}(m+n))}$, with $m = 0, 1, \ldots, 7$, we get the transform coefficients

$$X(k,n) = \begin{bmatrix} 0 & 0 & 0 & 0 & 0 & 0 & 0 & 0 \\ 8e^{j\frac{2\pi}{8}0} & 8e^{j\frac{2\pi}{8}1} & 8e^{j\frac{2\pi}{8}2} & 8e^{j\frac{2\pi}{8}3} & 8e^{j\frac{2\pi}{8}4} & 8e^{j\frac{2\pi}{8}5} & 8e^{j\frac{2\pi}{8}6} & 8e^{j\frac{2\pi}{8}7} \\ 0 & 0 & 0 & 0 & 0 & 0 & 0 & 0 \\ 0 & 0 & 0 & 0 & 0 & 0 & 0 & 0 \\ 0 & 0 & 0 & 0 & 0 & 0 & 0 & 0 \\ 0 & 0 & 0 & 0 & 0 & 0 & 0 & 0 \\ 0 & 0 & 0 & 0 & 0 & 0 & 0 & 0 \\ 0 & 0 & 0 & 0 & 0 & 0 & 0 & 0 \end{bmatrix}$$

Taking the 1-D DFT of the partial transform $X(k,n)$ along the rows, we get the 2-D DFT coefficients of $e^{j(\frac{2\pi}{8}(m+n))}$ in the standard and center-zero formats, respectively, as

$$X(k,l) = \begin{bmatrix} \mathbf{0} & 0 & 0 & 0 & 0 & 0 & 0 & 0 \\ 0 & 64 & 0 & 0 & 0 & 0 & 0 & 0 \\ 0 & 0 & 0 & 0 & 0 & 0 & 0 & 0 \\ 0 & 0 & 0 & 0 & 0 & 0 & 0 & 0 \\ 0 & 0 & 0 & 0 & 0 & 0 & 0 & 0 \\ 0 & 0 & 0 & 0 & 0 & 0 & 0 & 0 \\ 0 & 0 & 0 & 0 & 0 & 0 & 0 & 0 \\ 0 & 0 & 0 & 0 & 0 & 0 & 0 & 0 \end{bmatrix} \qquad X(k,l) = \begin{bmatrix} 0 & 0 & 0 & 0 & 0 & 0 & 0 & 0 \\ 0 & 0 & 0 & 0 & 0 & 0 & 0 & 0 \\ 0 & 0 & 0 & 0 & 0 & 0 & 0 & 0 \\ 0 & 0 & 0 & 0 & 0 & 0 & 0 & 0 \\ 0 & 0 & 0 & 0 & \mathbf{0} & 0 & 0 & 0 \\ 0 & 0 & 0 & 0 & 0 & 64 & 0 & 0 \\ 0 & 0 & 0 & 0 & 0 & 0 & 0 & 0 \\ 0 & 0 & 0 & 0 & 0 & 0 & 0 & 0 \end{bmatrix}$$

By swapping the quadrants of the spectrum, we get one format from the other. The nonzero DFT coefficient $X(1,1) = 64$ is the scaled amplitude of the exponential in the frequency domain. Taking the 1-D IDFT of $X(k,l)$ along the rows, we get $X(k,n)$. Now, taking the 1-D IDFT of $X(k,n)$ along the columns, we get back $x(m,n)$.

Since

$$\cos\left(\frac{2\pi}{8}(m+n)\right) = \frac{e^{j\left(\frac{2\pi}{8}(m+n)\right)} + e^{-j\left(\frac{2\pi}{8}(m+n)\right)}}{2},$$

the 2-D transform coefficients of $\cos(\frac{2\pi}{8}(m+n))$ in the standard and center-zero formats, respectively, are

$$X(k,l) = \begin{bmatrix} \mathbf{0} & 0 & 0 & 0 & 0 & 0 & 0 & 0 \\ 0 & 32 & 0 & 0 & 0 & 0 & 0 & 0 \\ 0 & 0 & 0 & 0 & 0 & 0 & 0 & 0 \\ 0 & 0 & 0 & 0 & 0 & 0 & 0 & 0 \\ 0 & 0 & 0 & 0 & 0 & 0 & 0 & 0 \\ 0 & 0 & 0 & 0 & 0 & 0 & 0 & 0 \\ 0 & 0 & 0 & 0 & 0 & 0 & 0 & 0 \\ 0 & 0 & 0 & 0 & 0 & 0 & 0 & 32 \end{bmatrix} \qquad X(k,l) = \begin{bmatrix} 0 & 0 & 0 & 0 & 0 & 0 & 0 & 0 \\ 0 & 0 & 0 & 0 & 0 & 0 & 0 & 0 \\ 0 & 0 & 0 & 0 & 0 & 0 & 0 & 0 \\ 0 & 0 & 0 & 32 & 0 & 0 & 0 & 0 \\ 0 & 0 & 0 & 0 & \mathbf{0} & 0 & 0 & 0 \\ 0 & 0 & 0 & 0 & 0 & 32 & 0 & 0 \\ 0 & 0 & 0 & 0 & 0 & 0 & 0 & 0 \\ 0 & 0 & 0 & 0 & 0 & 0 & 0 & 0 \end{bmatrix}$$

Since

$$\sin\left(\frac{2\pi}{8}(m+n)\right) = \frac{e^{j\left(\frac{2\pi}{8}(m+n)\right)} - e^{-j\left(\frac{2\pi}{8}(m+n)\right)}}{j2} = \frac{-je^{j\left(\frac{2\pi}{8}(m+n)\right)} + je^{-j\left(\frac{2\pi}{8}(m+n)\right)}}{2},$$

the 2-D transform coefficients of $\sin(\frac{2\pi}{8}(m+n)))$ in the standard and center-zero formats, respectively, are

$$X(k,l) = \begin{bmatrix} 0 & 0 & 0 & 0 & 0 & 0 & 0 & 0 \\ 0 & -j32 & 0 & 0 & 0 & 0 & 0 & 0 \\ 0 & 0 & 0 & 0 & 0 & 0 & 0 & 0 \\ 0 & 0 & 0 & 0 & 0 & 0 & 0 & 0 \\ 0 & 0 & 0 & 0 & 0 & 0 & 0 & 0 \\ 0 & 0 & 0 & 0 & 0 & 0 & 0 & 0 \\ 0 & 0 & 0 & 0 & 0 & 0 & 0 & 0 \\ 0 & 0 & 0 & 0 & 0 & 0 & 0 & j32 \end{bmatrix} \quad X(k,l) = \begin{bmatrix} 0 & 0 & 0 & 0 & 0 & 0 & 0 & 0 \\ 0 & 0 & 0 & 0 & 0 & 0 & 0 & 0 \\ 0 & 0 & 0 & 0 & 0 & 0 & 0 & 0 \\ 0 & 0 & 0 & j32 & 0 & 0 & 0 & 0 \\ 0 & 0 & 0 & 0 & \mathbf{0} & 0 & 0 & 0 \\ 0 & 0 & 0 & 0 & 0 & -j32 & 0 & 0 \\ 0 & 0 & 0 & 0 & 0 & 0 & 0 & 0 \\ 0 & 0 & 0 & 0 & 0 & 0 & 0 & 0 \end{bmatrix}$$

The general form of the 2-D complex exponential with a complex coefficient is

$$x(m, n) = Ae^{j\theta}e^{j\left(\frac{2\pi}{N}(km+ln)\right)} = Ae^{j\left(\frac{2\pi}{N}(km+ln)+\theta\right)}$$

An arbitrary 2-D surface is expressed as a linear combination of the sinusoidal surfaces of all the constituent complex sinusoids. The 2-D DFT determines the coefficients of the surfaces and the 2-D IDFT reconstructs the 2-D signal back. So far, we considered surfaces with coefficient 1 and zero phase. Let us consider some other sinusoids.

Figure 4.3a, b show the 64×64 sinusoidal surface $x(m, n) = \cos(\frac{2\pi}{64}(2m + n))$ and its spectrum. The frequency along the m direction is 2 and that in the n direction is 1, as can be seen in (a). As its phase is zero, the coefficients are $X(2, 1) = (64)(64)/2 = 2048$ and $X(62, 63) = 2048$. The 64×64 sinusoidal surface $x(m, n) = 2\cos(\frac{2\pi}{64}(2m + n) + \frac{\pi}{2})$ and its spectrum are shown in Fig. 4.3c, d. Figure 4.3e, f show the 64×64 sinusoidal surface $x(m, n) = 3\cos(\frac{2\pi}{64}(2m + n) + \frac{\pi}{4})$ and its spectrum. As its amplitude is 3 and phase $\pi/4$, the coefficients are

$$X(2, 1) = 3(64)(64)\left(\cos\left(\frac{\pi}{4}\right) + j\sin\left(\frac{\pi}{4}\right)\right)/2$$
$$= 4344.5 + j4344.5 \text{ and } X(62, 63) = 4344.5 - j4344.5$$

Figure 4.4a shows a 32×32 square pulse $x(m, n)$, $m = 0, 1, \ldots, 31$, $n = 0, 1, \ldots, 31$ with $x(8 : 23, 8 : 23) = 1$ and the rest of the values zero. Figure 4.4b shows its 32×32 log magnitude spectrum in the center-zero format, $\log_{10}(1 + |X(k, l)|)$. Figure 4.4c shows the Fourier reconstructed pulse with just the DC component. The DC value is the average of $x(m, n)$, which is $256/1024 = 0.25$. The spectral value $X(0, 0)$ in log magnitude format

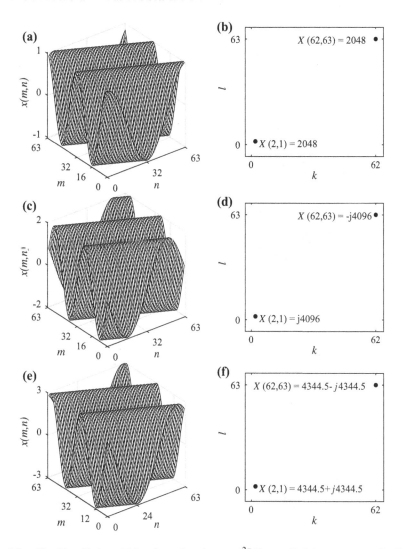

Fig. 4.3 **a** The 64 × 64 sinusoidal surface $x(m, n) = \cos(\frac{2\pi}{64}(2m + n))$; **b** its spectrum; **c** the 64 × 64 sinusoidal surface $x(m, n) = 2\cos(\frac{2\pi}{64}(2m + n) + \frac{\pi}{2})$; **d** its spectrum; **e** the 64 × 64 sinusoidal surface $x(m, n) = 3\cos(\frac{2\pi}{64}(2m + n) + \frac{\pi}{4})$; **f** its spectrum

$$X(0, 0) = \log_{10}(1 + 256) = \log_{10}(257) = 2.4099$$

is shown in (b). Figure 4.4d shows the Fourier reconstructed pulse with the first 4 × 4 frequency components around $X(0, 0)$. Figure 4.4e shows the Fourier reconstructed pulse with the first 8 × 8 frequency components around $X(0, 0)$. Figure 4.4f shows the Fourier reconstructed pulse with all the frequency components. As more and

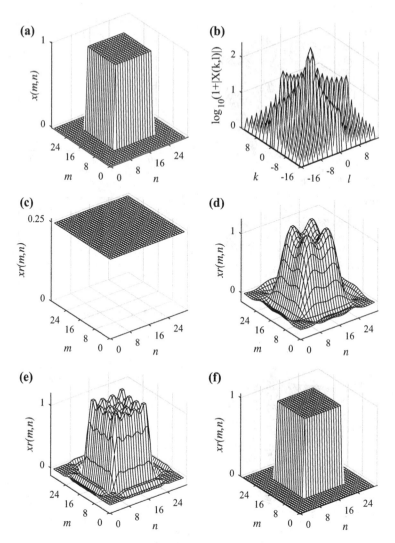

Fig. 4.4 **a** A A 32 × 32 square pulse; **b** its log magnitude spectrum in the center-zero format; **c** its Fourier reconstruction with just the DC component; **d** with the first 4 × 4 frequency components; **e** with the first 8 × 8 frequency components; **f** with all the frequency components

more of frequency components are used in the reconstruction, the signal approaches its original version through constructive and destructive interference of the frequency components.

Let us compute a 2-D DFT similar to this problem but of size 4 × 4. Let the input signal $x(m, n)$ be

$$x(m, n) = \begin{bmatrix} 0 & 0 & 0 & 0 \\ 0 & 1 & 1 & 0 \\ 0 & 1 & 1 & 0 \\ 0 & 0 & 0 & 0 \end{bmatrix} = \begin{bmatrix} 0 \\ 1 \\ 1 \\ 0 \end{bmatrix} \begin{bmatrix} 0 & 1 & 1 & 0 \end{bmatrix}$$

This signal can be expressed as the product of a column vector and a row vector. This type of signals is called separable signals. The 2-D DFT can be computed using 1-D DFT. The 1-D DFT of $\{0, 1, 1, 0\}$ is $\{2, -1 - j, 0, -1 + j\}$. The 2-D DFT of $x(m, n)$ is

$$\begin{bmatrix} 2 \\ -1-j \\ 0 \\ -1+j \end{bmatrix} \begin{bmatrix} 2 & -1-j & 0 & -1+j \end{bmatrix} = \begin{bmatrix} 4 & -2-j2 & 0 & -2+j2 \\ -2-j2 & j2 & 0 & 2 \\ 0 & 0 & 0 & 0 \\ -2+j2 & 2 & 0 & -j2 \end{bmatrix} = X(k, l)$$

It is to be noted that 2-D DFT is computed using 1-D DFTs, which becomes even simpler in the case of separable signals.

Example 4.1 Using the row–column method, find the 2-D DFT, $X(k, l)$, of the 4×4 signal $x(m, n)$.

$$n \rightarrow$$

$$\begin{matrix} m \\ \downarrow \end{matrix} \quad x(m, n) = \begin{bmatrix} 1 & -1 & 3 & 2 \\ 2 & 1 & 2 & 4 \\ 1 & -1 & 2 & -2 \\ 3 & 1 & 2 & 2 \end{bmatrix}$$

Find the 2-D IDFT of $X(k, l)$ to get back $x(m, n)$.

Solution
The 2-D DFT computation is decomposable. Therefore, it can be expressed as the product the 4×4 1-D DFT transform matrix, $x(m, n)$, and the 4×4 1-D DFT transform matrix.

$$\begin{bmatrix} 1 & 1 & 1 & 1 \\ 1 & -j & -1 & j \\ 1 & -1 & 1 & -1 \\ 1 & j & -1 & -j \end{bmatrix} \begin{bmatrix} 1 & -1 & 3 & 2 \\ 2 & 1 & 2 & 4 \\ 1 & -1 & 2 & -2 \\ 3 & 1 & 2 & 2 \end{bmatrix} \begin{bmatrix} 1 & 1 & 1 & 1 \\ 1 & -j & -1 & j \\ 1 & -1 & 1 & -1 \\ 1 & j & -1 & -j \end{bmatrix}$$

Multiplying $x(m, n)$ by the left transform matrix yields the 1-D DFT of the columns.

$$\begin{bmatrix} 7+j0 & 0+j0 & 9+j0 & 6+j0 \\ 0+j1 & 0+j0 & 1+j0 & 4-j2 \\ -3+j0 & -4+j0 & 1+j0 & -6+j0 \\ 0-j1 & 0+j0 & 1+j0 & 4+j2 \end{bmatrix} \begin{bmatrix} 1 & 1 & 1 & 1 \\ 1 & -j & -1 & j \\ 1 & -1 & 1 & -1 \\ 1 & j & -1 & -j \end{bmatrix}$$

Multiplying partially transformed signal matrix by the right transform matrix yields the 1-D DFT of the rows, and hence, the 2-D DFT of $x(m, n)$.

$$
\begin{array}{c}
k \\
\downarrow
\end{array}
\quad X(k, l) =
\begin{array}{c}
l \rightarrow \\
\begin{bmatrix}
22 + j0 & -2 + j6 & 10 + j0 & -2 - j6 \\
5 - j1 & 1 + j5 & -3 + j3 & -3 - j3 \\
-12 + j0 & -4 - j2 & 8 + j0 & -4 + j2 \\
5 + j1 & -3 + j3 & -3 - j3 & 1 - j5
\end{bmatrix}
\end{array}
$$

Computing the 2-D IDFT of $X(k, l)$ using the row–column method yields $x(m, n)$. However, the 1-D DFT algorithm can be used to reconstruct $x(m, n)$. One of the methods is to swap the real and imaginary parts of $X(k, l)$ and compute the 2-D DFT of the resulting matrix using the row–column method. By swapping the real and imaginary parts of the 2-D DFT and dividing by 16 yields $x(m, n)$. The advantage is that no separate IDFT algorithm is required.

Swapping the real and imaginary parts of $X(k, l)$, we get

$$
\begin{bmatrix}
0 + j22 & 6 - j2 & 0 + j10 & -6 - j2 \\
-1 + j5 & 5 + j1 & 3 - j3 & -3 - j3 \\
0 - j12 & -2 - j4 & 0 + j8 & 2 - j4 \\
1 + j5 & 3 - j3 & -3 - j3 & -5 + j1
\end{bmatrix}
$$

By computing the column 1-D DFTs, we get

$$
\begin{bmatrix}
0 + j20 & 12 - j8 & 0 + j12 & -12 - j8 \\
0 + j36 & 12 + j0 & 0 - j4 & -12 + j0 \\
0 + j0 & -4 - j4 & 0 + j24 & 4 - j4 \\
0 + j32 & 4 + j4 & 0 + j8 & -4 + j4
\end{bmatrix}
$$

By computing the row 1-D DFTs, we get

$$
j
\begin{bmatrix}
16 & -16 & 48 & 32 \\
32 & 16 & 32 & 64 \\
16 & -16 & 32 & -32 \\
48 & 16 & 32 & 32
\end{bmatrix}
$$

Swapping the real and imaginary parts and dividing by 16, we get back $x(m, n)$. ∎

To reduce the dynamic range of $X(k, l)$, the display of $\log_{10}(1 + |X(k, l)|)$ is used instead of $|X(k, l)|$. This improves the contrast of the spectrum. For the example signal, the magnitude of the spectrum, $|X(k, l)|$, in the center-zero format, is

$$\begin{bmatrix} 8 & 4.4721 & 12 & 4.4721 \\ 4.2426 & 5.0990 & 5.0990 & 4.2426 \\ 10 & 6.3246 & 22 & 6.3246 \\ 4.2426 & 4.2426 & 5.0990 & 5.0990 \end{bmatrix}$$

whereas $\log_{10}(1 + |X(k, l)|)$ is

$$\begin{bmatrix} 0.9542 & 0.7382 & 1.1139 & 0.7382 \\ 0.7196 & 0.7853 & 0.7853 & 0.7196 \\ 1.0414 & 0.8648 & 1.3617 & 0.8648 \\ 0.7196 & 0.7196 & 0.7853 & 0.7853 \end{bmatrix}$$

The log scale reduces the range of $|X(k, l)|$.

Either the case of computing the 1-D DFT or the 2-D DFT, computing the DFT of data with few numbers is important to understand and get proficient in the indispensable Fourier analysis. Problems, such as Example 4.1, is suited for manual computation. Equally important is to compute the DFT of larger data by programming. Now, we present an example of computing a 2-D 8×8 DFT. The 8×8 spatial domain data $x(m, n)$ is shown in Table 4.1. Using the row–column method, let us compute the 1-D DFT of the rows of $x(m, n)$ first. The result is the partial transform, shown in Table 4.2. Computing the 1-D DFT of the columns of the partial transform yields the 2-D DFT $X(k, l)$ of $x(m, n)$, shown in Table 4.4.

Equally valid procedure is to compute the 1-D DFT of the columns of $x(m, n)$ first, resulting the partial transform, shown in Table 4.3. Computing the 1-D DFT of the rows of the partial transform yields the 2-D DFT $X(k, l)$ of $x(m, n)$, shown in Table 4.4.

The 2-D DFT $X(k, l)$ of $x(m, n)$ in the center-zero format is shown in Table 4.5. The magnitude of the 2-D DFT $X(k, l)$ of $x(m, n)$ in the center-zero format is shown in Table 4.6. The log magnitude of the 2-D DFT $X(k, l)$ of $x(m, n)$ in the center-zero format is shown in Table 4.7.

Table 4.1 A 8×8 input signal $x(m, n)$

0	−2	2	1	2	0	4	3
1	0	1	3	3	2	3	5
0	−2	1	−3	2	0	3	−1
2	0	1	1	4	2	3	3
1	−1	3	2	1	2	1	3
2	1	2	4	−1	1	−1	1
1	−1	2	−2	3	2	2	2
3	1	2	2	2	4	−2	2

Table 4.2 1-D DFT of the rows of $x(m, n)$

10.00+j 0.00	−2.00+j 4.83	−4.00+j 6.00	−2.00+j 0.83	6.00+j 0.00	−2.00−j 0.83	−4.00−j 6.00	−2.00−j 4.83
18.00+j 0.00	−2.00+j 4.83	0.00+j 6.00	−2.00+j 0.83	−2.00+j 0.00	−2.00−j 0.83	0.00−j 6.00	−2.00−j 4.83
0.00+j 0.00	−2.00+j 4.83	−2.00−j 2.00	−2.00+j 0.83	12.00+j 0.00	−2.00−j 0.83	−2.00+j 2.00	−2.00−j 4.83
16.00+j 0.00	−2.00+j 4.83	2.00+j 2.00	−2.00+j 0.83	4.00+j 0.00	−2.00−j 0.83	2.00−j 2.00	−2.00−j 4.83
12.00+j 0.00	−1.41+j 0.83	−2.00+j 4.00	1.41+j 4.83	0.00+j 0.00	1.41−j 4.83	−2.00−j 4.00	−1.41−j 0.83
9.00+j 0.00	0.88−j 5.12	0.00+j 3.00	5.12+j 0.88	−5.00+j 0.00	5.12−j 0.88	0.00−j 3.00	0.88+j 5.12
9.00+j 0.00	−1.29+j 4.95	0.00−j 1.00	−2.71+j 4.95	7.00+j 0.00	−2.71−j 4.95	0.00+j 1.00	−1.29−j 4.95
14.00+j 0.00	−1.12−j 1.88	5.00−j 1.00	3.12+j 6.12	−4.00+j 0.00	3.12−j 6.12	5.00+j 1.00	−1.12+j 1.88

Table 4.3 1-D DFT of the columns of $x(m, n)$

10.00+j 0.00	-4.00+j 0.00	14.00+j 0.00	8.00+j 0.00	16.00+j 0.00	13.00+j 3.00	13.00+j 0.00	18.00+j 0.00
-1.00+j 2.41	-1.00+j 2.41	-1.00+j 2.41	-1.00+j 2.41	2.41-j 3.24	0.12+j 2.71	2.29-j 7.36	2.12-j 0.54
0.00+j 2.00	0.00+j 0.00	2.00+j 0.00	8.00-j 4.00	-2.00+j 4.00	0.00+j 3.00	0.00-j 1.00	5.00-j 1.00
-1.00+j 0.41	-1.00+j 0.41	-1.00+j 0.41	-1.00+j 0.41	-0.41-j 5.24	-4.12-j 1.29	3.71-j 5.36	-2.12-j 6.54
-6.00+j 0.00	-8.00+j 0.00	2.00+j 0.00	-12.00+j 0.00	0.00+j 0.00	-5.00+j 0.00	7.00+j 0.00	-4.00+j 0.00
-1.00-j 0.41	-1.00-j 0.41	-1.00-j 0.41	-1.00-j 0.41	-0.41+j 5.24	-4.12+j 1.29	3.71+j 5.36	-2.12+j 6.54
0.00-j 2.00	0.00+j 0.00	2.00+j 0.00	8.00+j 4.00	-2.00-j 4.00	0.00-j 3.00	0.00+j 1.00	5.00+j 1.00
-1.00-j 2.41	-1.00-j 2.41	-1.00-j 2.41	-1.00-j 2.41	2.41+j 3.24	0.12-j 2.71	2.29+j 7.36	2.12+j 0.54

Table 4.4 The 2-D DFT $X(k, l)$ of $x(m, n)$

88.00+j0.00	-10.95+j18.09	-1.00+j17.00	-1.05+j20.09	18.00+j0.00	-1.05-j20.09	-1.00-j17.00	-10.95-j18.09
2.95+j1.22	9.66+j9.66	3.36+j6.12	-12.73+j7.66	2.46-j12.78	3.07+j8.24	-3.12+j2.12	-13.66-j2.93
13.00+j3.00	-3.36-j6.12	4.00+j20.00	-1.12-j2.12	-13.00+j7.00	9.36-j1.88	-12.00-j6.00	3.12+j2.12
-6.95-j16.78	12.73+j3.66	1.12+j2.12	-1.66+j1.66	9.54-j2.78	-2.34+j17.07	-9.36-j1.88	-11.07+j0.24
-26.00+j0.00	-2.46+j12.78	-15.00-j3.00	-9.54-j2.78	32.00+j0.00	-9.54-j2.78	-15.00+j3.00	-2.46-j12.78
-6.95+j16.78	-11.07-j0.24	-9.36+j1.88	-2.34-j17.07	9.54+j2.78	-1.66-j1.66	1.12-j2.12	12.73-j3.66
13.00-j3.00	3.12-j2.12	-12.00+j6.00	9.36+j1.88	-13.00-j7.00	-1.12+j2.12	4.00-j20.00	-3.36+j6.12
2.95-j1.22	-13.66+j2.93	-3.12-j2.12	3.07-j8.24	2.46+j12.78	-12.73-j7.66	3.36-j6.12	9.66-j9.66

Table 4.5 The 2-D DFT $X(k, l)$ in the center-zero format

32.00+j0.00	−9.54−j2.78	−15.00+j3.00	−2.46−j12.78	−26.00+j0.00	−2.46+j12.78	−15.00−j3.00	−9.54+j2.78
9.54+j2.78	−1.66−j1.66	1.12−j2.12	12.73−j3.66	−6.95+j16.78	−11.07−j0.24	−9.36+j1.88	−2.34−j17.07
−13.00−j7.00	−1.12+j2.12	4.00−j20.00	−3.36+j6.12	13.00−j3.00	3.12−j2.12	−12.00+j6.00	9.36+j1.88
2.46+j12.78	−12.73−j7.66	3.36−j6.12	9.66−j9.66	2.95−j1.22	−13.66+j2.93	−3.12−j2.12	3.07−j8.24
18.00+j0.00	−1.05−j20.09	−1.00−j17.00	−10.95−j18.09	88.00+j0.00	−10.95+j18.09	−1.00+j17.00	−1.05+j20.09
2.46−j12.78	3.07+j8.24	−3.12+j2.12	−13.66−j2.93	2.95+j1.22	9.66+j9.65	3.36+j6.12	−12.73+j7.66
−13.00+j7.00	9.36−j1.88	−12.00−j6.00	3.12+j2.12	13.00+j3.00	−3.36−j6.12	4.00+j20.00	−1.12−j2.12
9.54−j2.78	−2.34+j17.07	−9.36−j1.88	−11.07+j0.24	−6.95−j16.78	12.73+j3.66	1.12+j2.12	−1.66+j1.66

Table 4.6 The magnitude of the 2-D DFT, $|X(k, l)|$, of $x(m, n)$ in the center-zero format

32	9.9320	15.2971	13.0137	26	13.0137	15.2971	9.9320
9.9320	2.3431	2.3994	13.2428	18.1606	11.0737	9.5506	17.2311
14.7648	2.3994	20.3961	6.9848	13.3417	3.7739	13.4164	9.5506
13.0137	14.8535	6.9848	13.6569	3.1928	13.9674	3.7739	8.7962
18	20.1193	17.0294	21.1474	88	21.1474	17.0294	20.1193
13.0137	8.7962	3.7739	13.9674	3.1928	13.6569	6.9848	14.8535
14.7648	9.5506	13.4164	3.7739	13.3417	6.9848	20.3961	2.3994
9.9320	17.2311	9.5506	11.0737	18.1606	13.2428	2.3994	2.3431

Table 4.7 The magnitude of the 2-D DFT of $x(m, n)$ in the center-zero format and in the log scale, $\log_{10}(1 + |X(k, l)|)$

1.5185	1.0387	1.2121	1.1466	1.4314	1.1466	1.2121	1.0387
1.0387	0.5242	0.5314	1.1536	1.2824	1.0818	1.0233	1.2608
1.1977	0.5314	1.3303	0.9023	1.1566	0.6789	1.1589	1.0233
1.1466	1.2001	0.9023	1.1660	0.6225	1.1751	0.6789	0.9911
1.2788	1.3247	1.2560	1.3453	1.9494	1.3453	1.2560	1.3247
1.1466	0.9911	0.6789	1.1751	0.6225	1.1660	0.9023	1.2001
1.1977	1.0233	1.1589	0.6789	1.1566	0.9023	1.3303	0.5314
1.0387	1.2608	1.0233	1.0818	1.2824	1.1536	0.5314	0.5242

4.2   The 2-D DFT and IDFT

It is easier to interpret the 2-DFT as two sets of 1-D DFTs and it is usually computed using 1-D DFTs. Formally, the 2-D DFT of a $N \times N$ signal $x(m, n)$ is defined as

$$X(k, l) = \sum_{m=0}^{N-1} \sum_{n=0}^{N-1} x(m, n) e^{-j\frac{2\pi}{N}(mk+nl)}, \quad k, l = 0, 1, \ldots, N - 1. \qquad (4.2)$$

As in the case of the 1-D IDFT, the 2-D IDFT synthesizes the 2-D signal by multiplying the basis signals with the respective coefficients and summing the samples at all points (m, n). The 2-D IDFT is given by

$$x(m, n) = \frac{1}{N^2} \sum_{k=0}^{N-1} \sum_{l=0}^{N-1} X(k, l) e^{j\frac{2\pi}{N}(mk+nl)}, \quad m, n = 0, 1, \ldots, N - 1. \qquad (4.3)$$

In this definition, the DC coefficient $X(0, 0)$ is placed in the top left-hand corner of the coefficient matrix. While this format is mostly used for computational purposes, placing $X(0, 0)$ in the center of the coefficient matrix is desired for better visualization of the spectrum. Further, it is easier to derive some derivations using this form. This form, called the center-zero format, with N even, is given as

$$X(k, l) = \sum_{m=-\frac{N}{2}}^{\frac{N}{2}-1} \sum_{n=-\frac{N}{2}}^{\frac{N}{2}-1} x(m, n) e^{-j\frac{2\pi}{N}(mk+nl)}, \quad k, l = -\frac{N}{2}, -\frac{N}{2}+1, \ldots, \frac{N}{2}-1$$

The corresponding 2-D IDFT is given by

$$x(m, n) = \frac{1}{N^2} \sum_{k=-\frac{N}{2}}^{\frac{N}{2}-1} \sum_{l=-\frac{N}{2}}^{\frac{N}{2}-1} X(k, l) e^{j\frac{2\pi}{N}(mk+nl)}, \quad m, n = -\frac{N}{2}, -\frac{N}{2}+1, \ldots, \frac{N}{2}-1$$

One format of the signal or the spectrum can be obtained from the other by swapping the quadrants of the signal or spectrum in the other format.

The 2-D DFT and IDFT definitions for a $M \times N$ signal are

$$X(k, l) = \sum_{m=0}^{M-1} \sum_{n=0}^{N-1} x(m, n) e^{-j2\pi\left(m\frac{k}{M}+n\frac{l}{N}\right)}, \quad k = 0, 1, \ldots, M-1, \, l = 0, 1, \ldots, N-1$$

$$(4.4)$$

$$x(m, n) = \frac{1}{MN} \sum_{k=0}^{M-1} \sum_{l=0}^{N-1} X(k, l) e^{j2\pi\left(m\frac{k}{M}+n\frac{l}{N}\right)}, \quad m = 0, 1, \ldots, M-1, \, n = 0, 1, \ldots, N-1$$

$$(4.5)$$

Let us compute the 2-D DFT of the 2×4 signal $x(m, n)$,

$$x(m, n) = \begin{bmatrix} 1 & 3 & 2 & 4 \\ 2 & 1 & 3 & 4 \end{bmatrix}$$

The row and column 1-D DFTs are

$$X(k, n) = \begin{bmatrix} 10 & -1+j1 & -4 & -1-j1 \\ 10 & -1+j3 & 0 & -1-j3 \end{bmatrix} \quad X(m, l) = \begin{bmatrix} 3 & 4 & 5 & 8 \\ -1 & 2 & -1 & 0 \end{bmatrix}$$

The 2-D DFT is obtained by computing the 1-D DFT of either of the partial transforms in the other direction.

$$X(k, l) = \begin{bmatrix} 20 & -2+j4 & -4 & -2-j4 \\ 0 & -j2 & -4 & j2 \end{bmatrix}$$

4.3 The 2-D DFT of Real-Valued Signals

The DFT, with complex exponential basis signals, is inherently designed for complex-valued signals. However, in practice, most of the signals are real-valued. For real-valued signals, the DFT coefficients always occur as complex conjugate pairs or real values. Coefficients

$$X(0,0), X\left(\frac{N}{2},0\right), X\left(0,\frac{N}{2}\right), X\left(\frac{N}{2},\frac{N}{2}\right)$$

of a $N \times N$ real-valued signal are real-valued, as the basis functions are of the form 1 and $(-1)^n$. The rest are complex conjugate pairs. For example,

$$2|X(k,l)| \cos\left(\frac{2\pi}{N}(mk+nl) + \angle(X(k,l))\right)$$
$$= X(k,l)e^{j\frac{2\pi}{N}(mk+nl)} + X^*(k,l)e^{-j\frac{2\pi}{N}(mk+nl)}$$
$$= X(k,l)e^{j\frac{2\pi}{N}(mk+nl)} + X^*(k,l)e^{j\frac{2\pi}{N}(m(N-k)+n(N-l))}$$

With $X(k,l) = X_r(k,l) + jX_i(k,l)$, the magnitude is

$$|X(k,l)| = \sqrt{X_r^2(k,l) + X_i^2(k,l)}$$

and the phase is

$$\angle X(k,l) = \tan^{-1}\frac{X_i(k,l)}{X_r(k,l)}$$

Using Eq. (4.3), with N even, a real-valued $N \times N$ signal can be expressed as a sum its constituent sinusoidal surfaces.

$$x(m,n) = \frac{1}{N^2}\left(X(0,0) + X\left(\frac{N}{2},0\right)\cos(\pi m)\right.$$
$$+ X\left(0,\frac{N}{2}\right)\cos(\pi n) + X\left(\frac{N}{2},\frac{N}{2}\right)\cos(\pi(m+n))$$
$$+ 2\sum_{k=1}^{\frac{N}{2}-1}\left(|X(k,0)|\cos\left(\frac{2\pi}{N}mk + \angle(X(k,0))\right)\right)$$
$$+ 2\sum_{l=1}^{\frac{N}{2}-1}\left(|X(0,l)|\cos\left(\frac{2\pi}{N}nl + \angle(X(0,l))\right)\right)$$
$$+ 2\sum_{l=1}^{\frac{N}{2}-1}\left(|X\left(\frac{N}{2},l\right)|\cos\left(\frac{2\pi}{N}\left(m\frac{N}{2}+nl\right) + \angle\left(X\left(\frac{N}{2},l\right)\right)\right)\right)$$

$$+ 2 \sum_{k=1}^{\frac{N}{2}-1} \sum_{l=1}^{N-1} \left(|X(k,l)| \cos \left(\frac{2\pi}{N}(mk+nl) + \angle(X(k,l)) \right) \right),$$

$$m, n = 0, 1, \ldots, N-1 \qquad (4.6)$$

Therefore,

$$\frac{N^2}{2} + 2$$

different sinusoidal surfaces constitute a $N \times N$ real-valued signal. The computational complexity of computing the 2-D DFT from the definition is $O(N^4)$. By computing the 2-D DFT using the row–column method, the complexity is reduced to $O(N^3)$. Fast algorithms are available for computing the 1-D DFT for lengths those are an integral power of 2. Using these algorithms, along with the row–column method, the computational complexity is reduced to $O(N^2 \log_2 N)$. For real-valued 2-D signals, the computational complexity can be further reduced, as presented in Chap. 10.

4.4 Properties of the 2-D DFT

Properties of the 2-D DFT relate the characteristics of signals in the spatial and frequency domains. The DFT of signals can be computed easily using the properties and the DFT of related simpler signals.

Linearity

Let $x(m,n) \leftrightarrow X(k,l)$ and $y(m,n) \leftrightarrow Y(k,l)$. Then,

$$z(m,n) = ax(m,n) + by(m,n) \leftrightarrow Z(k,l) = aX(k,l) + bY(k,l)$$

where a and b are real or complex constants. The 2-D DFT of a linear combination of a set of discrete signals is equal to the same linear combination of their individual DFTs. The dimensions of the signals must be same. If necessary, zero padding can be employed to meet this constraint. Linearity holds in both the spatial and frequency domains.

Example 4.2 Compute the DFT of $x(m,n)$ and $y(m,n)$. Using the linearity property, deduce the DFT of $z(m,n) = 2x(m,n) + 3y(m,n)$ from those of $x(m,n)$ and $y(m,n)$.

$$x(m,n) = \begin{bmatrix} 2 & 3 \\ 1 & 2 \end{bmatrix}, \quad y(m,n) = \begin{bmatrix} 1 & 3 \\ 2 & 1 \end{bmatrix}$$

Solution
The individual DFTs are

$$X(k, l) = \begin{bmatrix} 8 & -2 \\ 2 & 0 \end{bmatrix}, \quad Y(k, l) = \begin{bmatrix} 7 & -1 \\ 1 & -3 \end{bmatrix}$$

The DFT of

$$z(m, n) = 2x(m, n) + 3y(m, n) = \begin{bmatrix} 7 & 15 \\ 8 & 7 \end{bmatrix}$$

is

$$Z(k, l) = 2X(k, l) + 3Y(k, l) = \begin{bmatrix} 37 & -7 \\ 7 & -9 \end{bmatrix}$$

The DFT $Z(k, l)$ can be verified by directly computing that of $z(m, n)$.   ∎

Periodicity

If $x(m, n)$ and $X(k, l)$ are a $M \times N$-point DFT pair, then

$$x(m, n) = x(m + aM, n + bN), \quad \text{for all } m, n \text{ and}$$
$$X(k, l) = X(k + aM, l + bN), \quad \text{for all } k, l$$

where a and b are arbitrary integers and M and N are the periods. For example, the periodic extension of signal $x(m, n)$ in Example 4.2 is

$$\begin{bmatrix} & & & \vdots & & & \\ & 2 & 3 & 2 & 3 & 2 & 3 \\ & 1 & 2 & 1 & 2 & 1 & 2 \\ \cdots & 2 & 3 & 2 & 3 & 2 & 3 & \cdots \\ & 1 & 2 & 1 & 2 & 1 & 2 \\ & 2 & 3 & 2 & 3 & 2 & 3 \\ & 1 & 2 & 1 & 2 & 1 & 2 \\ & & & \vdots & & & \end{bmatrix}$$

The bottom and top edges are considered adjacent and so are the left and right edges.

Circular Shift of a Signal

A right (left) shift of a sinusoid results in adding a negative (positive) phase angle. The change in phase is proportional to its frequency. Its magnitude is unaffected. For a $N \times N$ signal,

$$x(m, n) \leftrightarrow X(k, l) \rightarrow x(m - m_0, n - n_0) \leftrightarrow X(k, l)e^{-j\frac{2\pi}{N}(km_0 + ln_0)}$$

The DFT of $x(m - m_0, n - n_0)$ is

$$= \sum_{m=0}^{N-1} \sum_{n=0}^{N-1} x(m - m_0, n - n_0) e^{-j\frac{2\pi}{N}(mk+nl)}$$

$$= e^{-j\frac{2\pi}{N}(km_0+ln_0)} \sum_{m=0}^{N-1} \sum_{n=0}^{N-1} x(m - m_0, n - n_0) e^{-j\frac{2\pi}{N}((m-m_0)k+(n-n_0)l)}$$

$$= e^{-j\frac{2\pi}{N}(km_0+ln_0)} X(k, l)$$

Example 4.3 Using the shift theorem, find the 2-D DFT of the shifted version $x(m - 2, n + 1)$ of the signal $x(m, n)$ in Example 4.1. Verify that by computing the DFT of the shifted signal.

Solution
For a 4×4 signal, a right shift by 2 sample intervals of the signal with frequency index 1 contributes a phase of $-180°$. A left shift by one sample interval contributes a phase of $90°$. Therefore, the 2-D DFT of the shifted signal

$$x(m, n) = \begin{bmatrix} 1 & -1 & 3 & 2 \\ 2 & 1 & 2 & 4 \\ 1 & -1 & 2 & -2 \\ 3 & 1 & 2 & 2 \end{bmatrix} \quad x(m - 2, n + 1) = \begin{bmatrix} -1 & 2 & -2 & 1 \\ 1 & 2 & 2 & 3 \\ -1 & 3 & 2 & 1 \\ 1 & 2 & 4 & 2 \end{bmatrix}$$

is

$$(-1)^k (j)^l X(k, l) = \begin{bmatrix} 22 & -6 - j2 & -10 & -6 + j2 \\ -5 + j1 & 5 - j1 & -3 + j3 & 3 - j3 \\ -12 & 2 - j4 & -8 & 2 + j4 \\ -5 - j1 & 3 + j3 & -3 - j3 & 5 + j1 \end{bmatrix}, \ k = 0, 1, 2, 3, \ l = 0, 1, 2, 3$$

∎

Circular Shift of a Spectrum
For a $N \times N$ signal,

$$x(m, n) \leftrightarrow X(k, l) \rightarrow x(m, n) e^{j\frac{2\pi}{N}(k_0 m + l_0 n)} \leftrightarrow X(k - k_0, l - l_0)$$

With N even and $k_0 = l_0 = \frac{N}{2}$, the center-zero spectrum of $x(m, n)$ can be obtained using this theorem.

Example 4.4 By multiplying the signal of Example 4.1 with $(-1)^{(m+n)}$, we get

$$(-1)^{(m+n)} x(m, n) = \begin{bmatrix} 1 & 1 & 3 & -2 \\ -2 & 1 & -2 & 4 \\ 1 & 1 & 2 & 2 \\ -3 & 1 & -2 & 2 \end{bmatrix}$$

By computing the DFT of this signal, the center-zero spectrum of $x(m, n)$ is obtained.

$$X(k-2, l-2) = \begin{bmatrix} 8 & -4+j2 & -12 & -4-j2 \\ -3-j3 & 1-j5 & 5+j1 & -3+j3 \\ 10 & -2-j6 & 22 & -2+j6 \\ -3+j3 & -3-j3 & 5-j1 & 1+j5 \end{bmatrix}$$

∎

Circular Convolution in the Spatial Time-Domain

The circular convolution of $x(m, n)$ and $h(m, n)$ is given by

$$y(m, n) = \sum_{p=0}^{N-1} \sum_{q=0}^{N-1} x(p, q)h(m-p, n-q), \quad m, n = 0, 1, \ldots, N-1$$

Let $x(m, n) \leftrightarrow X(k, l)$ and $h(m, n) \leftrightarrow H(k, l)$, $m, n, k, l = 0, 1, \ldots, N-1$. Then,

$$x(m, n) \circledast h(m, n) \leftrightarrow X(k, l)H(k, l)$$

Convolution of two 2-D signals, as in the case of 1-D signals, in the spatial-domain becomes pointwise multiplication of their DFT coefficients in the frequency-domain. This is due to the fact that convolution is a linear operation and the frequency of a signal is unaffected from the input to the output of a LTI system.

Example 4.5 Convolve

$$x(m, n) = \begin{bmatrix} 1 & -1 & 3 & 2 \\ 2 & 1 & 2 & 4 \\ 1 & -1 & 2 & -2 \\ 3 & 1 & 2 & 2 \end{bmatrix} \quad h(m, n) = \begin{bmatrix} 1 & 2 & 1 & 3 \\ 2 & 1 & 3 & 4 \\ 3 & 1 & 2 & 1 \\ 2 & 2 & 1 & 0 \end{bmatrix}$$

Solution
The DFT of $x(m, n)$ is

$$X(k, l) = \begin{bmatrix} 22+j0 & -2+j6 & 10+j0 & -2-j6 \\ 5-j1 & 1+j5 & -3+j3 & -3-j3 \\ -12+j0 & -4-j2 & 8+j0 & -4+j2 \\ 5+j1 & -3+j3 & -3-j3 & 1-j5 \end{bmatrix}$$

The DFT of $h(m, n)$ is

$$H(k, l) = \begin{bmatrix} 29 + j0 & 1 + j2 & 1 + j0 & 1 - j2 \\ 0 - j5 & 4 + j3 & -6 + j1 & -6 + j1 \\ -1 + j0 & 1 + j0 & -1 + j0 & 1 + j0 \\ 0 + j5 & -6 - j1 & -6 - j1 & 4 - j3 \end{bmatrix}$$

The pointwise product $Y(k, l) = X(k, l)H(k, l)$ is

$$Y(k, l) = X(k, l)H(k, l) = \begin{bmatrix} 638 + j0 & -14 + j2 & 10 + j0 & -14 - j2 \\ -5 - j25 & -11 + j23 & 15 - j21 & 21 + j15 \\ 12 + j0 & -4 - j2 & -8 + j0 & -4 + j2 \\ -5 + j25 & 21 - j15 & 15 + j21 & -11 - j23 \end{bmatrix}$$

The IDFT of $Y(k, l)$ is the convolution output in the spatial domain.

$$y(m, n) = \begin{bmatrix} 41 & 37 & 43 & 39 \\ 40 & 42 & 52 & 35 \\ 36 & 44 & 43 & 42 \\ 38 & 33 & 31 & 42 \end{bmatrix}$$

∎

Circular Convolution in the Frequency-Domain

$$x(m, n)h(m, n) \leftrightarrow \frac{1}{N^2} \sum_{p=0}^{N-1} \sum_{q=0}^{N-1} X(p, q)H(k - p, l - q)$$

Consider the two 2×2 signals.

$$x(m, n) = \begin{bmatrix} 2 & -3 \\ 1 & 2 \end{bmatrix}, \quad y(m, n) = \begin{bmatrix} -1 & 3 \\ 2 & 1 \end{bmatrix}$$

The respective DFTs are

$$X(k, l) = \begin{bmatrix} 2 & 4 \\ -4 & 6 \end{bmatrix}, \quad Y(k, l) = \begin{bmatrix} 5 & -3 \\ -1 & -5 \end{bmatrix}$$

The pointwise product of $x(m, n)$ and $y(m, n)$ is

$$c(m, n) = x(m, n)y(m, n) = \begin{bmatrix} -2 & -9 \\ 2 & 2 \end{bmatrix}$$

To find the circular convolution of $X(k, l)$ and $Y(k, l)$, we compute the respective DFTs, find their pointwise product, and compute the IDFT. The results are

$$\begin{bmatrix} 8 & -12 \\ 4 & 8 \end{bmatrix}, \begin{bmatrix} -4 & 12 \\ 8 & 4 \end{bmatrix}, \begin{bmatrix} -32 & -144 \\ 32 & 32 \end{bmatrix}, 4\begin{bmatrix} -7 & 7 \\ -15 & 7 \end{bmatrix}$$

The DFT of the pointwise product of $x(m, n)$ and $y(m, n)$

$$\begin{bmatrix} -2 & -9 \\ 2 & 2 \end{bmatrix} \text{ is } \begin{bmatrix} -7 & 7 \\ -15 & 7 \end{bmatrix}$$

Circular Cross-Correlation in the Spatial-Domain

The circular cross-correlation of two real-valued 2-D signals $x(m, n)$ and $h(m, n)$ is given by

$$r_{xh}(m, n) = \sum_{p=0}^{N-1} \sum_{q=0}^{N-1} x(p, q)h(p - m, q - n) \leftrightarrow H^*(k, l)X(k, l)$$

Since $h(N - m, N - n) \leftrightarrow H^*(k, l)$, this operation can also be interpreted as the convolution of $x(m, n)$ and $h(N - m, N - n)$.

$$r_{hx}(m, n) = r_{xh}(N - m, N - n) = \text{IDFT}(X^*(k, l)H(k, l))$$

The conjugate of the DFT of $x(m, n)$ in Example 4.5 is

$$X^*(k, l) = \begin{bmatrix} 22 & -2 - j6 & 10 & -2 + j6 \\ 5 + j1 & 1 - j5 & -3 - j3 & -3 + j3 \\ -12 & -4 + j2 & 8 & -4 - j2 \\ 5 - j1 & -3 - j3 & -3 + j3 & 1 + j5 \end{bmatrix}$$

The conjugate of the DFT of $h(m, n)$ is

$$H^*(k, l) = \begin{bmatrix} 29 & 1 - j2 & 1 & 1 + j2 \\ 0 + j5 & 4 - j3 & -6 - j1 & -6 - j1 \\ -1 & 1 & -1 & 1 \\ 0 - j5 & -6 + j1 & -6 + j1 & 4 + j3 \end{bmatrix}$$

The pointwise product $Y(k, l) = X(k, l)H^*(k, l)$ is

$$Y(k, l) = X(k, l)H^*(k, l)$$
$$= 100\begin{bmatrix} 6.38 & 0.10 + j0.10 & 0.10 & 0.10 - j0.10 \\ 0.05 + j0.25 & 0.19 + j0.17 & 0.21 - j0.15 & 0.15 + j0.21 \\ 0.12 & -0.04 - j0.02 & -0.08 & -0.04 + j0.02 \\ 0.05 - 0.25 & 0.15 - j0.21 & 0.21 + j0.15 & 0.19 - j0.17 \end{bmatrix}$$

The IDFT of $Y(k, l)$ is the correlation output in the spatial domain.

$$r_{xh}(m, n) = \begin{bmatrix} 49 & 38 & 39 & 39 \\ 36 & 31 & 42 & 35 \\ 34 & 41 & 41 & 44 \\ 48 & 42 & 35 & 44 \end{bmatrix}$$

The pointwise product $Y(k, l) = H(k, l)X^*(k, l)$ is

$$Y(k, l) = H(k, l)X^*(k, l)$$

$$= 100 \begin{bmatrix} 6.38 & 0.10 - j0.10 & 0.10 & 0.10 + j0.10 \\ 0.05 - j0.25 & 0.19 - j0.17 & 0.21 + j0.15 & 0.15 - j0.21 \\ 0.12 & -0.04 + j0.020 & -0.08 & -0.04 - j0.02 \\ 0.05 + j0.25 & 0.15 + j0.21 & 0.21 - j0.15 & 0.19 + j0.17 \end{bmatrix}$$

The IDFT of $Y(k, l)$ is the correlation output in the spatial domain.

$$r_{hx}(m, n) = \begin{bmatrix} 49 & 39 & 39 & 38 \\ 48 & 44 & 35 & 42 \\ 34 & 44 & 41 & 41 \\ 36 & 35 & 42 & 31 \end{bmatrix}$$

∎

Cross-correlation of a signal $x(m, n)$ with itself is the autocorrelation operation.

$$r_{xx}(m, n) = \text{IDFT}(|X(k, l)|^2)$$

$$|X(k, l)|^2 = \begin{bmatrix} 484 & 40 & 100 & 40 \\ 26 & 26 & 18 & 18 \\ 144 & 20 & 64 & 20 \\ 26 & 18 & 18 & 26 \end{bmatrix}$$

The IDFT of $|X(k, l)|^2$ is

$$r_{xx}(m, n) = \begin{bmatrix} 68 & 30 & 42 & 30 \\ 26 & 18 & 21 & 20 \\ 46 & 28 & 42 & 28 \\ 26 & 20 & 21 & 18 \end{bmatrix}$$

Sum and Difference of Sequences

Due to the fact that the twiddle factors are 1 and -1, the evaluation of coefficients $X(0, 0)$ and $X(\frac{N}{2}, \frac{N}{2})$ is relatively simple and these values can be used to check the DFT and IDFT computation.

$$X(0, 0) = \sum_{m=0}^{N-1} \sum_{n=0}^{N-1} x(m, n)$$

With N even,

$$X\left(\frac{N}{2}, \frac{N}{2}\right) = \sum_{m=0}^{N-1} \sum_{n=0}^{N-1} x(m, n)(-1)^{(m+n)}$$

$$x(0, 0) = \frac{1}{N^2} \sum_{k=0}^{N-1} \sum_{l=0}^{N-1} X(k, l)$$

With N even,

$$x\left(\frac{N}{2}, \frac{N}{2}\right) = \frac{1}{N^2} \sum_{k=0}^{N-1} \sum_{l=0}^{N-1} X(k, l)(-1)^{(k+l)}$$

For the signal in Example 4.1,

$$\{x(0, 0) = 1, x(2, 2) = 2, X(0, 0) = 22, X(2, 2) = 8\}$$

The Difference

$$x(m, n) - x(m - 1, n) \leftrightarrow X(k, l)(1 - e^{-j\frac{2\pi}{N}k})$$

$$x(m, n) - x(m, n - 1) \leftrightarrow X(k, l)(1 - e^{-j\frac{2\pi}{N}l})$$

Reversal Property

If $x(m, n)$ and $X(k, l)$ form a transform pair, then so are $x(N - m, N - n)$ and $X(N - k, N - l)$.

$$x(m, n) \leftrightarrow X(k, l) \rightarrow x(N - m, N - n) \leftrightarrow X(N - k, N - l)$$

Example 4.6 For the signal in Example 4.1

$$x(4 - m, 4 - n) = \begin{bmatrix} 1 & 2 & 3 & -1 \\ 3 & 2 & 2 & 1 \\ 1 & -2 & 2 & -1 \\ 2 & 4 & 2 & 1 \end{bmatrix}$$

$$X(4 - k, 4 - l) = \begin{bmatrix} 22 & -2 - j6 & 10 & -2 + j6 \\ 5 + j1 & 1 - j5 & -3 - j3 & -3 + j3 \\ -12 & -4 + j2 & 8 & -4 - j2 \\ 5 - j1 & -3 - j3 & -3 + j3 & 1 + j5 \end{bmatrix}$$

∎

Symmetry

As a pair of complex conjugate exponentials form a real sinusoid, the exponential Fourier representation of real-valued data is conjugate symmetric. The $N \times N$ DFT of a real-valued data can also have only $N \times N$ independent values. The redundancy in the transform neither increases the storage requirements nor the computational complexity significantly with an appropriate implementation of the transform.

The DFT values at diametrically opposite points form complex conjugate pairs.

$$X^*(N - k, N - l) = X(k, l)$$

An equivalent form of the symmetry is

$$X\left(\frac{N}{2} \pm k, \frac{N}{2} \pm l\right) = X^*\left(\frac{N}{2} \mp k, \frac{N}{2} \mp l\right)$$

Example 4.7 Underline the left-half of the nonredundant DFT values of the signal in Example 4.1.
Solution

$$
\begin{array}{c}
 \quad \quad l \rightarrow \\
\begin{array}{c} k \\ \downarrow \end{array}
\begin{bmatrix}
\underline{22} & -2 + j6 & \underline{10} & -2 - j6 \\
\underline{5 - j1} & \underline{1 + j5} & -3 + j3 & -3 - j3 \\
\underline{-12} & \underline{-4 - j2} & \underline{8} & -4 + j2 \\
5 + j1 & \underline{-3 + j3} & -3 - j3 & 1 - j5
\end{bmatrix}
\end{array}
$$

∎

Storing only the nonredundant part of the DFT coefficients, the storage requirement is the same as that of the input signal. The storage of the 2-D DFT values of columns (rows) 1 to $\frac{N}{2} - 1$ and the first $\frac{N}{2} + 1$ values of the zeroth and the $\frac{N}{2}$th columns (rows) is sufficient. The computation of the DFT requires the computation of $\frac{N}{2} - 1$ 1-D DFT of complex-valued data and $N + 2$ 1-D DFT of real-valued data.

Rotation

The rotation of a $N \times N$ signal by an angle θ, about its center, rotates its spectrum also by the same angle and in the same direction. Rotations other than multiples of 90° require interpolation. The rotation of a 2-D signal is a part of geometric transformations. The equations governing the counterclockwise rotation of a vector with coordinates (m, n) by an angle θ are

$$mr = m\cos(\theta) - n\sin(\theta) \tag{4.7}$$

$$nr = m\sin(\theta) + n\cos(\theta) \tag{4.8}$$

The length r of the vector remains the same. In matrix notation, we get

$$\begin{bmatrix} mr \\ nr \end{bmatrix} = \begin{bmatrix} \cos(\theta) & -\sin(\theta) \\ \sin(\theta) & \cos(\theta) \end{bmatrix} \begin{bmatrix} m \\ n \end{bmatrix}$$

where (mr, nr) are the new coordinates. For clockwise rotation, change the sign of the angle.

With $x(m, n) \leftrightarrow X(k, l)$, the rotated version of the 2-D DFT pair by θ is given by

$$x(m\cos(\theta) - n\sin(\theta), m\sin(\theta) + n\cos(\theta)) \leftrightarrow X(k\cos(\theta) - l\sin(\theta), k\sin(\theta) + l\cos(\theta))$$

Consider the following signal, its rotated version by $90°$ in the counterclockwise direction and their spectra.

$$x(m, n) = \begin{bmatrix} \check{1} & 2 \\ 3 & 4 \end{bmatrix} \leftrightarrow \begin{bmatrix} 10 & -2 \\ -4 & 0 \end{bmatrix}, \begin{bmatrix} 2 & 4 \\ \check{1} & 3 \end{bmatrix} \leftrightarrow \begin{bmatrix} -2 & 0 \\ 10 & -4 \end{bmatrix}$$

$$\begin{bmatrix} mr \\ nr \end{bmatrix} = \begin{bmatrix} \cos(90°) & -\sin(90°) \\ \sin(90°) & \cos(90°) \end{bmatrix} \begin{bmatrix} m \\ n \end{bmatrix} = \begin{bmatrix} 0 & -1 \\ 1 & 0 \end{bmatrix} \begin{bmatrix} m \\ n \end{bmatrix} = \begin{bmatrix} -n \\ m \end{bmatrix}$$

The periodic extension of signal $x(m, n)$ and its rotated version about $(0, 0)$ are

$$x_p(m, n) = \cdots \begin{bmatrix} & & \vdots & & & \\ 1 & 2 & 1 & 2 & 1 & 1 \\ 3 & 4 & 3 & 4 & 3 & 4 \\ 1 & 2 & \check{1} & 2 & 1 & 2 \\ 3 & 4 & 3 & 4 & 3 & 4 \\ 1 & 2 & 1 & 2 & 1 & 1 \\ 3 & 4 & 3 & 4 & 3 & 4 \\ & & \vdots & & & \end{bmatrix} \cdots \quad xr_p(m, n) = \cdots \begin{bmatrix} & & \vdots & & & \\ 1 & 3 & 1 & 3 & 1 & 3 \\ 2 & 4 & 2 & 4 & 2 & 4 \\ 1 & 3 & \check{1} & 3 & 1 & 3 \\ 2 & 4 & 2 & 4 & 2 & 4 \\ 1 & 3 & 1 & 3 & 1 & 3 \\ 2 & 4 & 2 & 4 & 2 & 4 \\ & & \vdots & & & \end{bmatrix} \cdots$$

The element at the origin is indicated by a check mark. For rotation other than about the center, translation operation can be used in addition.

Separable Signals

The 2-D DFT is a separable function in the variables m and n. Therefore, the DFT of a separable function $x(m, n) = x(m)x(n)$ is also separable. The product of the column vector with the row vector is equal to the 2-D function. That is,

$$x(m) \leftrightarrow X(k), x(n) \leftrightarrow X(l) \rightarrow X(k, l) = X(k)X(l)$$

$$X(k, l) = \sum_{m=0}^{N-1}\sum_{n=0}^{N-1} x(m, n)e^{-j\frac{2\pi}{N}mk}e^{-j\frac{2\pi}{N}nl} = \sum_{m=0}^{N-1}\sum_{n=0}^{N-1} x(m)x(n)e^{-j\frac{2\pi}{N}mk}e^{-j\frac{2\pi}{N}nl}$$

$$= \left(\sum_{m=0}^{N-1} x(m)e^{-j\frac{2\pi}{N}mk}\right)\left(\sum_{n=0}^{N-1} x(n)e^{-j\frac{2\pi}{N}nl}\right) = X(k)X(l)$$

Example 4.8 Compute the DFT of $x(m) = \{1, 3, 2, 1\}$ and $x(n) = \{2, 1, 1, 3\}$. Using the separability theorem, verify that the product $x(m, n) = x(m)x(n)$ of the column vector $x(m)$ and the row vector $x(n)$ in the time-domain and the 2-D IDFT of the product of their individual DFTs are the same.

Solution
The product $x(m, n) = x(m)x(n)$ is

$$\begin{bmatrix} 1 \\ 3 \\ 2 \\ 1 \end{bmatrix} \begin{bmatrix} 2 & 1 & 1 & 3 \end{bmatrix} = \begin{bmatrix} 2 & 1 & 1 & 3 \\ 6 & 3 & 3 & 9 \\ 4 & 2 & 2 & 6 \\ 2 & 1 & 1 & 3 \end{bmatrix}$$

$X(k) = \{7, -1 - j2, -1, -1 + j2\}$ and $X(l) = \{7, 1 + j2, -1, 1 - j2\}$.

$$X(k, l) = X(k)X(l) = \begin{bmatrix} 7 \\ -1 - j2 \\ -1 \\ -1 + j2 \end{bmatrix} \begin{bmatrix} 7 & 1 + j2 & -1 & 1 - j2 \end{bmatrix}$$

$$= \begin{bmatrix} 49 & 7 + j14 & -7 & 7 - j14 \\ -7 - j14 & 3 - j4 & 1 + j2 & -5 \\ -7 & -1 - j2 & 1 & -1 + j2 \\ -7 + j14 & -5 & 1 - j2 & 3 + j4 \end{bmatrix}$$

The 2-D IDFT is

$$\begin{bmatrix} 2 & 1 & 1 & 3 \\ 6 & 3 & 3 & 9 \\ 4 & 2 & 2 & 6 \\ 2 & 1 & 1 & 3 \end{bmatrix} = x(m, n) = x(m)x(n)$$

∎

Parseval's Theorem

This theorem implies that the signal power can also be computed from the DFT representation of the signal. Let $x(m, n) \leftrightarrow X(k, l)$ with the dimensions of the signal $N \times N$. Since the magnitude of the samples of the complex sinusoidal surface

$$e^{j\frac{2\pi}{N}(mk+nl)}, \quad m = 0, 1, \ldots, N - 1, \quad n = 0, 1, \ldots, N - 1$$

is one and the 2-D DFT coefficients are scaled by N^2, the power of a complex sinusoidal surface is

$$\left(\frac{|X(k,l)|^2}{N^4}\right)N^2 = \frac{|X(k,l)|^2}{N^2}$$

Therefore, the sum of the powers of all the components of a signal yields the power of the signal.

$$\sum_{m=0}^{N-1}\sum_{n=0}^{N-1}|x(m,n)|^2 = \frac{1}{N^2}\sum_{k=0}^{N-1}\sum_{l=0}^{N-1}|X(k,l)|^2$$

For the signal $x(m,n)$ in Example 4.2, the power computed in both the domains is 18.

The generalized form of this theorem holds for two different signals as given by

$$\sum_{m=0}^{N-1}\sum_{n=0}^{N-1}x(m,n)y^*(m,n) = \frac{1}{N^2}\sum_{k=0}^{N-1}\sum_{l=0}^{N-1}X(k,l)Y^*(k,l)$$

4.5   Summary

- The 2-D DFT is a straightforward extension of that of the 1-D DFT.
- As the 2-D signal is a surface, rather than a curve as in the case of 1-D signals, two frequencies, in two orthogonal directions, are required for its frequency-domain representation.
- The constituent frequency components of a 1-D signal are the sinusoidal curves while the constituent frequency components of a 2-D signal are the sinusoidal surfaces.
- In practice, the 2-D DFT of a $N \times N$ 2-D signal is computed using N 1-DFTs along the rows of the input and N 1-DFTs along the columns of the partial transform or vice versa.
- Using fast 1-D DFT algorithms, along with the row–column method, the computational complexity of computing the 2-D DFT is $O(N^2 \log_2 N)$.
- As in the case of the 1-D DFT, the center-zero format of the DFT spectrum is desired for better visual display and easier derivation of some results.
- As in the case of the 1-D DFT, due to large dynamic range of the spectrum, the logarithm of the magnitude spectrum is preferred for better visualization.
- As in the case of the 1-D DFT, the 2-D spectrum of a real-valued 2-D signal is conjugate symmetric.
- The 2-D DFT is extensively used in applications such as digital image processing.

Exercises

4.1 The 2-D discrete signal $x(m, n)$ is periodic with period 4 samples in m and n. Find its complex exponential form and thereby find its 2-D DFT coefficients $X(k, l)$. Verify that the samples are the same from the two forms.

* 4.1.1

$$x(m, n) = 2 - 3\cos\left(\frac{2\pi}{4}(m + n) + \frac{\pi}{6}\right) + \sin\left(\frac{2\pi}{4}(2m + n) + \frac{\pi}{3}\right)$$

4.1.2

$$x(m, n) = 1 + 2\sin\left(\frac{2\pi}{4}(m + n) - \frac{\pi}{3}\right) - \sin\left(\frac{2\pi}{4}(2m + n) - \frac{\pi}{4}\right)$$

4.1.3

$$x(m, n) = -3 - 2\sin\left(\frac{2\pi}{4}(m + n) + \frac{\pi}{3}\right) - 3\cos\left(2\frac{2\pi}{4}(m + n)\right)$$

4.2 Find the 2-D DFT $X(k, l)$ of the signal $x(m, n)$ using the row–column method. The origin of $x(m, n)$ is at the top left corner. Find the 2-D IDFT of $X(k, l)$ to get back $x(m, n)$. Verify Parseval's theorem.

* 4.2.1

$$x(m, n) = \begin{bmatrix} 2 & 4 & 2 & 3 \\ 0 & 2 & 3 & 4 \\ 4 & 2 & 3 & 1 \\ 2 & 1 & 3 & 4 \end{bmatrix}$$

4.2.2

$$x(m, n) = \begin{bmatrix} 3 & 1 & 4 & 2 \\ 1 & 1 & 2 & 2 \\ 3 & 3 & 2 & 2 \\ 4 & 1 & 2 & 3 \end{bmatrix}$$

4.2.3

$$x(m, n) = \begin{bmatrix} 1 & 2 & 3 & 4 \\ 4 & 2 & 1 & 3 \\ 4 & 3 & 1 & 2 \\ 3 & 3 & 0 & 1 \end{bmatrix}$$

4.3 Find the DFT of the 4×4 signal $x(m, n)$ using the row–column method. Each image is composed of a shifted impulse. Verify the DFT using the shift theorem in the spatial domain.

4.3.1

$$x(m,n) = \begin{bmatrix} 0 & 0 & 0 & 0 \\ 0 & 0 & 0 & 0 \\ 0 & 1 & 0 & 0 \\ 0 & 0 & 0 & 0 \end{bmatrix}$$

4.3.2

$$x(m,n) = \begin{bmatrix} 0 & 0 & 0 & 0 \\ 0 & 0 & 0 & 0 \\ 0 & 0 & 1 & 0 \\ 0 & 0 & 0 & 0 \end{bmatrix}$$

* 4.3.3

$$x(m,n) = \begin{bmatrix} 0 & 0 & 0 & 0 \\ 0 & 0 & 0 & 0 \\ 0 & 0 & 0 & 0 \\ 0 & 1 & 0 & 0 \end{bmatrix}$$

4.4 Using the DFT and IDFT, find: (a) the periodic convolution of $x(m,n)$ and $h(m,n)$, (b) the periodic correlation of $x(m,n)$ and $h(m,n)$, and $h(m,n)$ and $x(m,n)$, (c) the autocorrelation of $x(m,n)$.

4.4.1

$$x(m,n) = \begin{bmatrix} 2 & 0 & 1 & 3 \\ 3 & 0 & 4 & 2 \\ 3 & 1 & 0 & 1 \\ 2 & 0 & 0 & 2 \end{bmatrix} \quad h(m,n) = \begin{bmatrix} 0 & 1 & 3 & 2 \\ 1 & 3 & 1 & -2 \\ 3 & 0 & 2 & -1 \\ 2 & 0 & 2 & 2 \end{bmatrix}$$

* 4.4.2

$$x(m,n) = \begin{bmatrix} 1 & 2 & 3 & 4 \\ 3 & 0 & 1 & 2 \\ 1 & 1 & -1 & 3 \\ 0 & 1 & 2 & 3 \end{bmatrix} \quad h(m,n) = \begin{bmatrix} 0 & 3 & 1 & 2 \\ 1 & 1 & 1 & -1 \\ 2 & 1 & 0 & 0 \\ 3 & 1 & 1 & 2 \end{bmatrix}$$

4.4.3

$$x(m,n) = \begin{bmatrix} 1 & 1 & 3 & 3 \\ -1 & 0 & 0 & 4 \\ 1 & 1 & 0 & 2 \\ 2 & 1 & 0 & 4 \end{bmatrix} \quad h(m,n) = \begin{bmatrix} 1 & 0 & 2 & 4 \\ 3 & 1 & 0 & 2 \\ 1 & 1 & 0 & 2 \\ 0 & 1 & -2 & 1 \end{bmatrix}$$

Chapter 5
Convolution and Correlation

Practical systems are analyzed using mathematical models. Convolution and transfer function are two of the frequently used models in the study of LTI systems. In this chapter, we study the convolution operation and we study the transfer function in later chapters. Both the models are based on decomposing an arbitrary input signal in terms of well-defined basis signals, impulse in the case of convolution and complex exponential signals in the case of transfer function. After the decomposition of the input signal, the system output can be found using the response of the system to the basis signals and the linearity and time-invariance properties of the LTI systems. Convolution operation relates the input and output of a system through its impulse response. The impulse response of a system is its response to the unit-impulse signal, assuming that the system is initially relaxed (zero initial conditions). Convolution expresses the output of a system in terms of its input only.

The important concepts, such as convolution and Fourier analysis, are easier to understand and remember through their physical interpretations. Convolution operation is the same thing as finding the current balance for our deposits in a bank. Assuming compound interest, the interest is computed on the principal at regular intervals and it is added to the principal. Let the interval be one year and the interest rate per year be 10%. Let $n = 0$ be the starting time and the deposit made at that time is designated as $x(0)$. Then, $x(n)$ is the deposit made after the nth year. Assuming N number of years, the deposits are

$$\{x(0), x(1), x(2), \ldots, x(N)\}$$

The compound interest rates $h(n)$ for n years of deposit are

$$\{h(0), h(1), h(2), \ldots, h(N)\} = \{1, 1.1, 1.21, \ldots, (1.1)^N\}$$

Therefore, the balance in the deposit $y(N)$ at the Nth year is

© Springer Nature Singapore Pte Ltd. 2018
D. Sundararajan, *Fourier Analysis—A Signal Processing Approach*,
https://doi.org/10.1007/978-981-13-1693-7_5

$$y(N) = 1x(N) + 1.1x(N-1) + 1.21x(N-2)+, \cdots, +(1.1)^N x(0)$$

$$\{h(0)x(N) + h(1)x(N-1) + h(2)x(N-2)+, \cdots, +h(N)x(0)\}$$

$$= y(N) = \sum_{k=0}^{N} h(k)x(N-k)$$

Alternately,

$$\{x(0)h(N) + x(1)h(N-1) + x(2)h(N-2)+, \cdots, +x(N)h(0)\}$$

$$= y(N) = \sum_{k=0}^{N} x(k)h(N-k)$$

After time reversing and shifting one of the two sequences, the output is the sum of the product of the corresponding terms. The time-reversal is required, as each other's index is running in opposite directions. The generalization of this problem is the convolution operation. In system analysis, the interest rate is called the impulse response of the system. The deposits are called the input. The balance at intervals is the output.

5.1   Convolution

The impulse response, which characterizes the system in the time domain, is the response of a relaxed (initial conditions are zero) system for the unit-impulse $\delta(n)$. A discrete unit-impulse signal is defined as

$$\delta(n) = \begin{cases} 1, & \text{for } n = 0 \\ 0, & \text{for } n \neq 0 \end{cases}$$

It is an all-zero sequence, except that its value is one when its argument n is equal to zero. The input $x(n)$ is decomposed into a sum of scaled and delayed unit-impulses. The response to each impulse is found and the superposition summation of all the responses is the system output. Convolution can also be considered as the weighted average of sections of one of the inputs with the weighting sequence being the other input.

5.1.1   Linear Convolution

In the convolution operation, the input $x(n)$ is decomposed into scaled and delayed impulses. At each point, the contribution of all the impulses is summed to find the output $y(n)$. The input, in terms of impulses, is

$$
\begin{aligned}
x(n) &= \cdots + x(-2)\delta(n+2) + x(-1)\delta(n+1) \\
&\quad + x(0)\delta(n) + x(1)\delta(n-1) + x(2)\delta(n-2) + \cdots \\
&= \sum_{k=-\infty}^{\infty} x(k)\delta(n-k)
\end{aligned}
$$

Let the impulse response of the system be $h(n)$. Then, due to the time-invariance of the LTI system, the response to a delayed impulse $\delta(n-k)$ is $h(n-k)$. Due to the linearity of the LTI system, the response to $x(k)\delta(n-k)$ is $x(k)h(n-k)$ and the total response of the system is the sum of the contributions to all the scaled and shifted impulses. The 1-D linear convolution of two aperiodic sequences $x(n)$ and $h(n)$, again due to linearity, is defined as

$$
y(n) = \sum_{k=-\infty}^{\infty} x(k)h(n-k) = \sum_{k=-\infty}^{\infty} h(k)x(n-k) = x(n) * h(n) = h(n) * x(n)
$$

The convolution operation relates the input $x(n)$, the output $y(n)$, and the impulse response $h(n)$ of a system.

Figure 5.1 shows the convolution of the signal $x(n) = \{\check{2}, 1, 3, 4\}$ and $h(n) = \{\check{1}, -2, 3\}$.

The output $y(0)$, from the definition, is

$$
y(0) = x(k)h(0-k) = (2)(1) = 2,
$$

where $h(0-k)$ is the time-reversal of $h(k)$. Shifting $h(0-k)$ to the right, we get the remaining outputs as

Fig. 5.1 1-D linear convolution

k	0	1	2	3
$h(k)$	1	-2	3	
$x(k)$	2	1	3	4

$h(0-k)$	3	-2	1						
$h(1-k)$		3	-2	1					
$h(2-k)$			3	-2	1				
$h(3-k)$				3	-2	1			
$h(4-k)$					3	-2	1		
$h(5-k)$						3	-2	1	

n	0	1	2	3	4	5
$y(n)$	2	-3	7	1	1	12

$$y(1) = x(k)h(1 - k) = (2)(-2) + (1)(1) = -3$$
$$y(2) = x(k)h(2 - k) = (2)(3) + (1)(-2) + (3)(1) = 7$$
$$y(3) = x(k)h(3 - k) = (1)(3) + (3)(-2) + (4)(1) = 1$$
$$y(4) = x(k)h(4 - k) = (3)(3) + (4)(-2) = 1$$
$$y(5) = x(k)h(5 - k) = (4)(3) = 12$$

As can be seen from the figure, the convolution operation consists of repeatedly executing the following four operations. The first operation is the time-reversal of one of the sequences. The second operation is shifting. The time-reversal and shifting, resulting in $h(n - k)$, is relatively more difficult to visualize. It is easier to visualize the simple shifting operation resulting in $h(k - n)$. Since $(n - k)$ is the same as $-(k - n)$, it is the time-reversal of $h(k - n)$ about a vertical line at the point $k = n$. The third operation is to find the products of the overlapping values of $x(k)$ and $h(n - k)$. The fourth operation is to sum the products, which yields $y(n)$. There are two quick checks of the convolution output. The product of the sums of the two sequences convolved is equal to the sum of the output sequence.

$$(\check{2} + 1 + 3 + 4)(\check{1} - 2 + 3) = 20 = (\check{2} - 3 + 7 + 1 + 1 + 12)$$

The same test after changing the signs of the odd-indexed terms is

$$(\check{2} - 1 + 3 - 4)(\check{1} + 2 + 3) = 0 = (\check{2} + 3 + 7 - 1 + 1 - 12)$$

Convolution and Polynomial Multiplication

Let the coefficients of two polynomials be

$$x(n) = \{\check{2}, 1, 3, 4\} \quad \text{and} \quad h(n) = \{\check{1}, -2, 3\}$$

The polynomials are

$$2 + p + 3p^2 + 4p^3 \quad \text{and} \quad 1 - 2p + 3p^2$$

The product of the two polynomials is

$$(2 + p + 3p^2 + 4p^3)(1 - 2p + 3p^2) = 2 - 3p + 7p^2 + p^3 + p^4 + 12p^5$$

The coefficients of the product polynomial are the same as those obtained by convolving $x(n)$ and $h(n)$.

Linear Convolution Using the DFT

A measure of the comparison of two algorithms is the order of time complexity. For example, if the number of major operations required to execute an algorithm is proportional to N, the number of elements in the input, then its time complexity is $O(N)$. The time complexity of the direct convolution operation is $O(N^2)$, whereas that using the DFT is $O(N \log_2 N)$. The evaluation of the convolution of two sequences becomes faster for longer sequence lengths, if it is carried out in the frequency domain. The linear convolution output of convolving two sequences is of length that is equal to the sum of the lengths of the sequences minus one. Therefore, the sequences have to be zero padded so that their lengths satisfy this constraint to implement convolution using the DFT. Further, as practically used fast DFT algorithms are of lengths that is an integral power of 2, this becomes the second constraint. Another constraint is that the origin of the sequences must be aligned.

Let the sequences to be linearly convolved be

$$x(n) = \{\check{1}, 3, 2, 4\} \quad \text{and} \quad h(n) = \{3, \check{1}, 2\}$$

The linear convolution output of the sequences is

$$\{3, \check{1}0, 11, 20, 8, 8\}$$

The lengths of the sequences are 4 and 3. Therefore, the convolution output has to be of length $4 + 3 - 1 = 6$. As there are 6 independent values in the output, the length of the DFT must be at least of length 6. The nearest power of 2 is $2^3 = 8$. Therefore, the sequences are appended by zeros to make the sequences of length 8. This excess length produces $8 - 6 = 2$ zeros at the end of the output. Then, the sequences become

$$\{\check{1}, 3, 2, 4, 0, 0, 0, 0\} \quad \text{and} \quad \{3, \check{1}, 2, 0, 0, 0, 0, 0\}$$

The origins of the two sequences must be aligned. Circularly shifting the second sequence by one position left, we get

$$\{\check{1}, 2, 0, 0, 0, 0, 0, 3\}$$

The 8-point DFT of the aligned sequences, with a precision of 2 digits, is

$$X(k) = \{\check{1}0, 0.29 - j6.95, -1 + j1, 1.71 - j2.95, -4, 1.71 + j2.95, -1 - j1, 0.29 + j6.95\}$$

$$H(k) = \{\check{6}, 4.54 + j0.71, 1 + j1, -2.54 + j0.71, -4, -2.54 - j0.71, 1 - j1, 4.54 - j0.71\}$$

The term-by-term product $X(k)H(k)$ is

$$X(k)H(k) = \{\check{6}0, 6.24 - j31.31, -2, -2.24 + j8.69, 16, -2.24 - j8.69, -2, 6.24 + j31.31\}$$

The IDFT of $X(k)H(k)$ is

$$\{\check{1}0, 11, 20, 8, 8, 0, 0, 3\}$$

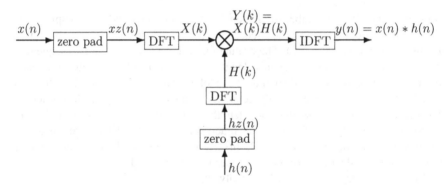

Fig. 5.2 Linear convolution using the DFT

Circularly shifting right by one sample interval, we get the linear convolution output appended by 2 zeros.

$$y(n) = \{3, \overset{\vee}{1}0, 11, 20, 8, 8, 0, 0\}$$

Figure 5.2 shows the block diagram for implementing linear convolution using the DFT. The sequences to be convolved, $x(n)$ and $h(n)$, are sufficiently zero padded according to the requirements given earlier to get $xz(n)$ and $hz(n)$. The DFT of the zero-padded sequences is computed to get, $X(k)$ and $H(k)$. The IDFT of the term-by-term product, $X(k)H(k)$, of the DFTs yields the linear convolution of $x(n)$ and $h(n)$ with some zeros appended. If necessary, alignment of the sequences must be taken care of, as already pointed out.

5.1.2 Circular Convolution

The circular convolution is also known as cyclic or periodic convolution. While the linear convolution is used most of the times in the analysis of LTI systems, the circular convolution is also important for the reason that signals are considered as periodic in DFT computation and the DFT is the tool to implement the linear convolution faster. The linear convolution is the periodic convolution in the limit, when the periods of the signals to be convolved become infinite. Both the circular and linear convolution are based on the same four operations (folding, shifting, multiplication, and summing). The difference is that the folding and shifting operations are carried out along a line in the linear convolution, whereas it is carried out around a circle in the circular convolution. Due to this difference, some number of output values of linear and circular convolutions, for the same inputs, are different at the borders.

The circular convolution of the two sequences $x(n)$ and $h(n)$, both of period N, is defined as

Fig. 5.3 1-D circular convolution

k				-3	-2	-1	0	1	2	3			
$h(k)$					2	-1	3	1	2	-1	3		
$x(k)$							2	1	-3	4			
$h(0-k)$			3	-1	2	1	3	-1	2				
$h(1-k)$				3	-1	2	1	3	-1	2			
$h(2-k)$					3	-1	2	1	3	-1	2		
$h(3-k)$						3	-1	2	1	3	-1	2	
n							0	1	2	3			
$y(n)$							16	-8	9	3			

$$y(n) = \sum_{m=0}^{N-1} x(m)h(n-m) = \sum_{m=0}^{N-1} h(m)x(n-m), \quad n = 0, 1, \ldots, N-1$$

resulting in the periodic output sequence $y(n)$ with the same period. Consider the circular convolution output $y(n)$ of the sequences

$$h(n) = \{\overset{\vee}{1}, 2, -1, 3\} \quad \text{and} \quad x(n) = \{\overset{\vee}{2}, 1, -3, 4\}$$

$$y(n) = \{\overset{\vee}{16}, -8, 9, 3\}$$

shown in Fig. 5.3. This is the same as the linear convolution with the sequences periodic. Consequently, the first three output values are different from that of the linear convolution and the fourth value is the same. The linear convolution output $y(n)$ of the same sequences is

$$y(n) = \{\overset{\vee}{2}, 5, -3, 3, 14, -13, 12\}$$

The last three values get added to the first three values to form the circular convolution output. The sum of the shifted, by 4 samples, copies of linear convolution output is the circular convolution output. In circular convolution, the periods of the two sequences to be convolved are assumed to be the same. Let $x(n)$ is a sequence of length N and $h(n)$ is a sequence of length M with $M \leq N$. Sequence $h(n)$ is zero padded to make its length also N. Then, the circular convolution of $x(n)$ and $h(n)$ yields N output values. The first $(M - 1)$ output values are not the corresponding linear convolution output values, while the rest of the $(N - M + 1)$ values are the same. In the last example, with $N = M = 4$, the last value $y(3) = 3$ only is the same in both the outputs.

Overlap–Save Method

In practical applications, the input sequence is often very long and the impulse response is relatively short. Even if the required memory is available, the output will be delayed too long. In these cases, due to the limited availability of the memory in digital systems and the desirability of fast output, the input signal is segmented into blocks to suit the memory availability and the speed of response. Each block is

Fig. 5.4 Convolution of
long sequences using the
overlap–save method

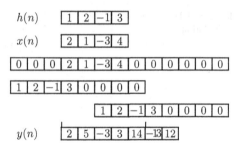

convolved with the impulse response and the convolution outputs of the successive
blocks are assembled to form the total convolution output. There are two equivalent
methods to carry out convolution in this way. One of it, called the overlap–save
method, is described.

The overlap–save method of convolution of long sequences is shown in Fig. 5.4.
Let the length of the input sequence $x(n)$ be N. Let the length of the impulse response
$h(n)$ be Q and the block length be B. Then, for efficient implementation of the method,
the following condition should be met.

$$N \gg B \gg Q$$

For illustrative purposes, short sequences are used in the example. Let $x(n)$ and $h(n)$
be

$$x(n) = \{\check{2}, 1, -3, 4\} \quad \text{and} \quad h(n) = \{\check{1}, 2, -1, 3\}$$

The output of linear convolution of $x(n)$ and $h(n)$ is

$$y(n) = \{\check{2}, 5, -3, 3, 14, -13, 12\}$$

Therefore, there are $N + Q - 1 = 7$ output values have to be computed. Let the block
length B be 8 and $N = Q = 4$. As first $Q - 1 = 3$ output values are corrupted, the
input data has to be prepended by 3 zeros. Since the block length B is 8, the data has to
be appended by one zero. The first block of extended $x(n)$ is $\{0, 0, 0, 2, 1, -3, 4, 0\}$.
The DFT of this block, with a precision of 2 digits, is

$$\{4, -0.29 + j0.46, -3 + j5, -1.71 - j7.54, 6, -1.71 + j7.54, -3 - j5, -0.29 - j0.46\}$$

The extended $h(n)$ is $\{1, 2, -1, 3, 0, 0, 0, 0\}$. The DFT of this data, which is computed only once and stored, is

$$\{5, 0.29 - j2.54, 2 + j1, 1.71 - j4.54, -5, 1.71 + j4.54, 2 - j1, 0.29 + j2.54\}$$

The pointwise product of the two DFTs is

$$\{20, 1.09 + j0.88, -11 + j7, -37.09 - j5.12, -30, -37.09 + j5.12, -11 - j7, 1.09 - j0.88\}$$

The IDFT of the product is

$$\{-13, 12, 0, 2, 5, -3, 3, 14\}$$

The last five values $\{2, 5, -3, 3, 14\}$ are the first five values of linearly convolving $x(n)$ and $h(n)$. The second block has to overlap the first block by 3 samples. Therefore, the second block is $\{-3, 4, 0, 0, 0, 0, 0, 0\}$. The DFT of this block, with a precision of 2 digits, is

$$\{1, -0.17 - j2.83, -3 - j4, -5.83 - j2.83, -7, -5.83 + j2.83, -3 + j4, -0.17 + j2.83\}$$

The pointwise product of this DFT with that of extended $h(n)$ is

$$\{5, -7.22 - j0.39, -2 - j11, -22.78 + j21.61, 35,$$
$$- 22.78 - j21.61, -2 + j11, -7.22 + j0.39\}$$

The IDFT of the product is

$$\{-3, -2, 11, -13, 12, 0, 0, 0\}$$

The last five values $\{-13, 12, 0, 0, 0\}$ are the last two values of linearly convolving $x(n)$ and $h(n)$ appended by three zeros.

5.1.3 2-D Linear Convolution

2-D convolution is a straightforward extension of the 1-D convolution. As the sequences involved are 2-D, folding, shifting, and zero-padding operations have to be carried out along the rows and columns of the 2-D data. If one of the sequences is separable, 2-D convolution can be implemented faster by a set of 1-D convolutions. The convolution of 2-D sequences $x(m, n)$ and $h(m, n)$ is defined as

$$y(m, n) = \sum_{k=-\infty}^{\infty} \sum_{l=-\infty}^{\infty} x(k, l)h(m - k, n - l)$$

$$= \sum_{k=-\infty}^{\infty} \sum_{l=-\infty}^{\infty} h(k, l)x(m - k, n - l) = h(m, n) * x(m, n)$$

The same four steps of the 1-D convolution (folding, shifting, multiplying, and summing) are repeatedly carried out in implementing the 2-D convolution in two dimensions.

1. Any one of the two sequences, say $h(k, l)$, is rotated in the (k, l) plane by 180° about the origin to get $h(-k, -l)$. Of course, the same result is achieved by folding

$h(k, l)$ about the k-axis to get $h(k, -l)$ and then folding $h(k, -l)$ about the l-axis to get $h(-k, -l)$ or vice versa.

2. The rotated sequence $h(-k, -l)$ is shifted by (m, n) to get $h(m - k, n - l)$ to find the convolution output at coordinates (m, n).
3. The term-by-term products, $x(k, l)h(m - k, n - l)$, of all the overlapping samples are computed.
4. Summing all the products is the convolution output $y(m, n)$ at (m, n).

Let us find the output of convolving the 3×3 sequence $h(k, l)$ and the 4×4 sequence $x(k, l)$

$$
h(k, l) = \begin{bmatrix} \check{2} & 1 & 3 \\ 1 & 2 & 2 \\ 3 & 2 & 1 \end{bmatrix} \quad \text{and} \quad x(k, l) = \begin{bmatrix} \check{3} & 1 & 3 & 2 \\ 2 & 1 & 3 & 4 \\ 2 & 1 & 2 & 3 \\ 1 & 1 & 2 & 2 \end{bmatrix}
$$

shown in Fig. 5.5. Four examples of computing the convolution output are shown. For example, with a shift of $(0 - k, 0 - l)$, there is only one overlapping pair $(3, 2)$. The product of these numbers is the output $y(0, 0) = 6$. The process is repeated to get the complete convolution output $y(m, n)$ shown in the figure.

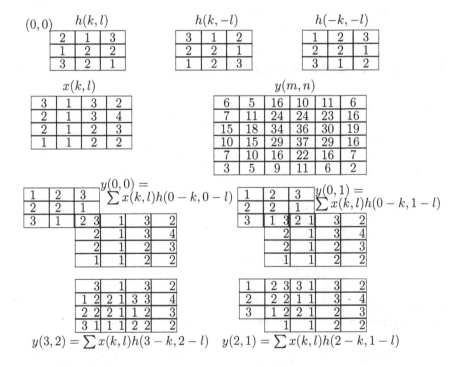

Fig. 5.5 2-D linear convolution

$$y(m, n) = \begin{bmatrix} 6 & 5 & 16 & 10 & 11 & 6 \\ 7 & 11 & 24 & 24 & 23 & 16 \\ 15 & 18 & 34 & 36 & 30 & 19 \\ 10 & 15 & 29 & 37 & 29 & 16 \\ 7 & 10 & 16 & 22 & 16 & 7 \\ 3 & 5 & 9 & 11 & 6 & 2 \end{bmatrix}$$

We assumed that the values outside the defined region of the sequence are zero. This assumption may or may not be suitable. Some other commonly used border extensions are based on periodicity, symmetry, or replication.

Let us compute the linear convolution of the last example using the DFT. As the convolution output is of size 6×6 and the nearest power of 2 is 8, the sequences are zero padded to get

$$xz(m, n) = \begin{bmatrix} \overset{\ast}{3} & 1 & 3 & 2 & 0 & 0 & 0 & 0 \\ 2 & 1 & 3 & 4 & 0 & 0 & 0 & 0 \\ 2 & 1 & 2 & 3 & 0 & 0 & 0 & 0 \\ 1 & 1 & 2 & 2 & 0 & 0 & 0 & 0 \\ 0 & 0 & 0 & 0 & 0 & 0 & 0 & 0 \\ 0 & 0 & 0 & 0 & 0 & 0 & 0 & 0 \\ 0 & 0 & 0 & 0 & 0 & 0 & 0 & 0 \\ 0 & 0 & 0 & 0 & 0 & 0 & 0 & 0 \end{bmatrix} \quad hz(m, n) = \begin{bmatrix} \overset{\ast}{2} & 1 & 3 & 0 & 0 & 0 & 0 & 0 \\ 1 & 2 & 2 & 0 & 0 & 0 & 0 & 0 \\ 3 & 2 & 1 & 0 & 0 & 0 & 0 & 0 \\ 0 & 0 & 0 & 0 & 0 & 0 & 0 & 0 \\ 0 & 0 & 0 & 0 & 0 & 0 & 0 & 0 \\ 0 & 0 & 0 & 0 & 0 & 0 & 0 & 0 \\ 0 & 0 & 0 & 0 & 0 & 0 & 0 & 0 \\ 0 & 0 & 0 & 0 & 0 & 0 & 0 & 0 \end{bmatrix}$$

The 2-D DFT of $xz(m, n)$, $X(k, l)$, is shown in Table 5.1. The 2-D DFT of $hz(m, n)$, $H(k, l)$, is shown in Table 5.2. The pointwise product of $X(k, l)$ and $H(k, l)$, $X(k, l)H(k, l)$ is shown in Table 5.3. The IDFT of $X(k, l)H(k, l)$ is the convolution output $y(m, n)$ appended by zeros in the last 2 rows and columns.

$$y(m, n) = \begin{bmatrix} 6 & 5 & 16 & 10 & 11 & 6 & 0 & 0 \\ 7 & 11 & 24 & 24 & 23 & 16 & 0 & 0 \\ 15 & 18 & 34 & 36 & 30 & 19 & 0 & 0 \\ 10 & 15 & 29 & 37 & 29 & 16 & 0 & 0 \\ 7 & 10 & 16 & 22 & 16 & 7 & 0 & 0 \\ 3 & 5 & 9 & 11 & 6 & 2 & 0 & 0 \\ 0 & 0 & 0 & 0 & 0 & 0 & 0 & 0 \\ 0 & 0 & 0 & 0 & 0 & 0 & 0 & 0 \end{bmatrix}$$

The top-left 6×6 of $y(m, n)$ are the output values.

2-D Convolution with Separable Filters

A 2-D separable filter can be expressed as the product of a column vector and a row vector. For convolving with separable filters, 1-D convolution is sufficient. Consider the separable filter $h(m, n)$ and the corresponding column and row filters.

Table 5.1 2-D DFT of zero-padded $x(m, n)$, $X(k, l)$

33.00+j 0.00	3.05−j20.61	−2.00+j 7.00	12.95−j 0.61	3.00+j 0.00	12.95+j 0.61	−2.00−j 7.00	3.05+j20.61
11.83−j19.31	−10.36−j 7.54	4.83+j 3.83	4.12−j 6.95	3.00+j 0.00	6.71−j 8.12	−4.83−j 1.00	14.36+j 6.12
1.00−j 4.00	−0.71+j 0.12	2.00−j 1.00	−0.12−j 0.71	3.00+j 0.00	0.71−j 4.12	−2.00+j 1.00	4.12+j 0.71
6.17−j 3.31	−0.12−j 2.95	0.83+j 1.00	2.36+j 0.46	3.00+j 0.00	1.64−j 1.88	−0.83+j 1.83	5.29+j 3.88
1.00+j 0.00	2.71+j 0.71	2.00−j 1.00	1.29+j 0.71	3.00+j 0.00	1.29−j 0.71	2.00+j 1.00	2.71−j 0.71
6.17+j 3.31	5.29−j 3.88	−0.83−j 1.83	1.64+j 1.88	3.00−j 0.00	2.36−j 0.46	0.83−j 1.00	−0.12+j 2.95
1.00+j 4.00	4.12−j 0.71	−2.00−j 1.00	0.71+j 4.12	3.00+j 0.00	−0.12+j 0.71	2.00+j 1.00	−0.71−j 0.12
11.83+j19.31	14.36−j 6.12	−4.83+j 1.00	6.71+j 8.12	3.00+j 0.00	4.12+j 6.95	4.83−j 3.83	−10.36+j 7.54

Table 5.2 2-D DFT of zero-padded $h(m, n)$, $H(k, l)$

17.00+j 0.00	9.54−j 9.54	0.00−j 5.00	2.46+j 2.46	7.00+j 0.00	2.46−j 2.46	0.00+j 5.00	9.54+j 9.54
9.54−j 9.54	−0.41−j12.24	−5.12−j 3.71	1.00+j 1.41	4.71−j 2.71	1.00−j 4.00	1.71+j 1.12	9.24+j 0.00
0.00−j 5.00	−5.12−j 3.71	−5.00+j 2.00	0.29+j 3.12	2.00−j 1.00	−0.88−j 2.29	−1.00+j 0.00	1.71−j 1.12
2.46+j 2.46	1.00+j 1.41	0.29+j 3.12	2.41+j 3.76	3.29+j 1.29	0.76+j 0.00	−0.88+j 2.29	1.00+j 4.00
7.00+j 0.00	4.71−j 2.71	2.00−j 1.00	3.29+j 1.29	5.00+j 0.00	3.29−j 1.29	2.00+j 1.00	4.71+j 2.71
2.46−j 2.46	1.00−j 4.00	−0.88−j 2.29	0.76−j 0.00	3.29−j 1.29	2.41−j 3.76	0.29−j 3.12	1.00−j 1.41
0.00+j 5.00	1.71+j 1.12	−1.00+j 0.00	−0.88+j 2.29	2.00+j 1.00	0.29−j 3.12	−5.00−j 2.00	−5.12+j 3.71
9.54+j 9.54	9.24+j 0.00	1.71−j 1.12	1.00+j 4.00	4.71+j 2.71	1.00−j 1.41	−5.12+j 3.71	−0.41+j12.24

$$h(m, n) = \begin{bmatrix} \check{1} & 1 & 1 \\ 1 & 1 & 1 \\ 1 & 1 & 1 \end{bmatrix} = \begin{bmatrix} 1 \\ 1 \\ 1 \end{bmatrix} \begin{bmatrix} 1 & 1 & 1 \end{bmatrix} = h(m)h(n)$$

The convolution of $x(m, n)$ given in the last example with this filter can be carried out using two 1-D filters, $\{h(0) = 1, h(1) = 1, h(2) = 1\}$. Since $xz(m, n)$ is of size 8×8, the zero-padded filter is

Table 5.3 Convolution output $Y(k, l) = X(k, l)H(k, l)$ in the frequency domain

561.00+j0.00	−167.41−j225.58	35.00+j10.00	33.41+j30.42	21.00+j0.00	33.41−j30.42	35.00−j10.00	−167.41+j225.58
−71.38−j296.96	−87.96+j130.00	−10.54−j37.51	13.95−j1.12	14.12−j8.12	−25.78−34.95	−7.12−j7.12	132.76+j56.58
−20.00−j5.00	4.07+j2.00	−8.00+j9.00	2.17−j0.59	6.00−j3.00	−10.07+j2.00	2.00−j1.00	7.83−j3.41
23.38+j7.04	4.05−j3.12	−2.88+j2.88	3.96+j10.00	9.88+j3.88	1.24−j1.42	−3.46−j3.51	−10.22+j25.05
7.00+j0.00	14.66−j4.00	3.00−j4.00	3.34+j4.00	15.00+j0.00	3.34−j4.00	3.00+j4.00	14.66+j4.00
23.38−j7.04	−10.22−j25.05	−3.46+j3.51	1.24+j1.42	9.88−j3.88	3.96−j10.00	−2.88−j2.88	4.05+j3.12
−20.00+j5.00	7.83+j3.41	2.00+j1.00	−10.07−j2.00	6.00+j3.00	2.17+j0.59	−8.00−j9.00	4.07−j2.00
−71.38+j296.96	132.76−j56.58	−7.12+j7.12	−25.78+j34.95	14.12+j8.12	13.95+j1 12	−10.54+j37.51	−87.96−j130.00

Table 5.4 Partial convolution output $P(k, l) = X(k, l)H(k)$ in the frequency domain

99	9.15−j61.82	−6.00+j21.00	38.85−j1.82	9.00+j0.00	38.85+j1.82	−6.00−j21.00	9.15+j61.82
−12.78−j53.16	−30.56+j4.83	14.78−j1.71	−4.83−j18.90	5.12−j5.12	−2.41−j25.31	−9.95+j6.54	34.97−j14.07
−4.00−j1.00	0.12+j0.71	−1.00−j2.00	−0.71+j0.12	0.00−j3.00	−4.12−j0.71	1.00+j2.00	0.71−j4.12
2.78+j0.84	0.83−j0.90	−0.05+j0.54	0.56+j0.83	0.88+j0.88	1.03−j0.07	−0.78+j0.29	0.41+j2.69
1.00+j0.00	2.71+j0.71	2.00−j1.00	1.29+j0.71	3.00+j0.00	1.29−j0.71	2.00+j1.00	2.71−j0.71
2.78−j0.84	0.41−j2.69	−0.78−j0.29	1.03+j0.07	0.88−j0.88	0.56−j0.83	−0.05−j0.54	0.83+j0.90
−4.00+j1.00	0.71+j4.12	1.00−j2.00	−4.12+j0.71	0.00+j3.00	−0.71−j0.12	−1.00+j2.00	0.12−j0.71
−12.78+j53.16	34.97+j14.07	−9.95−j6.54	−2.41+j25.31	5.12+j5.12	−4.83+j18.90	14.78+j1.71	−30.56−j4.83

Table 5.5 Convolution output $Y(k, l) = P(k, l)H(l)$ in the frequency domain

297	−89.91−j121.15	21.00+j6.00	11.91+j10.85	9.00+j0.00	11.91−j10.85	21.00−j6.00	−89.91+j121.15
−38.33−j159.49	−43.92+j60.41	−1.71−j14.78	4.12−j6.95	5.12−j5.12	−8.12−j6.71	−6.54−j9.95	83.72+j35.68
−12.00−j3.00	1.41+j1.00	−2.00+j1.00	−0.24−j0.17	0.00−j3.00	−1.41+j1.00	−2.00+j1.00	8.24−j5.83
8.33+j2.51	−0.12−j2.95	0.54+j0.05	−0.08+j0.41	0.88+j0.88	0.28−j0.32	−0.29−j0.78	−3.88+j5.29
3.00+j0.00	5.83−j3.41	−1.00−j2.00	0.17+j0.59	3.00+j0.00	0.17−j0.59	−1.00+j2.00	5.83+j3.41
8.33−j2.51	−3.88−j5.29	−0.29+j0.78	0.28+j0.32	0.88−j0.88	−0.08−j0.41	0.54−j0.05	−0.12+j2.95
−12.00+j3.00	8.24+j5.83	−2.00−j1.00	−1.41−j1.00	0.00+j3.00	−0.24+j0.17	−2.00−j1.00	1.41−j1.00
−38.33+j159.49	83.72−j35.68	−6.54+j9.95	−8.12+j6.71	5.12+j5.12	4.12+j6.95	−1.71+j14.78	−43.92−j60.41

$$\{h(0) = 1, h(1) = 1, h(2) = 1, h(3) = 0, h(4) = 0, h(5) = 0, h(6) = 0, h(7) = 0\}$$

The row and column filters have the same coefficients. Taking the 1-D DFT of this filter, we get

$$H(k) = \{3, 1.7071 - j1.7071, -j, 0.2929 + j0.2929, 1, 0.2929$$
$$- j0.2929, j, 1.7071 + j1.7071\}$$

The 2-D DFT, $X(k, l)$, of $xz(m, n)$ is shown in Table 5.1. The partial convolution output, $P(k, l) = X(k, l)H(k)$ shown in Table 5.4, in the frequency domain is obtained by pointwise multiplication of each column of $X(k, l)$ by $H(k)$. The convolution output, $Y(k, l) = P(k, l)H(l)$ shown in Table 5.5, in the frequency domain is obtained by pointwise multiplication of each row of $P(k, l)$ by $H(l)$. The output is the 2-D IDFT of $Y(k, l)$.

$$y(m, n) = \begin{bmatrix} 3 & 4 & 7 & 6 & 5 & 2 & 0 & 0 \\ 5 & 7 & 13 & 14 & 12 & 6 & 0 & 0 \\ 7 & 10 & 18 & 20 & 17 & 9 & 0 & 0 \\ 5 & 8 & 15 & 19 & 16 & 9 & 0 & 0 \\ 3 & 5 & 9 & 11 & 9 & 5 & 0 & 0 \\ 1 & 2 & 4 & 5 & 4 & 2 & 0 & 0 \\ 0 & 0 & 0 & 0 & 0 & 0 & 0 & 0 \\ 0 & 0 & 0 & 0 & 0 & 0 & 0 & 0 \end{bmatrix}$$

The top-left 6×6 of $y(m, n)$ are the output values. Figure 5.6 shows the block diagram for 2-D convolution with separable filters. The filter is decomposed in to row and column filters with sufficient zero padding and their DFTs $H(l)$ and $H(k)$ are computed. The input $x(m, n)$ is zero padded to get $xz(m, n)$ and its DFT $X(k, l)$ is computed. The columns of $X(k, l)$ are multiplied by $H(k)$ to get the partial convolution output

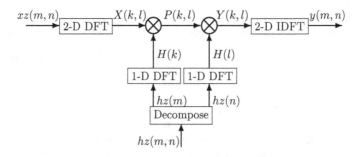

Fig. 5.6 2-D linear convolution using the DFT with separable filters

$P(k, l) = X(k, l)H(k)$ in the frequency domain. The rows of partial convolution output $P(k, l)$ are multiplied by $H(l)$ to get the convolution output $Y(k, l) = P(k, l)H(l)$ in the frequency domain. The 2-D IDFT of $Y(k, l)$ is the convolution output $y(m, n)$.

5.2 Correlation

5.2.1 The Linear Correlation

Correlation is a similarity measure between two signals. The correlation output indicates the strength of the relationships between the signals, which may be negative, zero or positive. If the signals are positively correlated, then both increase or decrease together. The more time we walk, the more calories we burn. If the signals are negatively correlated, then one increases and the other decreases. It is an inverse relationship. An increase in the amount of physical effort results in weight loss. Zero correlation implies no discernible relationship between the two variables. If a signal increases with the other remaining constant or increasing half the time and decreasing half the time, it indicates no correlation. In signal processing, object recognition and estimation are typical examples of correlation operation. The most famous and important example, of course, is the determination of the amplitudes of the signal components by correlating the given signal with each of its components in Fourier analysis.

The cross-correlation of two signals $x(n)$ and $y(n)$ is defined as

$$r_{xy}(m) = \sum_{n=-\infty}^{\infty} x(n)y^*(n-m) = \sum_{n=-\infty}^{\infty} x(n+m)y^*(n), \quad m = 0, \pm 1, \pm 2, \ldots$$

Equivalent but alternate definition is also used. The asterisk in the definition indicates complex conjugation operation, which has no effect for real-valued signals. The

Fig. 5.7 Linear correlation operation

n			0	1	2	3		
$y(n)$			2	1	−3			
$x(n)$			2	1	3	4		
$y(n+2)$		2	1	−3				
$y(n+1)$			2	1	−3			
$y(n)$				2	1	−3		
$y(n-1)$					2	1	−3	
$y(n-2)$						2	1	−3
$y(n-3)$							2	1	−3
m		−2	−1	0	1	2	3	
$r_{xy}(m)$		−6	−1	−4	−7	10	8	

output is the sum of products of two signals, with one of them shifted. The number of shifts is the independent variable and the sum is the dependent variable.

The correlation of $y(n) = \{\check{2}, 1, -3\}$ and $x(n) = \{\check{2}, 1, 3, 4\}$ is shown in Fig. 5.7. The output is $r_{xy}(n) = \{-6, -1, -4, -7, 10, 8\}$. The convolution operation without time-reversal is the correlation operation. Convolution of the time-reversed version of the sequence $y(n)$ with $x(n)$ is the same as correlation of $x(n)$ and $y(n)$. Two real signals $x(n)$ and $y(n)$ are said to be orthogonal over the entire time interval if

$$\sum_{n=-\infty}^{\infty} x(n)y(n) = 0$$

When two signals to be correlated are the same, the operation is called autocorrelation. The autocorrelation function of real-valued signals is even-symmetric. Unlike convolution, correlation operation, in general, is not commutative. The correlation of a function with an impulse shifts the time-reversed version of the function to the location of the impulse.

Let us implement the correlation using the DFT. The zero-padded 8-point DFTs, $X(k)$ and $Y(k)$, of $x(n)$ and $y(n)$ are, respectively,

$$\{10, -0.1213 - j6.5355, -1 + j3, 4.1213 - j0.5355,$$
$$0, 4.1213 + j0.5355, -1 - j3, -0.1213 + j6.5355\}$$

$$\{0, 2.7071 + j2.2929, 5 - j, 1.2929 - j3.7071,$$
$$- 2, 1.2929 + j3.7071, 5 + j, 2.7071 - j2.2929\}$$

The product $X(k)Y^*(k)$ is

$$\{0, -15.3137 - j17.4142, -8 + j14, 7.3137 + j14.5858,$$
$$0, 7.3137 - j14.5858, -8 - j14, -15.3137 + j17.4142\}$$

The IDFT of $X(k)Y^*(k)$ is

$$\{-4, -7, 10, 8, 0, 0, -6, -1\}$$

Right shifting it cyclically by 2 sample intervals is $r_{xy}(n)$ appended by 2 zeros.

5.3 Applications

Large number of applications are based on Fourier analysis, correlation, and convolution. In this section, we present some samples of the applications.

5.3.1 Lowpass Filtering of Images

Filter is a system that removes something from whatever passes through it. A coffee filter removes the coffee grounds and allows the decoction to flow through. A water filter removes salt and bacteria. An ultraviolet filter blocks or absorbs ultraviolet light.

In signal and image processing, the spectra of signals are altered in a desired way. If the low-frequency components are allowed and high-frequency components suppressed, it is called a lowpass filter. If the high-frequency components are allowed and low-frequency components suppressed, it is called a highpass filter. If we find the running averages of a set of numbers in a neighborhood of a number of a long data sequence, the bumpiness of the sequence is reduced. On the other hand, if we find the differences, the bumpiness will be enhanced and the smoothness reduced. A purely high frequency signal such as $\{1, -1\}$ will become zero, if averaged. A purely low-frequency signal such as $\{1, 1\}$ will become zero, if differenced. The filtering operation, in the time domain, is modeled by convolution. The impulse responses of the various types of filters characterize the filtering action.

The filtering operation is easier to visualize in the frequency domain, since convolution in time domain becomes multiplication in the frequency domain. The spectrum of a signal is a display of its ordered frequency components versus their respective amplitudes. Therefore, we can simply alter any part of the spectrum in a desired way. If we make the coefficients of the high-frequency components of the signal small or zero, the signal becomes smoother. If we make the coefficients of the low-frequency components of the signal small or zero, the signal becomes bumpier. In addition to easier visualization of the filtering operation, the implementation of the filtering operation in the frequency domain is faster for longer filters. Let us study the filtering operation in both the time domain and the frequency domain.

Lowpass filters with different characteristics are available. The impulse response of the simplest and widely used 3×3 2-D lowpass filter, called the averaging filter, is

$$h(m, n) = \begin{bmatrix} h(-1, -1) & h(-1, 0) & h(-1, 1) \\ h(0, -1) & \boldsymbol{h(0.0)} & h(0, 1) \\ h(1, -1) & h(1, 0) & h(1, 1) \end{bmatrix}$$

$$= \frac{1}{9} \begin{bmatrix} 1 & 1 & 1 \\ 1 & 1 & 1 \\ 1 & 1 & 1 \end{bmatrix}, \quad m = -1, 0, 1, \quad n = -1, 0, 1$$

The origin of the filter is shown in boldface. All the coefficient values are $1/9$. The filter outputs the average of the pixel values of the image in each neighborhood. Variable weighting is applied in other lowpass filters. When an image is passed through this filter, the output at each point is given by

$$y(m, n) = \frac{1}{9}(x(m - 1, n - 1) + x(m - 1, n) + x(m - 1, n + 1) + x(m, n - 1) + x(m, n)$$
$$+ x(m, n + 1) + x(m + 1, n - 1) + x(m + 1, n) + x(m + 1, n + 1))$$

The output image becomes smoother. The smoothing effect increases with larger filters.

When a 2-D filter is decomposable in terms of 1-D filters, it is always advantageous to use 1-D convolution to filter an image. The averaging filter is decomposable. This filter can be expressed as the product of a the 3×1 column filter $h_c(m) = \{1, 1, 1\}^T/3$ and the 1×3 row filter $h_r(n) = \{1, 1, 1\}/3$, which is the transpose of the column filter.

$$h(m, n) = \frac{1}{9} \begin{bmatrix} 1 & 1 & 1 \\ 1 & 1 & 1 \\ 1 & 1 & 1 \end{bmatrix} = \frac{1}{3} \begin{bmatrix} 1 \\ 1 \\ 1 \end{bmatrix} \frac{1}{3} \begin{bmatrix} 1 & 1 & 1 \end{bmatrix} = h_c(m)h_r(n)$$

The filtered output is obtained by convolving the image with the column filter first and then convolving the partial output by the row filter or vice versa, as convolution operation is commutative. With the 2-D filter $h(m, n)$ separable, $h(m, n) = h_c(m)h_r(n)$ and, with input $x(m, n)$,

$$h(m, n) * x(m, n) = (h_c(m)h_r(n)) * x(m, n)$$
$$= (h_c(m) * x(m, n)) * h_r(n) = h_c(m) * (x(m, n) * h_r(n))$$

$$y(k, l) = \sum_m h_c(m) \sum_n h_r(n)x(k - m, l - n) = \sum_n h_r(n) \sum_m h_c(m)x(k - m, l - n)$$

Let the input be

$$x(m, n) = \begin{bmatrix} 2 & 1 & 4 & 3 \\ 1 & -2 & 3 & 1 \\ 2 & -2 & 1 & -1 \\ 1 & 1 & -2 & 2 \end{bmatrix}$$

In filtering images, suitable border extensions are assumed. Assuming zero padding at the borders, the output of 1-D filtering of the columns of the input and the output of 1-D filtering of the rows of the partial output are, respectively,

$$yc(m, n) = \frac{1}{3}\begin{bmatrix} 3 & -1 & 7 & 4 \\ 5 & -3 & 8 & 3 \\ 4 & -3 & 2 & 2 \\ 3 & -1 & -1 & 1 \end{bmatrix} \qquad y(m, n) = \frac{1}{9}\begin{bmatrix} 2 & 9 & 10 & 11 \\ 2 & 10 & 8 & 11 \\ 1 & 3 & 1 & 4 \\ 2 & 1 & -1 & 0 \end{bmatrix}$$

Note that, only the central part of the output, of the same size as the input, is shown. Assuming periodicity at the borders, the extended input and the output are, respectively,

$$xe(m, n) = \begin{bmatrix} 2 & 1 & 1 & -2 & 2 & 1 \\ 3 & 2 & 1 & 4 & 3 & 2 \\ 1 & 1 & -2 & 3 & 1 & 1 \\ -1 & 2 & -2 & 1 & -1 & 2 \\ 2 & 1 & 1 & -2 & 2 & 1 \\ 3 & 2 & 1 & 4 & 3 & 2 \end{bmatrix} \qquad y(m, n) = \frac{1}{9}\begin{bmatrix} 10 & 9 & 11 & 15 \\ 5 & 10 & 8 & 16 \\ 3 & 3 & 1 & 8 \\ 9 & 8 & 7 & 12 \end{bmatrix}$$

With different border extensions, the outputs differ only at the borders.

Figure 5.8a shows a 256×256 8-bit gray level image. Figure 5.8b–d shows the filtered images with 5×5, 9×9 and 13×13 averaging filters, respectively. The blurring of the image is more with the larger filters.

Lowpass Filtering with the DFT

For a 2-D signal, we have to zero pad in two directions. Figure 5.9 shows the 8×8 zero-padded image $xz(m, n)$, obtained from the 4×4 image $x(m, n)$ used in the last example. The 3×3 filter $h(m, n)$, along with the corresponding zero-padded and shifted row and column filters, is also shown. Since the convolution output is of size is 6×6 and the the nearest power of 2 is 8, the size of the zero-padded image is 8×8.

As the averaging filter is separable. the convolution operation can be carried out using two 1-D filters, $\{h(-1) = 1, h(0) = 1, h(1) = 1\}/3$. Since $xz(m, n)$ is of size 8×8 and the origin is at the top-left corner, the zero-padded filter has to be of length 8 with $h(0)$ in the beginning. The zero-padded filter can be written as

$$\{h(-1) = 1, h(0) = 1, h(1) = 1, h(2) = 0, h(3) = 0, h(4) = 0, h(5) = 0, h(6) = 0\}/3$$

We get the zero-padded column filter

$$hz(m) = \frac{1}{3}\{1, 1, 0, 0, 0, 0, 0, 1\}$$

with the origin at the beginning, by circularly shifting left by one position, The row filter, shown in Fig. 5.9, is the transpose of the column filter. Computing the 1-D DFT

Fig. 5.8 **a** A 256 × 256 8-bit image; **b** filtered image with 5 × 5 averaging filter; **c** filtered image with 9 × 9 filter; **d** filtered image with 13 × 13 filter

$xz(m,n)$									$hz(m)$
2	1	4	3	0	0	0	0		1/3
1	−2	3	1	0	0	0	0		1/3
2	−2	1	−1	0	0	0	0		0
1	1	−2	2	0	0	0	0		0
0	0	0	0	0	0	0	0		0
0	0	0	0	0	0	0	0		0
0	0	0	0	0	0	0	0		0
0	0	0	0	0	0	0	0		1/3

$hz(n)$

1/3	1/3	0	0	0	0	0	1/3

$h(m,n)$

1/9	1/9	1/9
1/9	1/9	1/9
1/9	1/9	1/9

Fig. 5.9 Zero padding of images for implementing the 2-D linear convolution using the 2-D DFT

of this filter (division by 3 is deferred), we get

$$H(k) = \{3, 2.4142, 1, -0.4142, -1, -0.4142, 1, 2.4142\}$$

Since the filter is real-valued and even-symmetric in the time domain, its DFT is also real-valued and even-symmetric, The DFT of the column filter $H(k)$ is the transpose of that of the row filter $H(l)$. The 2-D DFT, $X(k, l)$, of $xz(m, n)$ in Fig. 5.9 is shown in Table 5.6. The partial convolution output, $3P(k, l) = X(k, l)H(l)$ shown in Table 5.7, in the frequency domain is obtained by pointwise multiplication of each row of $X(k, l)$ by $H(l)$. The convolution output, $9Y(k, l) = 3P(k, l)H(k)$ shown in Table 5.8, in the frequency domain is obtained by pointwise multiplication of each column of $3P(k, l)$ by $H(k)$.

The 2-D IDFT of $9Y(k, l)$ divided by 9 is the output, which is the same as that obtained by time-domain convolution, is

$$y(m, n) = \frac{1}{9} \begin{bmatrix} 2 & 9 & 10 & 11 & 4 & 0 & 0 & 3 \\ 2 & 10 & 8 & 11 & 3 & 0 & 0 & 5 \\ 1 & 3 & 1 & 4 & 2 & 0 & 0 & 4 \\ 2 & 1 & -1 & 0 & 1 & 0 & 0 & 3 \\ 2 & 0 & 1 & 0 & 2 & 0 & 0 & 1 \\ 0 & 0 & 0 & 0 & 0 & 0 & 0 & 0 \\ 0 & 0 & 0 & 0 & 0 & 0 & 0 & 0 \\ 3 & 7 & 8 & 7 & 3 & 0 & 0 & 2 \end{bmatrix}$$

The top left 4×4 of $y(m, n)$ are the center part of the output of the linear convolution of $x(m, n)$ and $h(m, n)$.

Figure 5.10a shows the magnitude spectrum in log scale, $\log_{10}(1 + |X(k, l)|)$ in the center-zero format, of the image in Fig. 5.8a. Figure 5.10b–d shows the magnitude spectra in log scale of the 5×5, 9×9 and 13×13 averaging filters, respectively. As with the characteristics of typical practical images, the spectrum of the image has high magnitude in the neighborhood of zero frequency and the magnitude decreases with increasing frequency. A small filter has a less blurring effect and is characterized by a high cutoff frequency. As the size of the filter increases, the cutoff frequency moves toward zero frequency and the blurring effect is more pronounced.

5.3.2　Highpass Filtering of Images

The Laplacian sharpening filter is defined as

$$h(m, n) = \begin{bmatrix} 0 & -1 & 0 \\ -1 & 5 & -1 \\ 0 & -1 & 0 \end{bmatrix} \tag{5.1}$$

Table 5.6 2-D DFT, $X(k, l)$, of $xz(m, n)$ in Fig. 5.9

15.00+j0.00	1.05−j8.12	0.00+j7.00	10.95+j3.88	9.00+j0.00	10.95−j3.88	0.00−j7.00	1.05+j8.12
10.71−j3.54	−1.00−j9.07	−1.71+j1.71	7.24+j0.59	8.36−j6.71	1.59−j12.83	−9.36−j5.12	0.17+j7.66
10.00−j1.00	−2.88−j6.54	−1.00+j6.00	8.54−j3.36	−4.00−j9.00	−7.12+j0.54	−5.00+j4.00	1.46+j9.36
9.29−j3.54	−1.24−j3.41	3.36+j0.88	−1.00−j5.07	−4.36+j5.29	5.83+j3.66	−0.29−j0.29	4.41+j7.17
5.00+j0.00	2.71−j3.29	−2.00−j1.00	1.29+j4.71	7.00+j0.00	1.29−j4.71	−2.00+j1.00	2.71+j3.29
9.29+j3.54	4.41−j7.17	−0.29+j0.29	5.83−j3.66	−4.36−j5.29	−1.00+j5.07	3.36−j0.88	−1.24+j3.41
10.00+j1.00	1.46−j9.36	−5.00−j4.00	−7.12−j0.54	−4.00+j9.00	8.54+j3.36	−1.00−j6.00	−2.88+j6.54
10.71+j3.54	0.17−j7.66	−9.36+j5.12	1.59+j12.83	8.36+j6.71	7.24−j0.59	−1.71−j1.71	−1.00+j9.07

Table 5.7 Partial convolution output $3P(k, l) = X(k, l)H(l)$ in the frequency domain

45.00+j0.00	2.54−j19.61	0.00+j7.00	−4.54−j1.61	−9.00+j0.00	−4.54+j1.61	0.00−j7.00	2.54+j19.61
32.12−j10.61	−2.41−j21.90	−1.71+j1.71	−3.00−j0.24	−8.36+j6.71	−0.66+j5.31	−9.36−j5.12	0.41+j18.49
30.00−j3.00	−6.95−j15.78	−1.00+j6.00	−3.54−j1.39	4.00+j9.00	2.95−j0.22	−5.00+j4.00	3.54+j22.61
27.88−j10.61	−3.00−j8.24	3.36+j0.88	0.41+j2.10	4.36−j5.29	−2.41−j1.51	−0.29−j0.29	10.66+j17.31
15.00+j0.00	6.54−j7.95	−2.00−j1.00	−0.54−j1.95	−7.00+j0.00	−0.54+j1.95	−2.00+j1.00	6.54+j7.95
27.88+j10.61	10.66−j17.31	−0.29+j0.29	−2.41+j1.51	4.36+j5.29	0.41−j2.10	3.36−j0.88	−3.00−j8.24
30.00+j3.00	3.54−j22.61	−5.00−j4.00	2.95+j0.22	4.00−j9.00	−3.54−j1.39	−1.00−j6.00	−6.95+j15.78
32.12+j10.61	0.41−j18.49	−9.36+j5.12	−0.66−j5.31	−8.36−j6.71	−3.00+j0.24	−1.71−j1.71	−2.41+j21.90

Table 5.8 Convolution output $9Y(k,l) = 3P(k,l)H(k)$ in the frequency domain

135.00+j0.00	7.61−j58.82	0.00+j21.00	−13.61−j4.82	−27.00+j0.00	−13.61+j4.82	0.00−j21.00	7.61+j58.82
77.55−j25.61	−5.83−j52.87	−4.12+j4.12	−7.24−j0.59	−20.19+j16.19	−1.59+j12.83	−22.61−j12.36	1.00+j44.63
30.00−j3.00	−6.95−j15.78	−1.00−j6.00	−3.54+j1.39	4.00+j9.00	2.95−j0.22	−5.00−j4.00	3.54+j22.61
−11.55+j4.39	1.24+j3.41	−1.39−j0.36	−0.17−j0.87	−1.81+j2.19	1.00+j0.63	0.12+j0.12	−4.41−j7.17
−15.00+j0.00	−6.54+j7.95	2.00+j1.00	0.54+j1.95	7.00+j0.00	0.54−j1.95	2.00−j1.00	−6.54−j7.95
−11.55−j4.39	−4.41+j7.17	0.12−j0.12	1.00−j0.63	−1.81−j2.19	−0.17+j0.87	−1.39+j0.36	1.24−j3.41
30.00+j3.00	3.54−j22.61	−5.00−j4.00	2.95+j0.22	4.00−j9.00	−3.54−j1.39	−1.00−j6.00	−6.95+j15.78
77.55+j25.61	1.00−j44.63	−22.61+j12.36	−1.59−j12.83	−20.19−j16.19	−7.24+j0.59	−4.12−j4.12	−5.83+j52.87

(a)   (b)

(c)   (d)

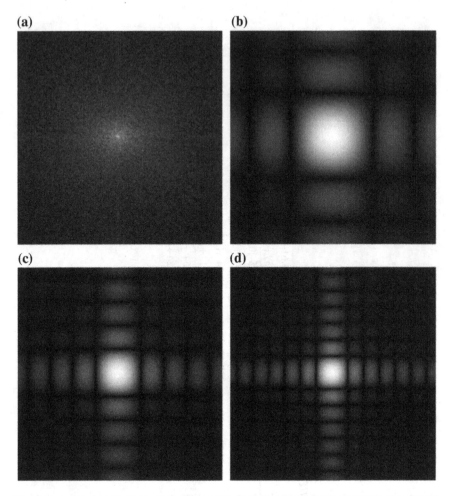

Fig. 5.10 **a** The magnitude spectrum in log scale, $\log_{10}(1 + |X(k, l)|)$ in the center-zero format, of the image in Fig. 5.8a; **b** the magnitude spectrum in log scale of the 5×5 averaging filter; **c** the magnitude spectrum in log scale, of the 9×9 averaging filter; and **d** the magnitude spectrum in log scale of the 13×13 averaging filter

Using this filter, with the same input used for lowpass filtering, the outputs with the input zero padded are

$$
y(m, n) = \begin{bmatrix}
8 & 1 & 13 & 10 \\
3 & -13 & 11 & 0 \\
10 & -12 & 7 & -9 \\
2 & 8 & -14 & 13
\end{bmatrix}
$$

(a)   **(b)**

Fig. 5.11 **a** Image in Fig. 5.8a after application of the 4 × 4 Laplacian sharpening filter (Eq. (5.1));
b image in Fig. 5.8a after application of a 8 × 8 Laplacian sharpening filter

Figure 5.11a shows the image in Fig. 5.8a after application of the 4 × 4 Laplacian filter (Eq. (5.1)). Figure 5.11b shows the image in Fig. 5.8a after application of the 8 × 8 Laplacian sharpening filter. The edges are sharper compared with Fig. 5.11a.

Image Sharpening with the DFT

High-frequency signals are characterized by the high difference between successive sample values. In this section, we present image sharpening with the DFT for the same input and the filter, assuming zero padding at the borders. The sharpening filter is inseparable with the center in the middle. By zero padding and shifting in two directions, we get

$$
hz(m, n) = \begin{bmatrix}
5 & -1 & 0 & 0 & 0 & 0 & 0 & -1 \\
-1 & 0 & 0 & 0 & 0 & 0 & 0 & 0 \\
0 & 0 & 0 & 0 & 0 & 0 & 0 & 0 \\
0 & 0 & 0 & 0 & 0 & 0 & 0 & 0 \\
0 & 0 & 0 & 0 & 0 & 0 & 0 & 0 \\
0 & 0 & 0 & 0 & 0 & 0 & 0 & 0 \\
0 & 0 & 0 & 0 & 0 & 0 & 0 & 0 \\
-1 & 0 & 0 & 0 & 0 & 0 & 0 & 0
\end{bmatrix}
$$

The 2-D DFT, $H(k, l)$, of $hz(m, n)$ is shown in Table 5.9. Since the filter is real-valued and even-symmetric, its DFT is also real-valued and even-symmetric. The 2-D DFT of the zero-padded input $X(k, l)$ is the same as in the last example. The convolution

Table 5.9 2-D DFT, $H(k, l)$, of zero-padded $hz(m, n)$

1.00	1.59	3.00	4.41	5.00	4.41	3.00	1.59
1.59	2.17	3.59	5.00	5.59	5.00	3.59	2.17
3.00	3.59	5.00	6.41	7.00	6.41	5.00	3.59
4.41	5.00	6.41	7.83	8.41	7.83	6.41	5.00
5.00	5.59	7.00	8.41	9.00	8.41	7.00	5.59
4.41	5.00	6.41	7.83	8.41	7.83	6.41	5.00
3.00	3.59	5.00	6.41	7.00	6.41	5.00	3.59
1.59	2.17	3.59	5.00	5.59	5.00	3.59	2.17

output $Y(k, l) = X(k, l)H(k, l)$, in the frequency domain, is obtained by pointwise multiplication and it is shown in Table 5.10. The output, which is the same as that obtained in in the spatial domain, is the top left 4×4 of the 8×8 2-D IDFT of $Y(k, l)$

$$
y(m, n) = \begin{bmatrix} 8 & 1 & 13 & 10 \\ 3 & -13 & 11 & 0 \\ 10 & -12 & 7 & -9 \\ 2 & 8 & -14 & 13 \end{bmatrix}
$$

5.3.3   Object Detection in Images

The normalized cross correlation (correlation coefficient) of images $x(m, n)$ and $y(m, n)$ is defined as

$$
rn_{xy}(m, n) = \frac{\sum_{k=-\infty}^{\infty} \sum_{l=-\infty}^{\infty} (x(k, l) - \bar{x}_l)(y(k - m, l - n) - \bar{y})}{\sqrt{\sum_{k=-\infty}^{\infty} \sum_{l=-\infty}^{\infty} (x(k, l) - \bar{x}_l)^2 \sum_{k=-\infty}^{\infty} \sum_{l=-\infty}^{\infty} (y(k - m, l - n) - \bar{y})^2}}
$$

The differences between this version and cross-correlation are: (i) the correlation is computed using the local mean-subtracted versions of the two inputs and (ii) the output is normalized to the range -1–1. One consequence is that the template cannot be composed of uniform values. The correlation coefficient is assigned the value zero, if the variance of the image over the overlapping portion with the template is zero. The higher the value of the coefficients, the better is the match between the template and the image. The fluctuating part of the values of the inputs is used in computing the correlation. The numerator is cross-correlation with the means subtracted. The denominator is a normalizing factor. It is the square root of the product of the variances of the overlapping samples of the inputs.

Consider the computation of the cross-correlation coefficients between $x(m, n)$ and $y(m, n)$.

Table 5.10 Convolution output $Y(k, l) = X(k, l)H(k, l)$ in the frequency domain

15.00+j0.00	1.67-j12.88	0.00+j21.00	48.33+j17.12	45.00+j0.00	48.33-j17.12	0.00-j21.00	1.67+j12.88
16.98-j5.61	-2.17-j19.70	-6.12+j6.12	36.21+j2.93	46.72-j37.46	7.93-j64.14	-33.58-j18.36	0.37+j16.63
30.00-j3.00	-10.32-j23.44	-5.00+j30.00	54.75-j21.58	-28.00-j63.00	-45.68+j3.44	-25.00+j20.00	5.25+j33.58
41.02-j15.61	-6.21-j17.07	21.58+j5.64	-7.83-j39.70	-36.72+j44.54	45.63+j28.63	-1.88-j1.88	22.07+j35.86
25.00+j0.00	15.12-j18.39	-14.00-j7.00	10.88+j39.61	63.00+j0.00	10.88-j39.61	-14.00+j7.00	15.12+j18.39
41.02+j15.61	22.07-j35.86	-1.88+j1.88	45.63-j28.63	-36.72-j44.54	-7.83+j39.70	21.58-j5.64	-6.21+j17.07
30.00+j3.00	5.25-j33.58	-25.00-j20.00	-45.68-j3.44	-28.00+j63.00	54.75+j21.58	-5.00-j30.00	-10.32+j23.44
16.98+j5.61	0.37-j16.63	-33.58+j18.36	7.93+j64.14	46.72+j37.46	36.21-j2.53	-6.12-j6.12	-2.17+j19.70

$$y(m, n) = \begin{bmatrix} 2 & 1 & 3 \\ 1 & 1 & -2 \\ 1 & 3 & 1 \end{bmatrix} \quad x(m, n) = \begin{bmatrix} 2 & 1 & 4 & 3 \\ 1 & -2 & 3 & 1 \\ 2 & -2 & 1 & -1 \\ 1 & 1 & -2 & 2 \end{bmatrix}$$

Subtracting the mean, 1.2222 from $y(m, n)$, we get

$$ym(m, n) = \begin{bmatrix} 2 & 1 & 3 \\ 1 & 1 & -2 \\ 1 & 3 & 1 \end{bmatrix} - \begin{bmatrix} 1,2222 & 1,2222 & 1,2222 \\ 1,2222 & 1,2222 & 1,2222 \\ 1,2222 & 1,2222 & 1,2222 \end{bmatrix}$$

$$= \begin{bmatrix} 0.7778 & -0.2222 & 1.7778 \\ -0.2222 & -0.2222 & -3.2222 \\ -0.2222 & 1.7778 & -0.2222 \end{bmatrix}$$

The variance of $ym(m, n)$ is 17.5556.

Subtracting the mean, 0.8889, from part of $x(m, n)$ for a neighborhood $x(1 : 3, 2 : 4)$, we get

$$xm(m, n) = \begin{bmatrix} 1 & 4 & 3 \\ -2 & 3 & 1 \\ -2 & 1 & -1 \end{bmatrix} - \begin{bmatrix} 0.8889 & 0.8889 & 0.8889 \\ 0.8889 & 0.8889 & 0.8889 \\ 0.8889 & 0.8889 & 0.8889 \end{bmatrix}$$

$$= \begin{bmatrix} 0.1111 & 3.1111 & 2.1111 \\ -2.8889 & 2.1111 & 0.1111 \\ -2.8889 & 0.1111 & -1.8889 \end{bmatrix}$$

The variance of $xm(m, n)$ is 38.8889. The sum of pointwise product of $ym(m, n)$ and $xm(m, n)$ is 4.2222. Now, $4.2222/\sqrt{(17.5556)(38.8889)} = 0.1616$.

Subtracting the mean, 1.2222, from part of $x(m, n)$ for a neighborhood $x(1 : 3, 1 : 3)$, we get

$$xm(m, n) = \begin{bmatrix} 2 & 1 & 4 \\ 1 & -2 & 3 \\ 2 & -2 & 1 \end{bmatrix} - \begin{bmatrix} 1.1111 & 1.1111 & 1.1111 \\ 1.1111 & 1.1111 & 1.1111 \\ 1.1111 & 1.1111 & 1.1111 \end{bmatrix}$$

$$= \begin{bmatrix} 0.8889 & -0.1111 & 2.8889 \\ -0.1111 & -3.1111 & 1.8889 \\ 0.8889 & -3.1111 & -0.1111 \end{bmatrix}$$

The variance of $xm(m, n)$ is 32.8889. The sum of pointwise product of $ym(m, n)$ and $xm(m, n)$ is -5.2222. Now, $-5.2222/\sqrt{(17.5556)(32.8889)} = -0.2173$. The complete correlation coefficients, for the example, are

$$\begin{bmatrix} -0.0563 & 0.3478 & 0.1143 & 0.3055 & 0.2615 & -0.0563 \\ -0.7316 & -0.2605 & -0.7976 & -0.3330 & 0.0573 & -0.0818 \\ -0.1566 & 0.4892 & -0.2173 & 0.1616 & 0.1014 & 0.1404 \\ -0.4102 & 0.1509 & 0.2406 & -0.0543 & 0.4013 & 0.0671 \\ -0.1379 & -0.5598 & 0.3448 & -0.6397 & 0.0671 & -0.0070 \\ 0.4500 & 0.2977 & 0.1063 & 0.5097 & -0.0563 & 0.1969 \end{bmatrix}$$

The location of an object can be precisely located by correlating the image with the template of the object. Figure 5.12a shows a 76×106 8-bit image composed of the text 'DISCRETE FOURIER TRANSFORM'. The problem is to locate the locations of the letter 'O'. The 10×10 8-bit template is shown in Fig. 5.12b. The two brightest points in Fig. 5.12c show the thresholded correlation coefficients image obtained by correlating the image with the template. The two points clearly indicate the locations of the letter 'O' in the image. The output is just 2 pixels. For clear visibility, they have been dilated. The un-normalized correlation output between the image and the template, shown in Fig. 5.12d, does not point out the locations of the letter 'O' in the image.

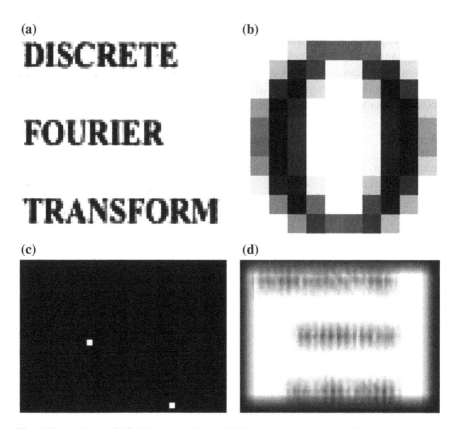

Fig. 5.12 **a** A 76×106 8-bit image; **b** 10×10 8-bit template of the letter 'O'; **c** the thresholded and dilated correlation coefficients image; **d** the un-normalized correlation output

5.3.4 Orthogonal Frequency Division Modulation

The purpose of a communication system is to transmit messages from source to destination. The source generates the message, such as human voice, an image, an e-mail message, or some kind of information. These signals are, as in other engineering applications, mostly nonelectrical (e.g., speech signal, image, pressure, vibration, and room temperature). These signals are converted by transducers to electrical signals for the purpose of efficient processing, storage, and transmission. Typical examples of transducers are microphone, thermocouple, strain gauge, computer keyboard, and CCD cameras.

The message signals are of lowpass nature and are also called baseband signals. Over short distances, the baseband signal from a source can reach the destination without much degradation. When we talk with people, this is what happens. However, it is not practically possible to transmit a baseband signal over long distances, a major reason being the requirement of an antenna of large size.

Therefore, the baseband signals are embedded (called modulation) in a carrier signal of suitable high frequency and that signal is transmitted. For example, we can walk to a place that is nearby. To travel long distances, we need a carrier, such as a car. Of course, a car can carry some number of people. Similarly, using a set of high-frequency carriers, we can transmit more than one baseband signal simultaneously over a channel. Therefore, modulation makes the transmission of baseband signals over long distances possible, in addition to the ability to transmit a set of signals at the same time. The transmitted signals pass through a channel, such as a pair of twisted copper wires, a coaxial cable, an optical fiber, or a radio link.

At the receiver, the distorted signal, in passing through the channel along with the added noise, is processed to reduce the distortion and noise to acceptable levels. Then, the modifications done to the message at the transmitter are reversed (called demodulation) to get the receiver output. This signal is converted to its original form by an appropriate transducer.

One of the ways a set of signals is transmitted through a channel, over nonoverlapping frequency bands, is called frequency division multiplexing (FDM). Multiplexing is combining several signals in to a single signal. The bandwidth of the channel of transmission is divided into nonoverlapping parts, with each part carrying a baseband signal. At the receiver, bandpass filters are used to separate the signals. Practical filters have a transition band. Therefore, sufficient frequency gap must be left between two modulated signals. A recent digital communication method that provides significant advantages is the orthogonal frequency division modulation (OFDM). The input to a digital communication system is a sequence of digits. The source of input could be inherently digital or digitized analog signal.

Fig. 5.13 DFT is a set of bandpass filters with a narrow passband

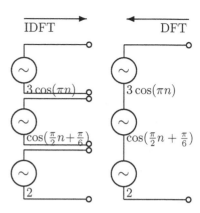

In OFDM, N subcarriers, each of which is independent due to orthogonality and having a constant gain, carry N digits of information. OFDM is one of the most successful signal transmission methods in digital communications, particularly in high data-rate applications. OFDM transforms an intersymbol interference (ISI) channel into N parallel subchannels without ISI. Additive white Gaussian noise is assumed. A longer sequence of data symbols is grouped into N subsequences and transmitted over N subcarriers. In effect, OFDM uses IDFT and cyclic prefix to realize multicarrier communications without actually generating and modulating the subcarriers. OFDM is basically a combination of Fourier analysis and convolution.

A signal is composed of frequency components, as shown in Fig. 5.13 (right side).

$$x(n) = 2 + \cos\left(\frac{\pi}{2}n + \frac{\pi}{6}\right) + 3\cos(\pi n)$$

The DFT decomposes the signal into its frequency components (left side).

$$\left\{ x_0(n) = 2, \quad x_1(n) = \cos\left(\frac{\pi}{2}n + \frac{\pi}{6}\right), \quad x_2(n) = 3\cos(\pi n) \right\}$$

The IDFT reconstructs the signal back from its components. In general, the DFT takes the N samples of an input signal $x(n)$ and outputs the N complex coefficients $X(k)$ of the individual frequency components. The IDFT reverses this process. The point is that the separation of the frequency components is much less than that required when the frequency components are not orthogonal, as in FDM. The orthogonality of the carriers leads to spectral efficiency giving a higher bit rate for the same bandwidth, in addition to several other advantages such as the elimination of the inter symbol interference, reduction of multipath effects, simplified equalization, and receiver design. Further, fast realization of the DFT and the IDFT using hardware and software modules is readily available. These advantages are obtained under the condition that frequency and time synchronization between the transmitter and receiver is accurate and the transmitter is designed to deliver the high peak power requirements, while the average power is low.

Let there be 2 complex coefficients, $X(0) = 1 + j2$ and $X(1) = 2 + j1$. These coefficients correspond to the frequency components

$$x_0(n) = (1 + j2)e^{\frac{2\pi}{2}0n} = 1 + j2 \quad \text{and} \quad x_1(n) = (2 + j1)e^{\frac{2\pi}{2}1n} = (2 + j1)\cos(\pi n)$$

With a complex coefficient, we are able to modulate the amplitude and phase of a sinusoid or, equivalently, the amplitudes of the in-phase and quadrature components of a sinusoid. As it is easier to implement, the quadrature form is used in practice. Note that, in OFDM, the subcarriers are orthogonal and their frequencies are fixed. With $X(0) = 2 + j1$ and $X(1) = 1 + j2$, these coefficients correspond to the frequency components

$$x_0(n) = (2 + j1)e^{\frac{2\pi}{2}0n} = 2 + j1 \quad \text{and} \quad x_1(n) = (1 + j2)e^{\frac{2\pi}{2}1n} = (1 + j2)\cos(\pi n)$$

Let us say we want to transmit 4 messages, AA, AB, BA, and BB. Let $1 + j2$ represent the symbol A and $2 + j1$ represent the symbol B. Then, the real and imaginary components of the modulated signals corresponding to the 4 messages, AA, AB, BA, and BB, are shown in Figs. 5.14 and 5.15, respectively.

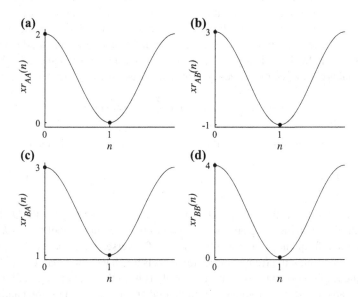

Fig. 5.14 Real components of the modulated subcarriers corresponding to different messages, **a** AA, **b** AB, **c** BA, and **d** BB

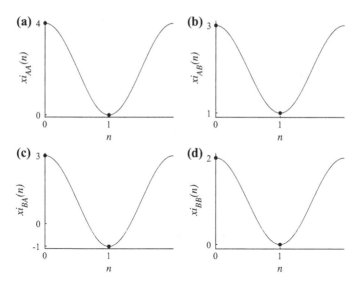

Fig. 5.15 Imaginary components of the modulated subcarriers corresponding to different messages, **a** AA, **b** AB, **c** BA, and **d** BB

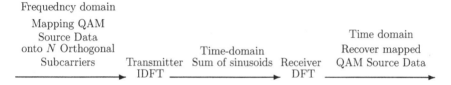

Fig. 5.16 Transmitter and receiver in OFDM

$$(1+j2)e^{\frac{2\pi}{2}0n} + (1+j2)e^{\frac{2\pi}{2}1n} = (1+j2) + (1+j2)\cos(\pi n) = \{2+j4, 0+j0\} \rightarrow AA$$

$$(1+j2)e^{\frac{2\pi}{2}0n} + (2+j1)e^{\frac{2\pi}{2}1n} = (1+j2) + (2+j1)\cos(\pi n) = \{3+j3, -1+j1\} \rightarrow AB$$

$$(2+j1)e^{\frac{2\pi}{2}0n} + (1+j2)e^{\frac{2\pi}{2}1n} = (2+j1) + (1+j2)\cos(\pi n) = \{3+j3, 1-j1\} \rightarrow BA$$

$$(2+j1)e^{\frac{2\pi}{2}0n} + (2+j1)e^{\frac{2\pi}{2}1n} = (2+j1) + (2+j1)\cos(\pi n) = \{4+j2, 0+j0\} \rightarrow BB$$

The DFT of these time-domain samples is the scaled frequency coefficients input by the transmitter.

These modulated discrete signals are converted to analog signals, also shown in Figs. 5.14 and 5.15, by a digital to analog converter. The analog signals are modulated by a high-frequency carrier for the purpose of transmission and transmitted. At the receiver, these signals are demodulated to get the baseband signal and the DFT yields the message. Figure 5.16 shows the essence of transmitting and receiving messages using OFDM. The input data is mapped to quadrature data, assumed in the frequency domain. The IDFT of this data is a sum of sinusoids in time domain, which is transmitted after converting to analog signal and modulating with a high

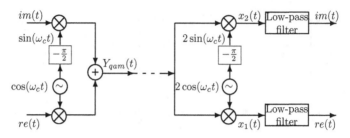

Fig. 5.17 Quadrature amplitude modulation, QAM

frequency. The DFT of the samples of the demodulated signal at the receiver is the source frequency-domain data transmitted. Reversing the mapping carried out at the transmitter, we get the input message.

Quadrature Amplitude Modulation

Quadrature amplitude modulation (QAM) transmits 2 signals using carriers of the same frequency but in phase quadrature. Figure 5.17 shows the quadrature amplitude modulation and demodulation. The carrier frequency is $\cos(\omega_c t)$. Phase shifters are used to delay the carrier frequency to get $\sin(\omega_c t)$. The in-phase component of the message signal is multiplied by $\cos(\omega_c t)$. The quadrature component is multiplied by $\sin(\omega_c(t)$. The sum of the two product signals is the modulator output

$$re(t)\cos(\omega_c t) + im(t)\sin(\omega_c t)$$

The modulated signal is transmitted using double sideband transmission. A similar process, in the demodulator at the right side, governed by the following equations produces the sum of the message signal and other high-frequency components.

$$
\begin{aligned}
x_1(t) &= 2y_{qam}\cos(\omega_c t) \\
&= 2(re(t)\cos(\omega_c t) + im(t)\sin(\omega_c t))\cos(\omega_c t) \\
&= re(t) + re(t)\cos(2\omega_c t) + im(t)\sin(2\omega_c t) \\
x_2(t) &= 2y_{qam}\sin(\omega_c t) \\
&= 2(re(t)\cos(\omega_c t) + im(t)\sin(\omega_c t))\sin(\omega_c t) \\
&= im(t) - im(t)\cos(2\omega_c t) + re(t)\sin(2\omega_c t)
\end{aligned}
$$

The message signal is obtained by lowpass filtering.

16-QAM Constellation

Figure 5.18 shows a 16-point QAM constellation, an arrangement of 16 distinct complex numbers. This constitutes the signal space. As each number is complex-valued, it has a magnitude and a phase or, equivalently, the amplitudes of the real and imaginary parts. Each number can be used to modulate any subcarrier sinusoid. Since

Fig. 5.18 16-point QAM
constellation

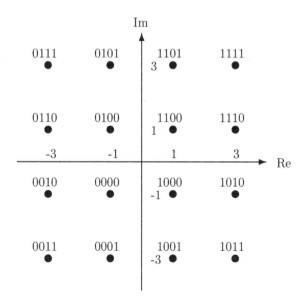

there are 16 numbers, each subcarrier can transmit the information of $\log_2 16 = 4$ bits. There are 16 distinct combinations with 4 bits and there are 16 distinct complex numbers. Therefore, a subcarrier, $a \cos(\omega_c t) + b \sin(\omega_c t)$ transmits 4 bits.

Example—Parallel code for the message 'Discrete Fourier Transform' with 4 subchannels each of length 8.

Table 5.11 shows the 16-point QAM assignment of codes to the distinct characters in the message. As there are less than 16 characters, the last few are assigned the code for space.

For example, the code for 'D' is $-3 - j1$, and it is the first entry in the first subchannel shown below. The code for 'I' is $-1 + j3$, which is the second entry in the first subchannel. Similarly, after assigning codes for all the characters in the message, we get

$$
\begin{bmatrix}
-3 - j & 1 + j3 & 1 + j3 & 1 - j3 \\
-1 + j3 & -1 + j & 3 - j3 & -3 + j1 \\
3 - j1 & 1 - j & 1 - j3 & 1 + j3 \\
-1 - j3 & 1 + j & -1 - j & 1 + j3 \\
1 - j3 & 1 - j3 & -3 + j3 & 1 + j3 \\
-3 - j3 & -1 + j3 & 3 - j & 1 + j3 \\
3 - j3 & -3 - j3 & -1 + j & 1 + j3 \\
-3 - j3 & 1 - j3 & 1 - j & 1 + j3
\end{bmatrix}
$$

These are the modulated signals in the frequency domain. We take the IDFT, which is a sum of all the modulated sinusoids in each subchannel. The result is

Table 5.11 16-point QAM constellation

Character	Symbol transmitted	Carrier		
		Phase	Amplitude/($3\sqrt{2}$)	Rect
A	0000	225°	0.33	$-1 - j1$
C	0001	255°	0.75	$-1 - j3$
D	0010	195°	0.75	$-3 - j1$
E	0011	225°	1.00	$-3 - j3$
F	0100	135°	0.33	$-1 + j1$
I	0101	105°	0.75	$-1 + j3$
M	0110	165°	0.75	$-3 + j1$
N	0111	135°	1.0	$-3 + j3$
O	1000	315°	0.33	$1 - j1$
R	1001	285°	0.75	$1 - j3$
S	1010	345°	0.75	$3 - j1$
T	1011	315°	1.00	$3 - j3$
U	1100	45°	0.33	$1 + j1$
	1101	75°	0.75	$1 + j3$
	1110	15°	0.75	$3 + j1$
	1111	45°	1.00	$3 + j3$

$$
\begin{bmatrix}
-0.5 - j1.8 & 0.0 - j0.3 & 0.5 - j0.3 & 0.5 + j2.0 \\
-1.3 + j1.1 & -0.4 + j0.7 & 1.4 - j0.1 & -0.2 - j1.3 \\
-1.8 + j0.0 & -0.3 + j0.0 & 0.0 + j1.8 & 0.3 - j1.3 \\
-0.8 + j0.1 & 0.1 + j0.8 & 0.0 - j0.2 & 0.5 - j0.9 \\
1.5 - j0.3 & 0.0 - j0.8 & -1.0 + j1.3 & 0.5 - j0.5 \\
-0.2 - j0.6 & -0.1 + j1.8 & 0.6 + j0.6 & 0.2 - j0.2 \\
-0.3 + j0.0 & 1.3 + j1.0 & -0.5 + j0.3 & -0.3 - j0.3 \\
0.3 + j0.4 & 0.4 - j0.3 & 0.0 - j0.3 & -0.5 - j0.6
\end{bmatrix}
$$

For example, the real part of the first entry -0.5 is obtained as the sum of the real part of the entries of the real parts of the first columns divided by 8.

$$
\frac{(-3 - 1 + 3 - 1 + 1 - 3 + 3 - 3)}{8} = \frac{-4}{8} = -0.5
$$

Passage of the Modulated Time-Domain Signal Through the Channel

The channel has an impulse response. For this example, let us assume that it is $h(n) = \{\breve{1}, 0.3, 0.1\}$. The output at the end of the channel is the linear convolution of the time-domain signal with $h(n)$. As the time-domain signal is periodic, circular convolution is required. Linear convolution of two sequences results in a sequence of length that is equal to the sum of the length of the two sequences minus 1. Circular convolution of two sequences of equal length results in a sequence of the same length. Circular convolution can be used to compute the linear convolution by sufficient zero padding of the sequences.

$$x(n) * h(n) = xz(n) \circledast hz(n)$$

$$\{1, 2\} * \{3, 2\} = \{1, 2, 0\} \circledast \{3, 2, 0\} = \{3, 8, 4\}$$

In OFDM, linear convolution is used to compute the circular convolution with a cyclic prefix.

$$xc(n) * h(n) = x(n) \circledast h(n)$$

$$\{2, 1, 2\} * \{3, 2\} = \{1, 2\} \oplus \{3, 2\} = \{7, 8\}$$

The partial periodic extension is called the cyclic prefix. Take the output from $n = 0$ to $n = N - 1$, where N is the length of the periodic sequences to be convolved. The necessity for this is because the data is periodic in OFDM, while the channel carries out linear convolution.

The periodically extended data is

$$
\begin{bmatrix}
-0.3 + j0.0 & 1.3 + j1.0 & -0.5 + j0.3 & -0.3 - j0.3 \\
0.3 + j0.4 & 0.4 - j0.3 & 0.0 - j0.3 & -0.5 - j0.6 \\
-0.5 - j1.8 & 0.0 - j0.3 & 0.5 - j0.3 & 0.5 + j2.0 \\
-1.3 + j1.1 & -0.4 + j0.7 & 1.4 - j0.1 & -0.2 - j1.3 \\
-1.8 + j0.0 & -0.3 + j0.0 & 0.0 + j1.8 & 0.3 - j1.3 \\
-0.8 + j0.1 & 0.1 + j0.8 & 0.0 - j0.2 & 0.5 - j0.9 \\
1.5 - j0.3 & 0.0 - j0.8 & -1.0 + j1.3 & 0.5 - j0.5 \\
-0.2 - j0.6 & -0.1 + j1.8 & 0.6 + j0.6 & 0.2 - j0.2 \\
-0.3 + j0.0 & 1.3 + j1.0 & -0.5 + j0.3 & -0.3 - j0.3 \\
0.3 + j0.4 & 0.4 - j0.3 & 0.0 - j0.3 & -0.5 - j0.6
\end{bmatrix}
$$

The last 2 rows of the last matrix have been prepended resulting in 8 rows becoming 10 rows. The modulated time-domain analog data, obtained using a digita-to-analog converter, with a prefix shown by the dotted line, is shown in Fig. 5.19. This prefixed data is linearly convolved withe the channel impulse response $h(n) = \{h(0) = 1, h(1) = 0.3, h(2) = 0.1\}$. The result is

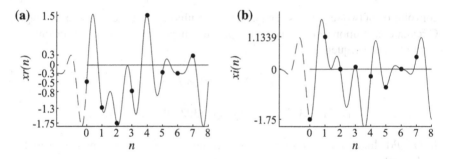

Fig. 5.19 Modulated time-domain analog data with a prefix, shown by the dotted line

$$\begin{bmatrix}
-0.3 + j0.0 & 1.3 + j1.1 & -0.2 + j0.3 & -0.3 - j0.3 \\
0.2 + j0.4 & 0.8 + j0.1 & -0.1 - j0.2 & -0.6 - j0.7 \\
-0.4 - j1.6 & 0.3 - j0.2 & 0.5 - j0.3 & 0.3 + j1.8 \\
-1.4 + j0.7 & -0.4 + j0.6 & 1.5 - j0.2 & -0.1 - j0.7 \\
-2.2 + j0.2 & -0.4 + j0.2 & 0.5 + j1.7 & 0.2 - j1.4 \\
-1.4 + j0.2 & -0.0 + j0.9 & 0.1 + j0.3 & 0.6 - j1.4 \\
1.1 - j0.2 & -0.0 - j0.5 & -1.0 + j1.4 & 0.7 - j0.9 \\
0.2 - j0.7 & -0.1 + j1.6 & 0.3 + j1.0 & 0.4 - j0.5 \\
-0.2 - j0.2 & 1.2 + j1.5 & -0.4 + j0.6 & -0.1 - j0.4 \\
0.2 + j0.4 & 0.8 + j0.2 & -0.1 - j0.1 & -0.6 - j0.7
\end{bmatrix}$$

For example,

$$(-0.5)(1) + (0.3)(0.3) + (-0.3)(0.1) = -0.4$$

is the real part value of the first value in the third row. The received analog signal is passed through an analog-to-digital converter to get the sampled data.

Now, the DFT of the columns of the last 8 rows is computed resulting in

$$\begin{bmatrix}
-4.2 - j1.4 & 1.4 + j4.2 & 1.4 + j4.2 & 1.4 - j4.2 \\
-0.3 + j3.9 & -0.9 + j1.5 & 2.7 - j4.6 & -3.3 + j2.1 \\
2.4 - j1.8 & 0.6 - j1.2 & -0.0 - j3.0 & 1.8 + j2.4 \\
-1.1 - j2.3 & 0.9 + j0.7 & -0.9 - j0.7 & 1.1 + j2.3 \\
0.8 - j2.4 & 0.8 - j2.4 & -2.4 + j2.4 & 0.8 + j2.4 \\
-2.0 - j2.7 & -1.1 + j2.3 & 2.5 - j0.5 & 0.5 + j2.5 \\
3.6 - j1.8 & -1.8 - j3.6 & -1.2 + j0.6 & 0.0 + j3.0 \\
-2.7 - j4.6 & 2.1 - j3.3 & 1.5 - j0.9 & 0.3 + j3.9
\end{bmatrix}$$

The frequency response of the channel, $H(k)$, is the 8-point DFT of its impulse response $h(n)$.

Fig. 5.20 OFDM scatter
plot. SNR = 20 dB

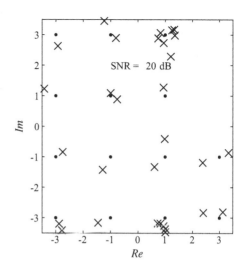

$$H(k) = \{1.4, 1.2121 - j0.3121, 0.9 - j0.3, 0.7879 - j0.1121, 0.8, 0.7879 + j0.1121, 0.9$$
$$+ j0.3, 1.2121 + j0.3121\}$$

The pointwise inverse is

$$\frac{1}{H(k)} = \{0.7, 0.8 + j0.2, 1.0 + j0.3, 1.2 + j0.2, 1.3, 1.2 - j0.2, 1.0 - j0.3, 0.8 - j0.2\}$$

The pointwise product of each column of the DFT values by these values yields the input message. The equalization is made simple due to orthogonality of the subcarriers. We are getting back the exact input message because of the assumption of zero noise.

With SNR = 20 dB, OFDM scatter plot is shown in Fig. 5.20. As noise is high, the received signals are scattered around the exact input message. For each received value, the input value nearest is assigned by computing the distances. To reduce the noise, Wiener filter can be used. Further error correcting codes may recover the exact data.

With SNR = 30 dB, OFDM scatter plot is shown in Fig. 5.21. Now, the received values are much closer to the actual values.

With SNR = 40 dB, OFDM scatter plot is shown in Fig. 5.22. With SNR high, the received and input values almost overlap.

The block diagram of OFDM transmission system is shown in Fig. 5.23. The QAM mapped data in the frequency domain, coming in serial form, is converted to parallel form. The IDFT of the parallel data is computed to get the time-domain samples. Now, the cyclic prefix is added. This data is converted to serial analog form, modulated by a carrier and transmitted. The distorted signal, due to the channel and noise, is demodulated, cyclic prefix removed and converted to parallel discrete form

Fig. 5.21 OFDM scatter
plot. SNR = 30 dB

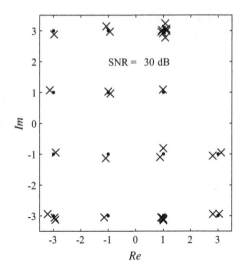

Fig. 5.22 OFDM scatter
plot. SNR = 40 dB

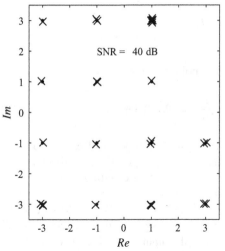

at the receiver. The analog processing is not shown in the diagram. The DFT of the
time-domain samples yields the frequency-domain data. This data is passed through
equalizer to compensate for the loss of gain in transmission through the channel. This
equalization is independently carried out for each subcarrier. The equalized signal
is passed through a detector to find the final frequency-domain data. One of the
ways of detection is to assign the data to the nearest, by Euclidean distance, QAM
constellation point. Further, scrambling for data security, error correction coding,
and coding for compression of data form part of the transmission system.

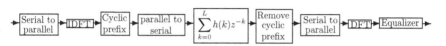

Fig. 5.23 Block diagram of OFDM transmission system

5.3.5 Hilbert Transform

A complex signal whose imaginary part is the Hilbert transform of its real part is called the analytic signal. An analytic signal has no negative frequency components. Hilbert transformer shifts the phase of every spectral component by $\frac{\pi}{2}$ radians. Typical applications are single sideband modulation in communication systems and sampling of bandpass signals.

The frequency response of the Hilbert transformer, for an even N, is

$$H(k) = \begin{cases} -j \text{ for } k = 1, 2, \ldots, \frac{N}{2} - 1 \\ 0 \text{ for } k = 0, \frac{N}{2} \\ j \text{ for } k = \frac{N}{2} + 1, \frac{N}{2} + 2, \ldots, N - 1 \end{cases}$$

Basically, it is an all-pass filter that imparts a $\pm 90°$ phase shift on the input signal. However, in practice, it is designed over the required bandwidth of a given input signal.

The IDFT of $H(k)$ is the impulse response. As the frequency response is imaginary and odd-symmetric, the impulse response is real and odd.

$$h(n) = \frac{1}{N} \sum_{k=0}^{N-1} H(k) e^{j\frac{2\pi}{N} kn} = \frac{j}{N} \sum_{k=0}^{N-1} H(k) \sin\left(\frac{2\pi}{N} kn\right)$$

For example,

$$x(n) = \cos\left(\frac{2\pi}{4} n\right) \leftrightarrow X(k) = \{0, 2, 0, 2\},$$

$$H(k) = \{0, -j, 0, j\}, \quad X(k)H(k) = \{0, -j2, 0, j2\}$$

The Hilbert transform $x_h(n)$ of $x(n)$ is the IDFT of $X(k)H(k)$.

$$x_h(n) = \sin\left(\frac{2\pi}{4} n\right) \leftrightarrow \{0, -j2, 0, j2\}$$

Now, the analytic signal $x_a(n)$ corresponding to $x(n)$ is the complex signal with $x(n)$ as its real part and $x_h(n)$ as its imaginary part.

$$x_a(n) = x(n) + jx_h(n) = \cos\left(\frac{2\pi}{4}n\right) + j\sin\left(\frac{2\pi}{4}n\right) = e^{j\frac{2\pi}{4}n} \leftrightarrow \{0, 4, 0, 0\},$$

which has a one-sided spectrum.

As always, the Hibert transform of a signal can also be found in the time domain by convolution.

$$h(n) = 0.5\sin\left(\frac{2\pi}{4}n\right) = \{0, 0.5, 0, -0.5\} \leftrightarrow \{0, -j, 0, j\}$$

$$(x(n) = \{1, 0, -1, 0\}) \circledast (\{0, 0.5, 0, -0.5\} = h(n)) = \{0, 1, 0, -1\} = x_h(n) = \sin\left(\frac{2\pi}{4}n\right)$$

5.4   Summary

- Convolution is one of the often used system models. It relates the input and output of a system through its impulse response. It can also be considered as the weighted average of a signal.
- The input signal is decomposed in terms of impulses and the superposition summation of the responses of the system to all the constituent impulses is the system output.
- In terms of computation, it is a sum of products of two sequences after time-reversal of one of them.
- In linear convolution, the sequences are located on a line. In circular convolution, the sequences are located on a circle. The output values differ only at the border.
- In applications, linear convolution is often required. However, with sufficient zero padding of the sequences, linear convolution is implemented using circular convolution efficiently.
- The computational complexity of the 1-D convolution of sequences of length N in the time domain is $O(N^2)$, whereas the complexity is reduced to $O(N \log_2 N)$ in the frequency domain. The reason is that convolution in the time domain becomes multiplication in the frequency domain with complexity $O(N)$. Mapping a time-domain sequence into a frequency-domain sequence requires a complexity of $O(N \log_2 N)$.
- Points to be noted in implementing the linear convolution operation using the DFT are: (i) the sequences must be zero padded so that their lengths are at the least equal to the sum of the lengths of the two sequences to be convolved minus one; (ii) the length of the zero-padded sequences is the nearest integral power of 2; and (iii) the origins of the sequences are aligned.

- 2-D convolution is a straightforward extension of the 1-D convolution. If one of the sequences is separable, 2-D convolution can be implemented faster by a set of 1-D convolutions.
- Correlation of two sequences is a similarity measure.
- Implementation of the correlation is the same as that of convolution with out the time-reversal operation.

Exercises

5.1 Find the linear convolution $y(n)$ of $x(n)$ and $h(n)$, using the DFT and the IDFT. Verify $y(n)$ by convolving $x(n)$ and $h(n)$ in the time domain.
5.1.1
$$x(n) = \{\check{2}, 3, 1, 4\} \quad \text{and} \quad h(n) = \{\check{1}, -3, 1, -4\}$$

5.1.2
$$x(n) = \{\check{1}, 1, -2, 4\} \quad \text{and} \quad h(n) = \{\check{1}, 2, 1, 4\}$$

* 5.1.3
$$x(n) = \{\check{2}, 1, 3, 4\} \quad \text{and} \quad h(n) = \{\check{1}, -2, 3, -4\}$$

5.2 Find the linear convolution $y(m, n)$ of $x(m, n)$ and $h(m, n)$, using the DFT and the IDFT. Verify $y(m, n)$ by convolving $x(m, n)$ and $h(m, n)$ in the time domain.
* 5.2.1
$$x(m, n) = \begin{bmatrix} 1 & -3 \\ 1 & 2 \end{bmatrix}, \quad h(m, n) = \begin{bmatrix} 2 & 3 \\ -2 & 1 \end{bmatrix}$$

5.2.2
$$x(m, n) = \begin{bmatrix} -2 & 1 \\ -4 & 2 \end{bmatrix}, \quad h(m, n) = \begin{bmatrix} 3 & -3 \\ -1 & 4 \end{bmatrix}$$

5.2.3
$$x(m, n) = \begin{bmatrix} 4 & -3 \\ 3 & -4 \end{bmatrix}, \quad h(m, n) = \begin{bmatrix} 2 & 2 \\ 1 & 3 \end{bmatrix}$$

5.3 Find the linear correlation $r_{xh}(n)$ of $x(n)$ and $h(n)$, using the DFT and the IDFT. Verify $r_{xh}(n)$ by correlating $x(n)$ and $h(n)$ in the time domain.
5.3.1
$$x(n) = \{\check{1}, 3, 1, 4\} \quad \text{and} \quad h(n) = \{\check{1}, -3, 3, -4\}$$

* 5.3.2
$$x(n) = \{\check{1}, 2, -2, 4\} \quad \text{and} \quad h(n) = \{\check{1}, 2, -1, 4\}$$

5.3.3
$$x(n) = \{\check{2}, 1, 3, 2\} \quad \text{and} \quad h(n) = \{\check{1}, 2, -3, 4\}$$

5.4 Find the linear correlation $r_{xh}(m, n)$ of $x(m, n)$ and $h(m, n)$, using the DFT and the IDFT. Verify $r_{xh}(m, n)$ by correlating $x(m, n)$ and $h(m, n)$ in the time domain.
* 5.4.1

$$x(m, n) = \begin{bmatrix} -1 & -3 \\ 3 & 2 \end{bmatrix}, \quad h(m, n) = \begin{bmatrix} 1 & 3 \\ -3 & 1 \end{bmatrix}$$

5.4.2

$$x(m, n) = \begin{bmatrix} -2 & 2 \\ -1 & 2 \end{bmatrix}, \quad h(m, n) = \begin{bmatrix} 4 & -3 \\ -1 & 3 \end{bmatrix}$$

5.4.3

$$x(m, n) = \begin{bmatrix} 2 & -3 \\ 3 & -1 \end{bmatrix}, \quad h(m, n) = \begin{bmatrix} 2 & 1 \\ 1 & -3 \end{bmatrix}$$

5.5 Find the linear Hilbert transform $x_h(n)$ of $x(n)$. Compute the DFT of $x(n) + jx_h(n)$ and verify that the spectrum is one-sided.
5.5.1 $x(n) = \{2, -3, 1, 4\}$.

* 5.5.2 $x(n) = \{2, 1, 1, 4\}$.

5.5.3 $x(n) = \{1, 3, 1, 4\}$.

Chapter 6
Aliasing and Leakage

There are four versions of the Fourier analysis. DFT is the only one that is discrete and finite in both the domains, and, hence, implementable using a digital system. In approximating other versions of the Fourier analysis, it has to be ensured that the data is adequately represented by the DFT in any one period. It is possible because physical devices can generate signals over a finite period only and can generate frequency components of a finite order only. Therefore, the sampling interval and the record length have to be carefully chosen. Then, for all practical purposes, all waveforms generated by physical devices can be represented by the DFT adequately. In this chapter, we learn how to determine the proper sampling interval and the record length.

6.1 Aliasing Effect

In unsigned binary number system, we can represent 2^N distinct numbers using N bits. For example, with 2 bits, we can represent 4 numbers $\{00, 01, 10, 11\}$. The range of numbers that can be represented uniquely depends on the number of bits used. If the number of bits is inadequate to represent a number, then it cannot be uniquely represented. Similarly, with N complex samples, we can represent only N complex exponentials. Since a real sinusoid needs two complex exponentials for its representation, only about $N/2$ real sinusoids can be uniquely represented with N complex samples.

With a periodic complex signal represented by 4 samples, the complex exponentials

$$\{e^{j\frac{2\pi}{4}0n}, e^{j\frac{2\pi}{4}1n}, e^{j\frac{2\pi}{4}2n}, e^{j\frac{2\pi}{4}3n}\}, \ n = 0, 1, 2, 3$$

© Springer Nature Singapore Pte Ltd. 2018

D. Sundararajan, *Fourier Analysis—A Signal Processing Approach*,
https://doi.org/10.1007/978-981-13-1693-7_6

can only be uniquely represented. For example,

$$e^{j\frac{2\pi}{4}5n} = e^{j\frac{2\pi}{4}(4+1)n} = e^{j\frac{2\pi}{4}4n}e^{j\frac{2\pi}{4}1n} = e^{j\frac{2\pi}{4}1n}$$

The impersonation of a higher frequency exponential $e^{j\frac{2\pi}{4}5n}$ as a lower-frequency exponential $e^{j\frac{2\pi}{4}1n}$, due to insufficient number of samples, is called the aliasing effect. For complex signals, with period N, the aliasing effect is characterized by

$$x(n) = e^{j\left(\frac{2\pi}{N}(k+lN)n+\phi\right)} = e^{j\left(\frac{2\pi}{N}kn+\phi\right)}, \quad k = 0, 1, \ldots, N-1$$

where l is any integer. Remember that periodic signals are defined over a circle. There are only N unique samples for a complex exponential with period N. Therefore,

$$e^{j\left(\frac{2\pi}{N}(k+lN)n+\phi\right)} = e^{j(2\pi ln)}e^{j\left(\frac{2\pi}{N}kn+\phi\right)} = e^{j\left(\frac{2\pi}{N}kn+\phi\right)}$$

since $e^{j(2\pi ln)} = 1$ for any integer values of l.

For real signals, with an even period N, the aliasing effect is characterized by three formulas, since each sinusoid is characterized by complex conjugate exponentials.

$$x(n) = \cos\left(\frac{2\pi}{N}(k+lN)n+\phi\right) = \cos\left(\frac{2\pi}{N}kn+\phi\right), \quad k = 0, 1, \ldots, \frac{N}{2}-1$$

$$x(n) = \cos\left(\frac{2\pi}{N}(k+lN)n+\phi\right) = \cos(\phi)\cos\left(\frac{2\pi}{N}kn\right), \quad k = \frac{N}{2}$$

$$x(n) = \cos\left(\frac{2\pi}{N}(lN-k)n+\phi\right) = \cos\left(\frac{2\pi}{N}kn-\phi\right), \quad k = 1, 2, \ldots, \frac{N}{2}-1$$

where N and index l are positive integers. Oscillations increase only up to $k = \frac{N}{2}$ with N even, decrease afterward, and cease at $k = N$ and this pattern repeats indefinitely. With frequency indices greater than $\frac{N}{2}$, frequency folding occurs. Therefore, sinusoids with frequency index up to $\frac{N}{2}$ can only be uniquely identified with N samples. Frequency with index $\frac{N}{2}$ is called the folding frequency. At this frequency, only cosine waveform can be represented. With the number of samples fixed (say N), the number of sinusoids those can be uniquely represented is $N/2$. For example, with 512 samples, the uniquely identifiable sinusoids are

$$x(n) = \cos\left(\frac{2\pi}{512}kn+\phi\right), \quad k = 0, 1, \ldots, 255$$

Figure 6.1 shows the aliasing effect with sinusoids $x(n) = \cos(\frac{2\pi}{8}n + \frac{\pi}{3})$ and $xa(n) = \cos(\frac{2\pi}{8}7n - \frac{\pi}{3})$ having the same set of samples. While their continuous versions have different amplitude profiles, as shown by the continuous and dashed lines, their discrete versions have the same amplitude profile.

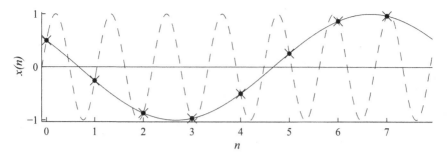

Fig. 6.1 Sinusoid $x(n) = \cos(\frac{2\pi}{8}n + \frac{\pi}{3})$. Sinusoid $xa(n) = \cos(\frac{2\pi}{8}7n - \frac{\pi}{3})$ in dashed line

Fig. 6.2 The aliasing effect of the frequency components of a periodic waveform with period 16 samples

$$\uparrow$$

$$
\begin{array}{cccccccc}
32 & 31 & 30 & 29 & 28 & 27 & 26 & 25 \\
17 & 18 & 19 & 20 & 21 & 22 & 23 & 24 \\
16 & 15 & 14 & 13 & 12 & 11 & 10 & 9 \\
0 & 1 & 2 & 3 & 4 & 5 & 6 & 7 & 8
\end{array}
$$

Frequency index, k

$$
xa(n) = \cos\left(\frac{2\pi}{8}(8-1)n - \frac{\pi}{3}\right) = \cos\left(\frac{2\pi}{8}(-1)n - \frac{\pi}{3}\right) = \cos\left(\frac{2\pi}{8}n + \frac{\pi}{3}\right) = x(n)
$$

If the sum or difference of the frequency indices of two discrete sinusoids is an integral multiple of the sampling period, then it is impossible to differentiate between them. For example, a discrete waveform with period 16 samples can have distinct frequency components with frequency indices $k = \{0, 1, 2, 3, 4, 5, 6, 7, 8\}$ only. Frequency component with index zero is DC and that with 8 is a cosine waveform. Frequency components with frequency indices $k = \{16, 15, 14, 13, 12, 11, 10, 9\}$ alias as $k = \{0, 1, 2, 3, 4, 5, 6, 7\}$. Frequency components with frequency indices $k = \{17, 18, 19, 20, 21, 22, 23, 24\}$ alias as $k = \{1, 2, 3, 4, 5, 6, 7, 8\}$. Figure 6.2 shows the aliasing effect, which is also shown in Table 6.1. For example, the sum of frequency indices 1 and 15 is 16 (the difference of frequencies 1 and 17 is also 16) and it is not possible differentiate between them with the sampling frequency 1/16 cycles/sample. In general, with a sampling frequency f_s cycles/sample, the formula to find the index of the alias f_a of any frequency component f is given by

Table 6.1 Frequency components with frequencies f and the corresponding aliased frequencies with sampling frequency $f_s = 1/16$ cycles/sample

f	0	1	2	3	4	5	6	7	8	9	10	11	12	13	14	15
f	16	17	18	19	20	21	22	23	24	25	26	27	28	29	30	31
f_a	0	1	2	3	4	5	6	7	8	7	6	5	4	3	2	1

$$f_a = \left| f - f_s \text{ round} \left(\frac{f}{f_s} \right) \right|$$

To avoid the aliasing effect, the sampling theorem states that the sampling frequency must be greater than twice that of the highest frequency component of a real-valued signal. Given a sampling frequency, the input signal has to be prefiltered to eliminate the frequency components with frequencies greater than or equal to half the sampling frequency. In practice, aliasing cannot be eliminated but can be reduced to negligible levels with appropriate choice of the sampling frequency.

6.2 Leakage Effect

In approximating signals by the DFT, due to the finite and discrete nature of the DFT, the sampling interval and the record length have to be selected appropriately. The aliasing effect due to insufficient number of samples over a period was presented in the last section. Now, we are going to study the importance of selecting the appropriate record length. The record length is usually very long and the frequency content is very high. However, for practical signals, it is possible to approximate them adequately due to the fact that the magnitude of signals becomes negligible in both the time domain and the frequency domain beyond some finite range. A time-limited signal cannot be band-limited. For practical purposes, we are able to assume that signals are both time-limited and band-limited with adequate accuracy. For example, while the exponential signal $e^{-at}u(t)$, $a > 0$ approaches zero asymptotically, its spectrum can be effectively approximated by the DFT using a relatively short record length. In digital filter design, everlasting impulse responses are suitably truncated for practical purposes. To eliminate aliasing, the signal is prefiltered, which is spectral truncation. The point is that truncation is inevitable and it distorts the data. However, it has to be ensured that the representation of the signal or spectrum is still adequate for practical purposes. The criterion for truncation is based on signal energy or amplitude. For example, the difference between the energy of the original signal and its truncated version is to be less than 5%.

The DFT computation assumes periodic extension of the given finite data. If a frequency component makes an integral number of cycles in one period, then its DFT coefficient will be an impulse at its frequency index. Otherwise, a discontinuity is created between the first and last samples in each period. To synthesize such a signal requires more number of frequency components. The energy of the signal is leaked to other frequency components. In practice, leakage cannot be eliminated. To reduce leakage, a more appropriate record length has to be selected. Once leakage occurs, the choice is to reduce the leakage error and increase the smearing effect or vice versa.

6.2.1 Modeling Data Truncation

Truncation of a N-point signal $x(n)$ to get a L-point signal $\hat{x}(n)$ $(L < N)$ may be considered as multiplying $x(n)$ by a rectangular window $w_r(n)$ defined as

$$w_r(n) = \begin{cases} 1 & \text{for } n = 0, 1, \ldots, L-1 \\ 0 & \text{for } n = L, L+1, \ldots, N-1 \end{cases}$$

For example, with $L = N = 4$,

$$w_r(n) = \{1, 1, 1, 1\}$$

With $L = N = 8$,
$$w_r(n) = \{1, 1, 1, 1, 1, 1, 1, 1\}$$

The objective is to relate the DFT of $x(n)$ and $\hat{x}(n)$. The truncated signal is the product of $x(n)$ and $w_r(n)$. Then, due to the DFT frequency-domain circular convolution theorem, we get

$$\hat{x}(n) = x(n)w_r(n) \leftrightarrow \hat{X}(k) = \frac{1}{N}X(k) \circledast W_r(k)$$

where $x(n) \leftrightarrow X(k)$ and

$$w_r(n) = \begin{cases} 1 & \text{for } n = 0, 1, \ldots, L-1 \\ 0 & \text{for } n = L, L+1, \ldots, N-1 \end{cases} \leftrightarrow W_r(k) = e^{(-j\frac{\pi}{N}(L-1)k)}\frac{\sin(\frac{\pi}{N}Lk)}{\sin(\frac{\pi}{N}k)} \tag{6.1}$$

For example, let

$$x(n) = \{\breve{1}, 0, -1, 0\} \leftrightarrow X(k) = \{\breve{0}, 2, 0, 2\}$$

and

$$w_r(n) = \{\breve{1}, 1, 0, 0\} \leftrightarrow W_r(k) = \{\breve{2}, 1 - j, 0, 1 + j\}$$

$$x(n)w_r(n) = \{\breve{1}, 0, 0, 0\} \leftrightarrow \{\breve{1}, 1, 1, 1\}$$

Let us find the circular convolution using the DFT.

$$\{\breve{0}, 2, 0, 2\} \leftrightarrow \{\breve{4}, 0, -4, 0\} \quad \text{and} \quad \{\breve{2}, 1 - j, 0, 1 + j\} \leftrightarrow \{\breve{4}, 0, 0, 4\}$$

$$\{\breve{4}, 0, -4, 0\}\{\breve{4}, 0, 0, 4\}/4 = \{\breve{4}, 0, 0, 0\}$$

The IDFT of $\{\breve{4}, 0, 0, 0\}$ divided by 4 is $\{\breve{1}, 1, 1, 1\}$. The IDFT of $\{\breve{1}, 1, 1, 1\}$ is $\{\breve{1}, 0, 0, 0\} = x(n)w_r(n)$.

The spectrum of the original signal is distorted due to truncation. First, more number of frequency components are required to reconstruct the signal. This is called leakage or spectral spreading. Since the resultant spectrum is the convolution of spectra of the signal and the window, due to a convolution property, the length of the nonzero spectral components increases. In the aliasing effect, a set of frequencies fold back on to a single frequency. Due to the leakage effect, a single frequency component produces a set of frequencies. This is also similar to signal compression and expansion. The amplitude of the spectrum is reduced due to smoothing (from $X(1) = X(3) = 2$ to $X(1) = X(3) = 1$), resulting in loss of detail. Similar to reducing the aliasing by decreasing the sampling interval, the leakage effect can be reduced by selecting a more appropriate record length. Once truncation has occurred, the only alternative is to reduce the leakage at the cost of increasing the smearing of the spectrum. To reduce the leakage, the signal has to be multiplied by a tapered window, which truncates the signal gradually. The characteristics of some of the often used windows are presented next.

6.2.2 Tapered Windows

The rectangular, triangular, Hann, and Hamming window functions are shown in Fig. 6.3 with $N = L = 32$.

Rectangular Window

The rectangular window is shown by dots and its DFT is given by Eq. (6.1). Let $x(n)$ be a N-point sequence. Retaining only the first L samples and making the rest equal to zero to get a truncated sequence $\hat{x}(n)$ are equivalent to multiplying it with the window $w_r(n)$. That is

$$\hat{x}(n) = x(n)w_r(n)$$

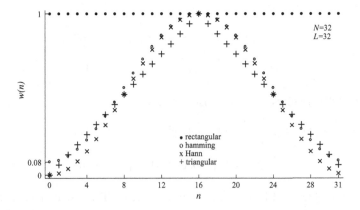

Fig. 6.3 Window functions in the time domain

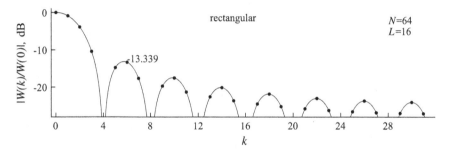

Fig. 6.4 The normalized magnitude in dB of the frequency response of the rectangular window

The normalized magnitude in dB of the frequency response of the rectangular window function, with $N = 64$ and $L = 16$, is shown in Fig. 6.4. The magnitude of the first side lobe is -13.339 dB, which is about 20% of that of the main lobe. At $k = 0$, the magnitude is $L = 16$ and $20 \log_{10}(16/16) = 0$ dB. At integral multiples of N/L, $X(k) = 0$. The main lobe width is $(2N)/L$. Other than at $k = 0$, peaks occur at $k = \frac{3N}{2L}, \frac{5N}{2L}, \ldots$ Since the magnitude of the numerator becomes 1 at these frequencies, the magnitudes are given by $1/\sin((\pi(2k + 1))/(2L)), k = 1, 2, \ldots$ For example, with $k = 1$,

$$20 \log 10 \left(\frac{1}{16} \left| \frac{1}{\sin((3\pi)/32)} \right| \right) = -13.339$$

Example 6.1 List the values of the rectangular window $w_r(n)$ with $N = 8$ and $L = 5$. Find the truncated version, $x_t(n)$, of one cycle of $x(n)$, starting from $n = 0$, by applying the window. Find the magnitude of the DFT of $x(n)$ and $x_t(n)$.

$$x(n) = e^{j\frac{2\pi}{8}n}, \quad n = 0, 1, 2, 3, 4, 5, 6, 7$$

$$w_r(n) = \{1, 1, 1, 1, 1, 0, 0, 0\}$$

$$x(n) = \{1, 0.7071 + j0.7071, j1, -0.7071 + j0.7071, -1, \\ -0.7071 - j0.7071, -j1, 0.7071 - j0.7071\}$$

$$X(k) = \{0, 8, 0, 0, 0, 0, 0, 0\}$$

$$x_t(n) = \{1, 0.7071 + j0.7071, j1, -0.7071 + j0.7071, -1, 0, 0, 0\}$$

$$|X_t(k)| = \{2.4142, 5, 2.4142, 1, 0.4142, 1, 0.4142, 1\}$$

The rectangular window is the best in terms of resolution and is the worst in terms of the magnitude of the side lobes. Therefore, the rectangular window is preferred when there is no leakage, the leakage is tolerable or if it is possible to select a more appropriate record length that results in tolerable leakage. Otherwise, a suitable tapered window is to be used. The truncated data is multiplied by the window so that the data is gradually reduced to zero or near zero at the ends. These windows provide the advantage of detection of frequency components with small magnitudes; those may be masked by the large side lobes of the rectangular window. Several windows are available and their suitability for specific applications is well known. The general approach in designing a window to reduce the side lobes is to make the transition of the signal more smoother from the middle toward the ends. The magnitude of the side lobes of the rectangular window is relatively large, due to its discontinuity at the ends. Therefore, its spectrum decays slowly at the rate of $1/k$, where k is the frequency index. By reducing or eliminating this discontinuity, the spectrum of the tapered windows decays faster resulting in smaller side lobes. This is achieved at the cost of increasing the main lobe width, which reduces resolution.

Triangular Window

The convolution of a function by itself in the time domain corresponds to the product of its spectrum by itself in the frequency domain. Therefore, the side lobes adjacent to the main lobes have considerably smaller magnitudes. From the time-domain viewpoint, as the resulting function becomes smoother, the spectrum is expected to decay at a faster rate resulting in smaller side lobes. However, the width of the function is increased in the time domain and the width of the main lobe of the spectrum is also increased.

The triangular window is defined as

$$w_t(n) = \begin{cases} \frac{2}{L}n & \text{for } n = 0, 1, \ldots, \frac{L}{2} \\ w_t(L-n) & \text{for } n = \frac{L}{2}+1, \frac{L}{2}+2, \ldots, L-1 \\ 0 & \text{for } n = L, L+1, \ldots, N-1 \end{cases}$$

The time-domain representation of this window, with $L = 32$ and $N = 32$, is shown in Fig. 6.3 by the $+$ symbol. For example, with $L = N = 4$,

$$w_t(n) = \{0, 0.5, 1, 0.5\}$$

With $L = N = 8$,

$$w_t(n) = \{0, 0.25, 0.5, 0.75, 1, 0.75, 0.5, 0.25\}$$

The circular convolution of a rectangular window

$$w_r(n) = \begin{cases} 1 \text{ for } n = 0, 1, \ldots, \frac{L}{2}-1 \\ 0 \text{ for } n = \frac{L}{2}, \frac{L}{2}+1, \ldots, N-1 \end{cases}$$

with itself followed by a circular right shift of one sample interval is the triangular window with a scale factor. For example, with $N = 4$ and $L = 2$,

$$w_r(n) = \{1, 1, 0, 0\} \leftrightarrow W_r(k) = \{2, 1 - j1, 0, 1 + j1\}$$

$$W_r^2(k) = \{4, -j2, 0, j2\} \leftrightarrow \frac{N}{2} w_t(n + 1) = \{1, 2, 1, 0\}$$

$$W_r^2(k)e^{-j\frac{2\pi}{4}k} = \frac{N}{2} W_t(k) = \{4, -2, 0, -2\} \leftrightarrow \frac{N}{2} w_t(n) = \{0, 1, 2, 1\}$$

Therefore, simplifying, we get the DFT of an even-length triangular window as

$$W_t(k) = \frac{2}{N} W_r^2(k)e^{-j\frac{2\pi}{N}k} = \frac{2}{N} \left(\frac{\sin(\frac{\pi}{2}k)}{\sin(\frac{\pi}{N}k)} \right)^2 e^{j\pi k}$$

As the time-domain function is real and even, the spectrum is also real and even.

The window is even-symmetric with the first sample value zero. For N even, the middle value is 1. Therefore, the DFT of an even-length triangular window of length N is also given by

$$W_t(k) = (-1)^k + \sum_{n=1}^{\frac{N}{2}-1} \frac{4n}{N} \cos \frac{2\pi}{N} nk$$

With $N = 4$, for example,

$$w_t(n) = \{0, 0.5, 1, 0.5\}, \quad W_t(k) = \{2, -1, 0, -1\}$$

For an odd length N,

$$W_t(k) = \sum_{n=1}^{\frac{N-1}{2}} \frac{4n}{N} \cos \frac{2\pi}{N} nk$$

With $N = 3$, for example,

$$w_t(n) = \left\{ 0, \frac{2}{3}, \frac{2}{3} \right\}, \quad W_t(k) = \left\{ \frac{4}{3}, -\frac{2}{3}, -\frac{2}{3} \right\}$$

Another formula defining the triangular window of length N is given by

$$w_t(n) = 1 - \frac{2\left|n - \frac{N}{2}\right|}{N}, \quad n = 0, 1, \ldots, N - 1$$

The normalized magnitude in dB of the DFT of this window is shown in Fig. 6.5 with $L = 16$ and $N = 64$. The magnitude of the largest side lobe is $-25.913\,\text{dB}$.

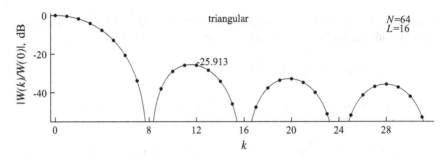

Fig. 6.5 The normalized magnitude in dB of the frequency response of the triangular window

The main lobe width is $\frac{2N}{L}$ for the rectangular window and $\frac{4N}{L}$ for the triangular window. Therefore, the main lobe width is 8 and 16 in Figs. 6.4 and 6.5, respectively.

Example 6.2 List the values of the triangular window $w_r(n)$ with $N = 8$ and $L = 5$. Find the truncated version, $x_t(n)$, of one cycle of $x(n)$, starting from $n = 0$, by applying the window. Find the magnitude of the DFT of $x(n)$ and $x_t(n)$.

$$x(n) = e^{j\frac{2\pi}{8}n}, \quad n = 0, 1, 2, 3, 4, 5, 6, 7$$

$$w_t(n) = \{0, 0.4, 0.8, 0.8, 0.4, 0, 0, 0\}$$

$$x(n) = \{1, 0.7071 + j0.7071, j1, -0.7071 + j0.7071, -1,$$
$$- 0.7071 - j0.7071, -j1, 0.7071 - j0.7071\}$$

$$X(k) = \{0, 8, 0, 0, 0, 0, 0, 0\}$$

$$x_t(n) = \{0, 0.2828 + j0.2828, j0.8, -0.5657 + j0.5657, -0.4, 0, 0, 0\}$$

$$|X_t(k)| = \{1.7844, 2.4, 1.7844, 0.5657, 0.1268, 0, 0.1268, 0.5657\}$$

6.2.3   Hann and Hamming Windows

An alternate approach to design tapered windows is to express their frequency responses as a linear combination of the scaled and shifted spectra of rectangular windows. The combination tends to reduce the large side lobes of the rectangular window at the cost of increasing the length of the main lobe. Shifting the spectrum

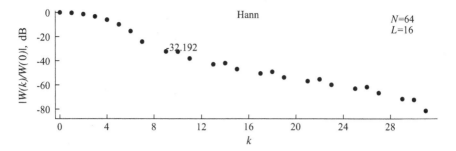

Fig. 6.6 The normalized magnitude in dB of the frequency response of the Hann window

in the frequency domain requires multiplication of the window in the time domain with a complex exponential (frequency shift theorem).

The Hann window is defined as

$$w_{han}(n) = \begin{cases} 0.5 - 0.5\cos\left(\frac{2\pi}{L}n\right) & \text{for } n = 0, 1, \ldots, L-1 \\ 0 & \text{for } n = L, L+1, \ldots, N-1 \end{cases}$$

The time-domain representation of this window is shown in Fig. 6.3 by the cross symbol, with $L = 32$ and $N = 32$. For example, with $L = N = 4$,

$$w_{han}(n) = \{0, 0.5, 1, 0.5\}$$

With $L = N = 8$,

$$w_{han}(n) = \{0, 0.1464, 0.5, 0.8536, 1, 0.8536, 0.5, 0.1464\}$$
$$\leftrightarrow W_{han}(k) = \{4, -2, 0, 0, 0, 0, 0, -2\}$$

Using Euler's formula, we get

$$\cos\left(\frac{2\pi}{L}n\right) = \frac{e^{j\frac{2\pi}{L}n} + e^{-j\frac{2\pi}{L}n}}{2}$$

Therefore, the frequency response is given in terms of that of the rectangular window as

$$W_{han}(k) = 0.5W_r(k) - 0.25W_r(k+1) - 0.25W_r(k-1)$$

The magnitude of the DFT in dB is shown in Fig. 6.6 with $L = 16$ and $N = 64$. The magnitude of the largest side lobe is $-32.192\,\text{dB}$.

Example 6.3 List the values of the Hann window $w_{han}(n)$ with $N = 8$ and $L = 5$. Find the truncated version, $x_t(n)$, of one cycle of $x(n)$, starting from $n = 0$, by applying the window. Find the magnitude of the DFT of $x(n)$ and $x_t(n)$.

$$x(n) = e^{j\frac{2\pi}{8}n}, \quad n = 0, 1, 2, 3, 4, 5, 6, 7$$

$$w_{han}(n) = \{0, 0.3455, 0.9045, 0.9045, 0.3455, 0, 0, 0\}$$

$$x(n) = \{1, 0.7071 + j0.7071, j1, -0.7071 + j0.7071, -1,$$
$$- 0.7071 - j0.7071, -j1, 0.7071 - j0.7071\}$$

$$X(k) = \{0, 8, 0, 0, 0, 0, 0, 0\}$$

$$x_t(n) = \{0, 0.2443 + j0.2443, j0.9045, -0.6396 + j0.6396, -0.3455, 0, 0, 0\}$$

$$|X_t(k)| = \{1.9357, 2.5, 1.9357, 0.7906, 0.0539, 0, 0.0539, 0.7906\}$$

The Hamming window is defined as

$$w_{ham}(n) = \begin{cases} 0.54 - 0.46\cos\left(\frac{2\pi}{L}n\right) & \text{for } n = 0, 1, \ldots, L-1 \\ 0 & \text{for } n = L, L+1, \ldots, N-1 \end{cases}$$

The time-domain representation of this window is shown in Fig. 6.3 by unfilled circles with $L = 32$ and $N = 32$. For example, with $L = N = 4$,

$$w_{ham}(n) = \{0.08.0.54, 1, 0.54\}$$

With $L = N = 8$,

$$w_{ham}(n) = \{0.08, 0.2147, 0.54, 0.8653, 1, 0.8653, 0.54, 0.2147\} \leftrightarrow$$

$$W_{ham}(k) = \{4.32, -1.84, 0, 0, 0, 0, 0, -1.84\}$$

The frequency response is given in terms of that of the rectangular window as

$$W_{ham}(k) = 0.54W_r(k) - 0.23W_r(k+1) - 0.23W_r(k-1)$$

The magnitude of the DFT in dB is shown in Fig. 6.7 with $L = 16$ and $N = 64$. The magnitude of the largest side lobe is $-40.160\,$dB.

Example 6.4 List the values of the Hamming window $w_r(n)$ with $N = 8$ and $L = 5$. Find the truncated version, $x_t(n)$, of one cycle of $x(n)$, starting from $n = 0$, by applying the window. Find the magnitude of the DFT of $x(n)$ and $x_t(n)$.

$$x(n) = e^{j\frac{2\pi}{8}n}, \quad n = 0, 1, 2, 3, 4, 5, 6, 7$$

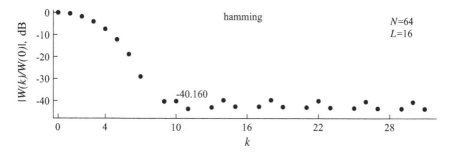

Fig. 6.7 The normalized magnitude in dB of the frequency response of the Hamming window

$$w_{ham}(n) = \{0.08, 0.3979, 0.9121, 0.9121, 0.3979, 0, 0, 0\}$$

$$x(n) = \{1, 0.7071 + j0.7071, j1, -0.7071 + j0.7071, -1,$$
$$- 0.7071 - j0.7071, -j1, 0.7071 - j0.7071\}$$

$$X(k) = \{0, 8, 0, 0, 0, 0, 0, 0\}$$

$$x_t(n) = \{0.08, 0.2813 + j0.2813, j0.9121, -0.6450 + j0.6450, -0.3979, 0, 0, 0\}$$

$$|X_t(k)| = \{1.9607, 2.7, 1.9607, 0.6731, 0.0479, 0.08, 0.0479, 0.6731\}$$

6.2.4 Reducing the Spectral Leakage

The ability to distinguish the different frequency components of a spectrum is called the resolution. The best resolution is obtained with no truncation of the time-domain signal. That means the record of the signal must be the whole nonzero part of an aperiodic signal or an integral number of periods of a periodic signal. With truncation, we have to use an appropriate window to ensure that the leakage and spectral resolutions are acceptable. Figure 6.8a, b show the sinusoid $x(n) = \cos(\frac{2\pi}{32}2n)$ and its spectrum, respectively. Since there is no truncation, the DFT spectral representation is distinct, clearly showing the two components. Figure 6.8c, d shows the truncated sinusoid using the rectangular window and its spectrum, respectively. Since there is truncation, both the resolution and the leakage of the DFT spectral representation have deteriorated significantly. Figure 6.8e, f shows the truncated sinusoid using the Hann window and its spectrum, respectively. The spectral resolution is low compared with that in Fig. 6.8d. But, the magnitude of the side lobes is much smaller than in Fig. 6.8d.

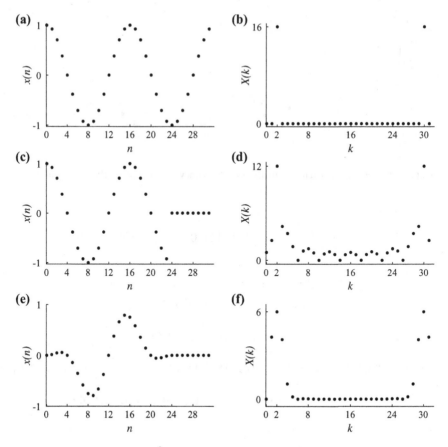

Fig. 6.8 **a** Sinusoid $x(n) = \cos(\frac{2\pi}{32}2n)$; **b** its spectrum; **c** the truncated version of $x(n)$ using the rectangular window; **d** its spectrum; **e** the truncated version of $x(n)$ using the Hann window; **f** its spectrum

Figure 6.9a shows the DFT magnitude spectrum of $x(n) = e^{j\frac{2\pi}{64}n} + e^{j\frac{2\pi}{64}2n}$. In (a), the two frequency components of the signal are clearly identified. In (b), there is only one peak, due to the truncation of the signal, and, clearly, the spectral representation is not a clear identification of the constituent frequency components of the signal. This is due to the convolution of the spectra of the signal and the rectangular window.

Figure 6.10a shows the DFT magnitude spectrum of $x(n) = e^{j\frac{2\pi}{64}n} + e^{j\frac{2\pi}{64}4n}$. In this case, the separation of the frequency components is more. In (a), the two frequency components of the signal are clearly identified. In (b) also, there are two peaks although not as distinct as in (a), due to the truncation of the signal.

Figure 6.11a shows the DFT magnitude spectrum of $x(n) = e^{j\frac{2\pi}{64}n} + 0.1e^{j\frac{2\pi}{64}14n}$. In this case, the amplitude of one of the frequency components is large compared with the other. In (a), the two frequency components of the signal are clearly identified.

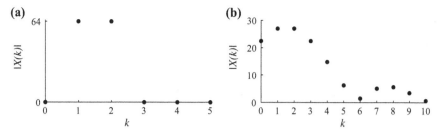

Fig. 6.9 **a** The DFT magnitude spectrum of $x(n) = e^{j\frac{2\pi}{64}n} + e^{j\frac{2\pi}{64}2n}$; **b** the DFT magnitude spectrum of truncated $x(n)$

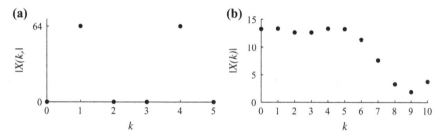

Fig. 6.10 **a** The DFT magnitude spectrum of $x(n) = e^{j\frac{2\pi}{64}n} + e^{j\frac{2\pi}{64}4n}$; **b** the DFT magnitude spectrum of truncated $x(n)$

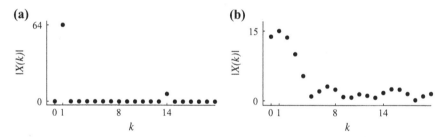

Fig. 6.11 **a** The DFT magnitude spectrum of $x(n) = e^{j\frac{2\pi}{64}n} + 0.1e^{j\frac{2\pi}{64}14n}$; **b** the DFT magnitude spectrum of truncated $x(n)$

In (b), due to the truncation of the signal, the clear identification of the second component is not possible due to the large side lobes of the rectangular window.

Figure 6.12a shows the DFT magnitude spectrum of $x(n) = e^{j\frac{2\pi}{64}n} + 0.1e^{j\frac{2\pi}{64}14n}$. In (a), the two frequency components of the signal are clearly identified. In (b), despite the truncation of the signal, the identification of the second component is possible due to the small side lobes of the Hann window.

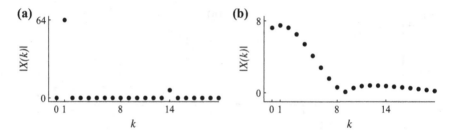

Fig. 6.12 **a** The DFT magnitude spectrum of $x(n) = e^{j\frac{2\pi}{64}n} + 0.1e^{j\frac{2\pi}{64}14n}$; **b** the DFT magnitude spectrum of truncated $x(n)$ using the Hann window

6.3 Picket-Fence Effect

As the DFT provides only the uniform samples of a spectrum, it is possible that features such as the peak of a spectrum are missed. As this is like viewing the spectrum through a picket-fence, this effect is called the picket-fence effect. The period in one domain is always fixed by the uniform sampling interval in the other. Therefore, the data record should be long enough to provide a sufficiently small sampling interval in the frequency domain. If necessary, zero padding can be employed in the time domain to reduce the spectral sampling interval.

Figure 6.13a shows the DFT samples of the continuous spectrum of a signal. The main lobe peak is missed. Figure 6.13b shows the DFT samples of the continuous spectrum of the zero-padded signal to make its length double. This reduces the sampling interval of the spectrum by one-half and the main lobe peak is located. In general, sufficient record length should be ensured by zero padding or otherwise so that all the important features of the spectrum are located.

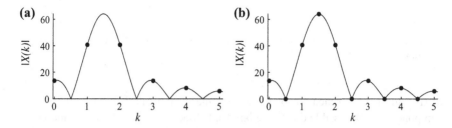

Fig. 6.13 **a** The DFT samples of the continuous spectrum of a signal; **b** the DFT samples of the zero-padded signal

6.4 Summary

- The DFT is used to approximate all other versions of Fourier analysis. While the DFT is finite and discrete in both the domains, signals may have infinite duration and/or infinite bandwidth. Further, most of the naturally occurring signals are continuous type. Therefore, signal analysis by the DFT is only approximate but can be made adequate.
- The approximation of the signals to suit the DFT analysis usually requires sampling, truncation, and quantization.
- The sampling theorem stipulates that a sinusoid with frequency f Hz requires more than $2f$ samples for its unambiguous representation in its sampled form.
- If the sampling theorem is not satisfied, then aliasing occurs. Aliasing is the impersonation by a low-frequency discrete sinusoid of a higher-frequency continuous sinusoid, due to a sampling interval not short enough.
- In order to avoid aliasing in the representation of a signal, the sampling interval must be sufficiently short. Given a sampling interval, the signal has to be prefiltered by an antialiasing filter so that the sampling theorem is satisfied. If aliasing is unavoidable, then it has to be ensured, by selecting a sampling interval that is as short as possible, that the error in the representation of the signal is negligible.
- The sampled version can be made sufficiently accurate, since signals generated by physical devices are characterized by a drooping spectrum with increasing frequency.
- As the duration of the signal may be very long, truncation is usually required.
- The DFT requires that all the frequency components of a signal complete an integral number of cycles in the record length for its accurate representation. This requirement may be violated due to truncation.
- Truncation reduces the resolution, which is the clear identification of the different frequency components. Further, the energy of the frequency components is leaked to neighboring frequencies indicating nonexisting frequency components in the spectrum.
- Truncation is modeled as multiplying the signal by a rectangular window. This window provides the best resolution of a truncated signal compared with other windows but results in large leakage of energy. This is due to the discontinuity created by this window at the borders, since a discontinuity makes the convergence of the spectrum slow.
- The tapered windows eliminate or reduce the discontinuity by gradually reducing the data towards the ends of the data record. This reduction in the discontinuity reduces the leakage at the cost of reducing the spectral resolution. The reduction in leakage improves the ability to detect frequency components with small magnitudes.
- Several tapered windows are available with different characteristics. The window suitable for a specific application should be selected.

- The truncated version can be made sufficiently accurate, since signals generated by physical devices are characterized by a drooping amplitude with increasing time.
- As the DFT provides only the samples of a spectrum, it is possible that features such as the peak of a spectrum are missed. This effect is called the picket-fence effect.
- The data record in the time domain should be long enough to provide a sufficiently small sampling interval in the frequency domain. If necessary, zero padding can be employed to reduce the spectral sampling interval.

Exercises

6.1 Given a periodic signal $x(n)$ with period $N = 4$. Increase the value of the frequency index from $k = 0$ to $k = 8$ and find the 4 samples for each k. Verify that the equation

$$Ae^{j\left(\frac{2\pi}{4}(k+4l)n+\phi\right)} = Ae^{j\left(\frac{2\pi}{4}kn+\phi\right)}, \quad n = 0, 1, 2, 3$$

holds, where l is an integer.

* 6.1.1

$$x(n) = 2e^{j\left(\frac{2\pi}{4}kn+\frac{\pi}{3}\right)}, \quad n = 0, 1, 2, 3$$

6.1.2

$$x(n) = e^{j\left(\frac{2\pi}{4}kn-\frac{\pi}{4}\right)}, \quad n = 0, 1, 2, 3$$

6.1.3

$$x(n) = -e^{j\left(\frac{2\pi}{4}kn-\frac{\pi}{3}\right)}, \quad n = 0, 1, 2, 3$$

6.1.4

$$x(n) = -2e^{j\left(\frac{2\pi}{4}kn+\frac{\pi}{6}\right)}, \quad n = 0, 1, 2, 3$$

6.1.5

$$x(n) = 3e^{j\left(\frac{2\pi}{4}kn+\frac{\pi}{8}\right)}, \quad n = 0, 1, 2, 3$$

6.2 Given a periodic signal $x(n)$ with period $N = 8$. Increase the value of the frequency index from $k = 0$ to $k = 12$ and find the 8 samples for each k. Verify that the equations

$$x(n) = \cos\left(\frac{2\pi}{N}(k+lN)n + \phi\right) = \cos\left(\frac{2\pi}{N}kn + \phi\right), \quad k = 0, 1, \ldots, \frac{N}{2} - 1$$

$$x(n) = \cos\left(\frac{2\pi}{N}(k+lN)n + \phi\right) = \cos(\phi)\cos\left(\frac{2\pi}{N}kn\right), \quad k = \frac{N}{2}$$

$$x(n) = \cos\left(\frac{2\pi}{N}(lN-k)n + \phi\right) = \cos\left(\frac{2\pi}{N}kn - \phi\right),$$

$$k = 1, 2, \ldots, \frac{N}{2} - 1, \quad n = 0, 1, 2, 3, 4, 5, 6, 7$$

hold, where l is an integer.

6.2.1
$$x(n) = \cos\left(\frac{2\pi}{8}kn - \frac{\pi}{6}\right)$$

* 6.2.2
$$x(n) = -2\cos\left(\frac{2\pi}{8}kn + \frac{\pi}{6}\right)$$

6.2.3
$$x(n) - -\cos\left(\frac{2\pi}{8}kn - \frac{\pi}{4}\right)$$

6.2.4
$$x(n) = \cos\left(\frac{2\pi}{8}kn - \frac{\pi}{8}\right)$$

6.2.5
$$3\cos\left(\frac{2\pi}{8}kn - \frac{\pi}{5}\right)$$

6.3 List the values of the rectangular window $w_r(n)$ with $N = 8$ and $L = 5$. Verify these values by finding the IDFT of its frequency-domain version. Find the truncated version, $x_t(n)$, of one cycle of $x(n)$ by applying the window with $L = 5$. Find the magnitude of the DFT, $X_t(k)$, of $x_t(n)$.

6.3.1
$$x(n) = \cos\left(\frac{2\pi}{8}n\right), \quad n = 0, 1, 2, 3, 4, 5, 6, 7$$

6.3.2
$$x(n) = \cos\left(\frac{2\pi}{8}2n\right), \quad n = 0, 1, 2, 3, 4, 5, 6, 7$$

* 6.3.3
$$x(n) = \cos\left(\frac{2\pi}{8}3n\right), \quad n = 0, 1, 2, 3, 4, 5, 6, 7$$

6.3.4
$$x(n) = \cos\left(\frac{2\pi}{8}n - \frac{\pi}{3}\right), \quad n = 0, 1, 2, 3, 4, 5, 6, 7$$

6.3.5

$$x(n) = \cos\left(\frac{2\pi}{8}n + \frac{\pi}{6}\right), \quad n = 0, 1, 2, 3, 4, 5, 6, 7$$

6.4 List the values of the Hann window $w_{han}(n)$ with $N = 8$ and $L = 5$. Verify these values by finding the IDFT of its frequency-domain version. Find the truncated version, $x_t(n)$, of one cycle of $x(n)$ by applying the window with $L = 5$. Find the magnitude of the DFT, $X_t(k)$, of $x_t(n)$.

6.4.1

$$x(n) = \cos\left(\frac{2\pi}{8}n\right), \quad n = 0, 1, 2, 3, 4, 5, 6, 7$$

6.4.2

$$x(n) = \cos\left(\frac{2\pi}{8}2n\right), \quad n = 0, 1, 2, 3, 4, 5, 6, 7$$

6.4.3

$$x(n) = \cos\left(\frac{2\pi}{8}3n\right), \quad n = 0, 1, 2, 3, 4, 5, 6, 7$$

* 6.4.4

$$x(n) = \cos\left(\frac{2\pi}{8}n - \frac{\pi}{3}\right), \quad n = 0, 1, 2, 3, 4, 5, 6, 7$$

6.4.5

$$x(n) = \cos\left(\frac{2\pi}{8}n + \frac{\pi}{6}\right), \quad n = 0, 1, 2, 3, 4, 5, 6, 7$$

Chapter 7
Fourier Series

Signals and spectrums can be continuous or discrete and periodic or aperiodic. There are four possibilities resulting in the four versions of the Fourier analysis, as shown in the Tables 7.1 and 7.2. In both the time and frequency domains, the DFT (discrete Fourier transform) version is discrete and periodic. In both the time and frequency domains, the FT (Fourier transform) version is continuous and aperiodic. The FS (Fourier series) version is continuous and periodic in the time domain and it is discrete and aperiodic in the frequency domain. The DTFT (discrete-time Fourier transform) version is continuous and periodic in the frequency domain and it is discrete and aperiodic in the time domain. Note the duality between the DTFT and FS. Fourier analysis is representation of an arbitrary waveform in terms of sinusoids. It enables to determine the sinusoidal content of the waveforms. The differences between the versions are that signals and their spectra being continuous or discrete and periodic or aperiodic.

Each of the four versions of the Fourier analysis can be independently developed using the orthogonal properties of the sinusoids. We have first derived the DFT version for two major reasons. One is that it is the only version that can be implemented using digital systems and all the other three versions are approximated, in practice, using the DFT. Another reason is that it is the easiest to visualize the analysis and synthesis of the waveforms. Therefore, we derive the three other versions of the Fourier analysis starting from the DFT.

In this chapter, we derive the FS version of the Fourier analysis. In the Fourier series representation, the time-domain waveform is continuous and periodic and the corresponding spectrum in the frequency domain is discrete and aperiodic. A continuous periodic waveform is represented as a sum of sinusoids whose frequencies are integral multiples of that of the smallest, called the fundamental. The other sinusoids are called the harmonics. The frequency of the fundamental component is called the fundamental frequency. The multiplicative inverse of the period of the periodic waveform being represented is the cyclic fundamental frequency. The cyclic frequency multiplied by 2π is the frequency in radians. If the independent variable

© Springer Nature Singapore Pte Ltd. 2018
D. Sundararajan, *Fourier Analysis—A Signal Processing Approach*,
https://doi.org/10.1007/978-981-13-1693-7_7

Table 7.1 Time-domain classification of Fourier analysis

	Periodic	Aperiodic
Continuous	FS	FT
Discrete	DFT	DTFT

Table 7.2 Frequency-domain classification of Fourier analysis

	Periodic	Aperiodic
Continuous	DTFT	FT
Discrete	DFT	FS

of the time-domain waveform is time, then the unit of the cyclic frequency, usually denoted by f, is cycles per second (Hz). The unit of the radian frequency, usually denoted by ω, is radians per second. However, it should be remembered that Fourier analysis is equally applicable even if the independent variable is other than time. For example, the periodicity of the rate of change of intensity of the pixels with respect to distance is of interest in image processing.

The frequency increment of the spectrum of a periodic function is equal to its fundamental frequency. If the number of samples in a period is increased in the DFT representation of a signal, then, due to a longer range of the spectrum, the number of frequency components is increased with the ability to represent a more rapidly varying periodic waveform. When the sampling interval approaches zero, the resulting waveform becomes continuous and periodic. The aperiodic nature of the spectrum can be considered as a periodic spectrum becoming aperiodic in the limit as the period becomes infinite.

7.1   Fourier Series

In the FS representation of signals, a continuous periodic signal $x(t)$ with period T and cyclic frequency $f_0 = 1/T$ is expressed as a sum of a constant and sinusoids with frequencies f_0, called the fundamental, and

$$\{2f_0, 3f_0, \ldots, \infty\}$$

called the harmonic frequencies. A sinusoid with frequency kf_0 is the kth harmonic of the fundamental sinusoid with frequency f_0. The corresponding radian frequencies are

$$\{\omega_0 = 2\pi f_0, 2\omega_0 = 2\pi(2f_0), 3\omega_0 = 2\pi(3f_0), \ldots, \infty\}$$

Then, $x(t)$ is represented in terms of sinusoids as

$$x(t) = X_p(0) + X_p(1)\cos(\omega_0 t + \theta_1)$$
$$+ X_p(2)\cos(2\omega_0 t + \theta_2) + \cdots + X_p(\infty)\cos(\infty\omega_0 t + \theta_\infty)$$
$$= X_p(0) + \sum_{k=1}^{\infty} X_p(k)\cos(k\omega_0 t + \theta_k), \quad \omega_0 = \frac{2\pi}{T} \tag{7.1}$$

In Eq. (7.1), $x(t)$ and the frequencies of the sinusoids are known. The Fourier analysis problem is the determination of the amplitudes and phases of the sinusoids so that the equation is satisfied in the least squares error sense. While, in theory, the frequency range of the sinusoids is infinite, as no physical device can generate a harmonic of infinite order, the number of harmonics used, in practice, is always finite.

Using trigonometric identities, Eq (7.1) can be equivalently expressed, in terms of cosine and sine waveforms, as

$$x(t) = X_c(0) + \sum_{k=1}^{\infty} (X_c(k)\cos(k\omega_0 t) + X_s(k)\sin(k\omega_0 t)), \quad \omega_0 = \frac{2\pi}{T} \tag{7.2}$$

Using the Euler's formula, Eq. (7.1) can also be equivalently expressed, in terms of complex exponentials with a pure imaginary exponent, as

$$x(t) = \sum_{k=-\infty}^{\infty} X_{fs}(k)e^{jk\omega_0 t}, \quad \omega_0 = \frac{2\pi}{T} \tag{7.3}$$

7.1.1 FS as a Limiting Case of the DFT

FS can be derived using the orthogonality property of the sinusoids, similar to the derivation of the DFT. In this section, we derive the FS as a limiting case of the DFT with the sampling interval of the time-domain sequence tending to zero. We use the center-zero format for convenience. Let the DFT of sequence $x(n), N \le n \le N$ be $X(k), N \le k \le N$. Then, the Fourier representation of $x(n)$ is given as

$$x(n) = \frac{1}{2N+1}\sum_{k=-N}^{N} X(k)e^{j\frac{2\pi}{(2N+1)}nk}, \quad n = 0, \pm1, \pm2, \ldots, \pm N \tag{7.4}$$

where

$$X(k) = \sum_{m=-N}^{N} x(m)e^{-j\frac{2\pi}{(2N+1)}mk} \tag{7.5}$$

Substituting for $X(k)$ in Eq. (7.4), we get

$$x(n) = \frac{1}{2N+1} \sum_{k=-N}^{N} \left(\sum_{m=-N}^{N} x(m)e^{-j\frac{2\pi}{(2N+1)}mk} \right) e^{j\frac{2\pi}{(2N+1)}nk} \tag{7.6}$$

Suppose the $2N+1$ samples are obtained by sampling a periodic signal, of period T s, over a period with a sampling interval of T_s s, then

$$\frac{T}{T_s} = 2N+1$$

The index of the time-domain samples has to be replaced by nT_s s. The fundamental frequency is

$$\omega_0 = \frac{2\pi}{T} = \frac{2\pi}{(2N+1)T_s}$$

radians per second. Substituting these changes in Eq. (7.6), we get

$$x(nT_s) = \sum_{k=-N}^{N} \left(\frac{1}{T} \sum_{m=-N}^{N} x(mT_s)e^{-j\frac{2\pi}{T}mT_sk}T_s \right) e^{j\frac{2\pi}{T}nT_sk}$$

$$= \sum_{k=-N}^{N} \left(\frac{1}{T} \sum_{m=-N}^{N} x(mT_s)e^{-j\omega_0 mT_sk}T_s \right) e^{j\omega_0 nT_sk} \tag{7.7}$$

For a given T, ω_0 is fixed irrespective of T_s. With T_s decreasing, the time-domain waveform is densely populated and the spectrum becomes longer. As $T_s \to 0$, mT_s and nT_s become continuous variables, designated, respectively, τ and t. The inner summation becomes an integral between limits $-T/2$ and $T/2$ with T_s becoming the differential $d\tau$. The number of harmonics $2N+1$ becomes infinite. With these changes, the discrete periodic waveform becomes a continuous periodic waveform and the discrete periodic spectrum becomes an aperiodic discrete spectrum. Equation (7.7) is transformed to

$$x(t) = \sum_{k=-\infty}^{\infty} \left(\frac{1}{T} \int_{-T/2}^{T/2} x(\tau)e^{-j\omega_0 \tau k}d\tau \right) e^{j\omega_0 tk} \tag{7.8}$$

Consequently, the FS representation of a continuous periodic waveform $x(t)$ with period T is given as

$$x(t) = \sum_{k=-\infty}^{\infty} X_{fs}(k)e^{jk\omega_0 t} \tag{7.9}$$

where

$$X_{fs}(k) = \frac{1}{T} \int_{-T/2}^{T/2} x(t) e^{-jk\omega_0 t} dt = \frac{1}{T} \int_0^T x(t) e^{-jk\omega_0 t} dt \tag{7.10}$$

As $x(t)$ is periodic with period T, the integral can be evaluated over any continuous interval of duration T. Two commonly used limits are shown in Eq. (7.10). Equations (7.9) and (7.10) are, respectively, the exponential form of the FS synthesis and analysis of $x(t)$.

A periodic waveform

$$x(t) = 0.2 + \cos\left(\frac{2\pi}{16} t\right)$$

with period $T = 16$ s is shown in Fig. 7.1a. The cyclic frequency is $f_0 = 1/16$ Hz and its radian frequency is $\omega_0 = 2\pi/16$ rad/s. The FS for $x(t)$ in exponential form, using Euler's formula, is

$$x(t) = 0.2 + \left(\frac{1}{2} e^{j\frac{2\pi}{16} t} + \frac{1}{2} e^{-j\frac{2\pi}{16} t}\right)$$

and its spectrum is shown in Fig. 7.1b. With 3 time-domain samples, we can represent the DC component and the fundamental harmonic. With the same period, the waveform, shown in Fig. 7.1c,

$$x(t) = 0.2 + \cos\left(\frac{2\pi}{16} t\right) + 0.3 \cos\left(\frac{2\pi}{16} 3t\right),$$

with a higher frequency content, requires a minimum of 7 samples. The FS for $x(t)$ in exponential form is

$$x(t) = 0.2 + \left(\frac{1}{2} e^{j\frac{2\pi}{16} t} + \frac{1}{2} e^{-j\frac{2\pi}{16} t}\right) + \left(\frac{0.3}{2} e^{j\frac{2\pi}{16} 3t} + \frac{0.3}{2} e^{-j\frac{2\pi}{16} 3t}\right)$$

Its spectrum is shown in Fig. 7.1d. With more number of samples, the time-domain waveform becomes more densely sampled and its spectrum becomes longer. With the same period, the waveform, shown in Fig. 7.1e,

$$x(t) = 0.2 + \cos\left(\frac{2\pi}{16} t\right) + 0.3 \cos\left(\frac{2\pi}{16} 3t\right) + 0.2 \cos\left(\frac{2\pi}{16} 5t\right),$$

with a still higher frequency content, requires 11 samples. The FS for $x(t)$ in exponential form is

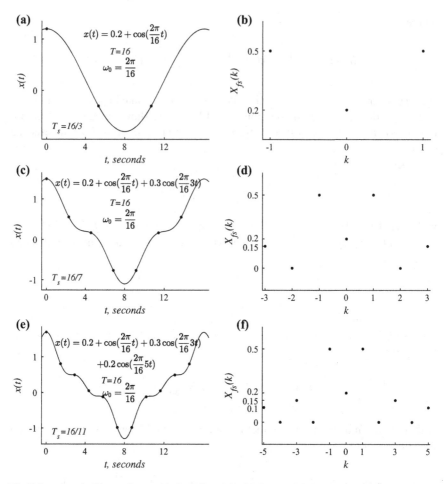

Fig. 7.1 **a** A periodic waveform with the DC and the fundamental; **b** its FS spectrum; **c** a periodic waveform with the DC, the fundamental and the 3rd harmonic; **d** its FS spectrum; **e** a periodic waveform with the DC, the fundamental and the 3rd and 5th harmonics; **f** its FS spectrum

$$
x(t) = 0.2 + \left(\frac{1}{2}e^{j\frac{2\pi}{16}t} + \frac{1}{2}e^{-j\frac{2\pi}{16}t}\right) + \left(\frac{0.3}{2}e^{j\frac{2\pi}{16}3t} + \frac{0.3}{2}e^{-j\frac{2\pi}{16}3t}\right)
$$
$$
+ \left(\frac{0.2}{2}e^{j\frac{2\pi}{16}5t} + \frac{0.2}{2}e^{-j\frac{2\pi}{16}5t}\right)
$$

Its spectrum is shown in Fig. 7.1f. The time-domain waveform becomes more densely sampled than those of (a) and (c) and its spectrum is longer than those of (b) and (d). In the limit, the sampling interval tends to zero. The time-domain waveform becomes continuous with the same period and the spectrum becomes aperiodic with the same harmonic spacing.

With the sinusoids represented in polar form, we get the compact trigonometric form of the FS. Equation (7.9) can be rewritten as

$$x(t) = X_{fs}(0) + \sum_{k=1}^{\infty}(X_{fs}(k)e^{jk\omega_0 t} + X_{fs}(-k)e^{-jk\omega_0 t})$$

All the terms, except the constant term $X_{fs}(0)$, are complex conjugate pairs. The pair with $k = \pm 1$ combines to form the fundamental sinusoid. The pair with $k = \pm 2$ combines to form the second harmonic and so on. Using Euler's formula, we get

$$x(t) = X_p(0) + \sum_{k=1}^{\infty} X_p(k)\cos(k\omega_0 t + \theta(k)), \qquad (7.11)$$

where

$$X_p(0) = X_{fs}(0), \quad X_p(k) = 2|X_{fs}(k)|, \quad \theta(k) = \angle(X_{fs}(k)), \quad k = 1, 2, \ldots, \infty$$

Comparing these two forms, the phase angles appear explicitly in the trigonometric form whereas the complex coefficients of the complex form contain the phase angles implicitly. While the complex form is the most suitable for analysis, the other form is easier to visualize and physical devices generate the waveforms in that form.

With the sinusoids represented in rectangular form, we get another version of the trigonometric form of the FS. Expressing the sinusoid in Eq. (7.11) in rectangular form, we get

$$x(t) = X_c(0) + \sum_{k=1}^{\infty}(X_c(k)\cos(k\omega_0 t) + X_s(k)\sin(k\omega_0 t)), \quad k = 1, 2, \ldots, \infty \quad (7.12)$$

where $X_c(0) = X_p(0)$, $X_c(k) = X_p(k)\cos(\theta(k))$, and $X_s(k) = -X_p(k)\sin(\theta(k))$.

Periodicity of the FS

The FS synthesized waveform is periodic with the same period as that of the fundamental, $T = \frac{2\pi}{\omega_0}$. Replacing t by $t + mT$ in Eq. (7.11) with m being any positive or negative integer, we get

$$x(t + mT) = X_p(0) + \sum_{k=1}^{\infty} X_p(k)\cos(k\omega_0(t + mT) + \theta_k)$$

$$= X_p(0) + \sum_{k=1}^{\infty} X_p(k)\cos(k\omega_0 t + 2km\pi + \theta_k)$$

$$= X_p(0) + \sum_{k=1}^{\infty} X_p(k)\cos(k\omega_0 t + \theta_k) = x(t)$$

If $x(t)$ is defined only over a finite duration, then one period of the FS represents the function. For a periodic $x(t)$, the FS representation is valid for all t.

Existence of the FS

Dirichlet conditions specify the sufficient conditions for the existence of a FS representation of a signal. As the coefficients are defined by an integral over a period, the first condition is that the signal $x(t)$ is absolutely integrable over one period. The function to be integrated is the product of $x(t)$ and the complex exponential with a pure imaginary exponent. As the magnitude of the complex exponential is unity, the condition is $\int_0^T |x(t)| dt < \infty$. From the definition of the FS, we get

$$|X_{fs}(k)| \leq \frac{1}{T} \int_{t_1}^{t_1+T} |x(t)e^{-jk\omega_0 t}| \, dt = \frac{1}{T} \int_{t_1}^{t_1+T} |x(t)||e^{-jk\omega_0 t}| \, dt = \frac{1}{T} \int_{t_1}^{t_1+T} |x(t)| \, dt$$

since $|e^{-jk\omega_0 t}| = 1$.

The second condition is that the number of finite maxima and minima in one period of the signal must be finite. The third condition is that the number of discontinuities in one period of the signal must be finite. All signals generated by physical devices satisfy these conditions.

Example 7.1 Find the three forms of the FS for the signal

$$x(t) = 1 - \cos\left(\frac{2\pi}{8}t - \frac{\pi}{3}\right) + \frac{1}{2}\sin\left(2\frac{2\pi}{8}t + \frac{\pi}{6}\right)$$
$$- \frac{1}{3\sqrt{2}}\cos\left(3\frac{2\pi}{8}t\right) + \frac{1}{3\sqrt{2}}\sin\left(3\frac{2\pi}{8}t\right)$$

Solution

As $x(t)$ can be rewritten in the form of the definitions easily, no evaluation of integral is required to find its FS. The fundamental frequency of the waveform is $\omega_0 = \frac{2\pi}{8}$, as the dc component is periodic with any period.

Compact Trigonometric Form

All the terms of $x(t)$ are rewritten using cosine waveform with positive amplitude for each frequency.

$$x(t) = 1 + \cos\left(\frac{2\pi}{8}t + \frac{2\pi}{3}\right) + \frac{1}{2}\cos\left(2\frac{2\pi}{8}t - \frac{\pi}{3}\right) + \frac{1}{3}\cos\left(3\frac{2\pi}{8}t - \frac{3\pi}{4}\right)$$

Comparing this expression with the definition, Eq. (7.11), we get the compact trigonometric form of the FS coefficients as

$$\left\{X_p(0) = 1, \quad X_p(1) = 1, \quad \theta(1) = \frac{2\pi}{3}, \quad X_p(2) = \frac{1}{2}, \quad \theta(2) = -\frac{\pi}{3}, \right.$$

$$\left. X_p(3) = \frac{1}{3}, \quad \theta(3) = -\frac{3\pi}{4}\right\}$$

Trigonometric Form

$$x(t) = 1 - \frac{1}{2}\cos\left(\frac{2\pi}{8}t\right) - \frac{\sqrt{3}}{2}\sin\left(\frac{2\pi}{8}t\right) + \frac{1}{4}\cos\left(2\frac{2\pi}{8}t\right) + \frac{\sqrt{3}}{4}\sin\left(2\frac{2\pi}{8}t\right)$$

$$- \frac{1}{3\sqrt{2}}\cos\left(3\frac{2\pi}{8}t\right) + \frac{1}{3\sqrt{2}}\sin\left(3\frac{2\pi}{8}t\right)$$

Comparing this expression with the definition, Eq. (7.12), we get the trigonometric form of the FS coefficients as

$$\left\{X_c(0) = 1, \quad X_c(1) = -\frac{1}{2}, \quad X_s(1) = -\frac{\sqrt{3}}{2}, \quad X_c(2) = \frac{1}{4}, \right.$$

$$\left. X_s(2) = \frac{\sqrt{3}}{4}, \quad X_c(3) = -\frac{1}{3\sqrt{2}}, \quad X_s(3) = \frac{1}{3\sqrt{2}}\right\}$$

Exponential Form

$$x(t) = 1 + \frac{1}{2}e^{j\left(\frac{2\pi}{8}t + \frac{2\pi}{3}\right)} + \frac{1}{2}e^{-j\left(\frac{2\pi}{8}t + \frac{2\pi}{3}\right)}$$

$$+ \frac{1}{4}e^{j\left(2\frac{2\pi}{8}t - \frac{\pi}{3}\right)} + \frac{1}{4}e^{-j\left(2\frac{2\pi}{8}t - \frac{\pi}{3}\right)} + \frac{1}{6}e^{j\left(3\frac{2\pi}{8}t - \frac{3\pi}{4}\right)} + \frac{1}{6}e^{-j\left(3\frac{2\pi}{8}t - \frac{3\pi}{4}\right)}$$

Comparing this expression with the definition, Eq. (7.9), we get the exponential form of the FS coefficients as

$$\left\{X_{fs}(0) = 1, \quad X_{fs}(1) = 0.5\angle\frac{2\pi}{3}, \quad X_{fs}(-1) = 0.5\angle-\frac{2\pi}{3}, \right.$$

$$X_{fs}(2) = 0.25\angle-\frac{\pi}{3}, \quad X_{fs}(-2) = 0.25\angle\frac{\pi}{3},$$

$$\left. X_{fs}(3) = \frac{1}{6}\angle-\frac{3\pi}{4}, \quad X_{fs}(-3) = \frac{1}{6}\angle\frac{3\pi}{4}\right\}$$

∎

The condition that a linear combination of sinusoids results in a periodic waveform is that their frequencies have a common measure. The frequencies are rational numbers or multiplied by the same irrational or transcendental numbers such as $\sqrt{3}$, π, or e. Then, the ratio becomes a rational number. The fundamental frequency of a combination of sinusoids is the greatest common divisor of the numerators of their

frequencies divided by the least common multiple of the denominators, after canceling any common factors of the numerators and denominators of each of them. Note
that a constant signal (DC) is periodic with any period.

Example 7.2 Find the fundamental frequency of the set of sinusoids

$$\left\{ \cos\left(\frac{\sqrt{3}}{7}t\right), \cos\left(\frac{4\sqrt{3}}{6}t\right), \cos\left(\frac{6\sqrt{3}}{3}t\right) \right\}$$

Solution
After canceling common factors of the frequencies

$$\left\{ \frac{\sqrt{3}}{7}, \frac{4\sqrt{3}}{6}, \frac{6\sqrt{3}}{3} \right\},$$

we get

$$\left\{ \frac{\sqrt{3}}{7}, \frac{2\sqrt{3}}{3}, \frac{2\sqrt{3}}{1} \right\},$$

The least common multiple of the denominators (7, 3, 1) is 21. The greatest com-
mon divisor of the numerators $(\sqrt{3}, 2\sqrt{3}, 2\sqrt{3})$ is $\sqrt{3}$. Therefore, the fundamental
frequency is $\omega_0 = \frac{\sqrt{3}}{21}$ rad/s. The fundamental period is $T = \frac{2\pi}{\omega_0} = \frac{2\pi 21}{\sqrt{3}} = 14\sqrt{3}\pi$ s.
One cycle of the sum of the three sinusoids is shown in Fig. 7.2a. The first sinusoid
(period $\frac{14\pi}{\sqrt{3}}$), shown in Fig. 7.2b, is the third harmonic and it completes 3 cycles during the period of $14\sqrt{3}\pi$ s. The second sinusoid (period $\sqrt{3}\pi$), shown in Fig. 7.2c,
is the 14th harmonic and it completes 14 cycles. The third sinusoid (period $\sqrt{3}$),
shown in Fig. 7.2d, is the 42nd harmonic and it completes 42 cycles (only a part of
it is shown).   ∎

Example 7.3 Find the FS for the periodic impulse train with period T s defined as

$$x(t) = \sum_{n=-\infty}^{\infty} \delta(t - nT)$$

Solution
The impulse train is shown in Fig. 7.3a. Each impulse is of continuous type with
strength 1, indicated by an upward pointing arrow. As the continuous unit-impulse
signal $\delta(t)$, located at $t = 0$, is defined, in terms of an integral

$$\int_{-\infty}^{\infty} x(t)\delta(t)\, dt = x(0)$$

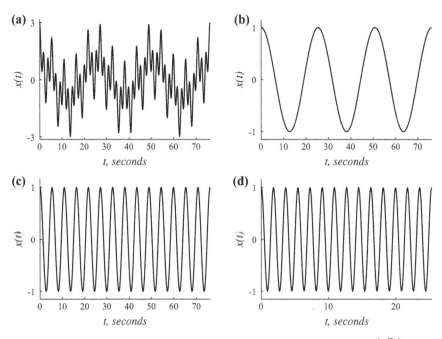

Fig. 7.2 **a** One cycle of the waveform $x(t)$ with period $14\sqrt{3}\pi$ s; **b** sinusoid $\cos\left(\frac{\sqrt{3}}{7}t\right)$ with 3 cycles in the period; **c** sinusoid $\cos\left(\frac{4\sqrt{3}}{6}t\right)$ with 14 cycles in the period; **d** sinusoid $\cos\left(\frac{6\sqrt{3}}{3}t\right)$ with 42 cycles in the period (only a part of it is shown)

and from the exponential FS definition, we get

$$X_{fs}(k) = \frac{1}{T}\int_{-\frac{T}{2}}^{\frac{T}{2}} \delta(t)e^{-jk\omega_0 t}\,dt = \frac{e^{-jk\omega_0 0}}{T} = \frac{1}{T}, \quad -\infty < k < \infty$$

The FS for the impulse train, in exponential form, is given by

$$x(t) = \sum_{k=-\infty}^{\infty} X_{fs}(k)e^{jk\omega_0 t} = \frac{1}{T}\sum_{k=-\infty}^{\infty} e^{jk\omega_0 t}, \quad \omega_0 = \frac{2\pi}{T}$$

The spectrum, shown in Fig. 7.3b, is also a periodic impulse train with period $\omega_0 = \frac{2\pi}{T}$ and amplitude $\frac{1}{T}$. Each impulse is of discrete type.

The FS coefficients, in compact trigonometric form, are

$$X_p(0) = X_{fs}(0) = \frac{1}{T}, \quad X_p(k) = 2|X_{fs}(k)| = \frac{2}{T}, \quad \theta(k) = 0, \ldots, \quad k = 1, 2, 3, \ldots$$

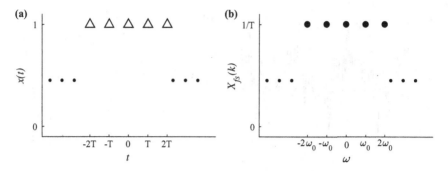

Fig. 7.3 **a** Impulse train with period T s and **b** its FS spectrum

The FS is given by

$$x(t) = \frac{1}{T}(1 + 2(\cos(\omega_0 t) + \cos(2\omega_0 t) + \cos(3\omega_0 t) + \cdots)), \quad \omega_0 = \frac{2\pi}{T} \quad (7.13)$$

As the impulse train is composed of cosine components only, the impulse is an even signal. ∎

The impulse $\delta(t)$ can be approximated by a unit-area rectangular pulse of width a and height $\frac{1}{a}$, located at $t = 0$. In the limiting case of $a \to 0$, the pulse degenerates into an impulse. Several other functions also degenerate into impulse in their limiting form. Consider the reconstructed signal $x(t)$ defined by Eq. (7.13) using a finite number of terms. Let $T = 2\pi$. Then, $\omega_0 = 1$. The partially reconstructed $x(t)$ with the first $N + 1$ terms is given by

$$x(t) \approx \frac{1}{2\pi}(2(1 + \cos(t) + \cos(2t) + \cos(3t) + \cdots + \cos(Nt)) - 1)$$
$$= \frac{1}{2\pi}\left(\frac{\sin(0.5(2N+1)t)}{\sin(0.5t)}\right)$$

Figure 7.4 shows the reconstructed impulse with (a) $N = 8$; (b) $N = 16$; (c) $N = 32$; (d) $N = 64$; (e) $N = 128$; (f) $N = 256$. The waveforms are composed of a large positive hump at the center and damped oscillations on either side. The total area of the oscillations is 1 for any N. The area of the oscillations is negative. Therefore, the area of the large hump must have an area greater than 1 by this amount so that the net area becomes 1. As the impulse encloses an area of 1 at $t = 0$, the reconstructed waveforms with a finite N are not ideal impulses. They are approximations of the impulse. In the limit $N \to \infty$, the large hump and the oscillations coincide at $t = 0$ and the reconstructed waveform attains its ideal form. Therefore, for finite values of N, however large, the reconstructed waveform is a deviation from its ideal form. In reconstructing waveforms with discontinuities, a suitable value for N has to be chosen so that the deviation is acceptable.

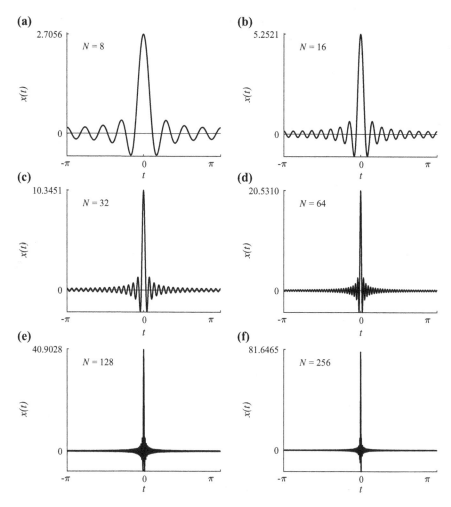

Fig. 7.4 Reconstructed impulse with **a** $N = 8$; **b** $N = 16$; **c** $N = 32$; **d** $N = 64$; **e** $N = 128$; **f** $N = 256$

Example 7.4 Find the FS for a square wave defined over one period as

$$x(t) = \begin{cases} 1 & \text{for } |t| < \frac{\pi}{2} \\ 0 & \text{for } \frac{\pi}{2} < |t| < \pi \end{cases}$$

Solution

The square wave is shown in Fig. 7.6. The period of the waveform is 2π and the fundamental frequency ω_0 is one. The waveform is even-symmetric and, therefore, the coefficients of all the sine components are zero. Further, subtracting the DC bias, the first-half and second-half of $x(t)$ are antisymmetric. This is called the odd half-wave symmetry. This implies that all the even-indexed, except the DC, components are absent. Therefore, the waveform is composed of odd-indexed cosine waves and a DC component.

$$X_c(0) = \frac{2}{2\pi} \int_0^{\frac{\pi}{2}} dt = \frac{1}{2}$$

$$X_c(k) = \frac{4}{2\pi} \int_0^{\frac{\pi}{2}} \cos(k\,t)dt = \begin{cases} \frac{2}{k\pi}\sin\left(\frac{\pi}{2}k\right) & \text{for } k \text{ odd} \\ 0 & \text{for } k \text{ even and } k \neq 0 \end{cases}$$

The coefficient $X_c(0)$ is the average value of $x(t)$.

$$x(t) = \frac{1}{2} + \frac{2}{\pi}\left(\cos(t) - \frac{1}{3}\cos(3t) + \frac{1}{5}\cos(5t) - \cdots\right) \qquad (7.14)$$

The FS magnitude spectrum and the phase spectrum of the signal in exponential form are shown, respectively, in Fig. 7.5a, b. ∎

As the spectrum is usually complex-valued, it requires two plots. A plot to show the magnitude and another to show the phase, as shown in Fig. 7.5. Of course, a plot to show the real part and another to show the imaginary part is also equivalent. The time-domain version of $x(t)$, as shown in Fig. 7.6, with t and $x(t)$ as the independent and dependent variables is its representation in the time domain. The FS spectrum, shown in Fig. 7.5, is another representation of $x(t)$, where k and $X_{fs}(k)$ are the independent and dependent variables. This representation is called the frequency-domain representation. Both the representations are complete and specific representation of the signal. The change of representation is called a transformation, since variable t is changed to k. The major reasons for the importance of this representation are: (i) the amplitude spectrum and power spectrum are convenient representations of signals in many application of signal analysis, (ii) the convolution operation becomes the much simpler multiplication, and (iii) effective signal compression is possible.

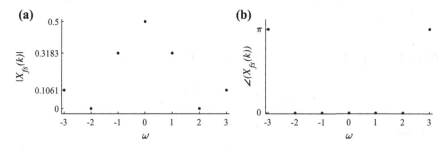

Fig. 7.5 **a** The FS magnitude spectrum and **b** the phase spectrum of the square wave in exponential form

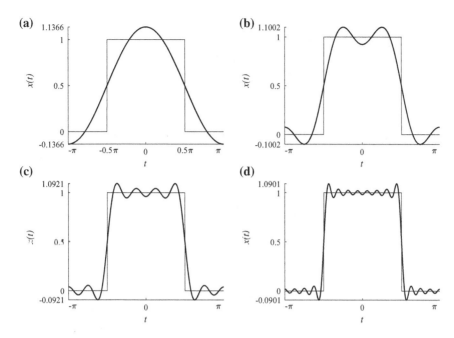

Fig. 7.6 The FS reconstructed square wave. **a** Using up to the first harmonic; **b** using up to the third harmonic; **c** using up to the seventh harmonic; **d** using up to the fifteenth harmonic

7.1.2 Gibbs Phenomenon

The reconstructed square waveforms using up to the first, third, seventh, and fifteenth harmonics are shown in Fig. 7.6a–d, respectively. The magnitude of the FS coefficients decreases as the order of the harmonics increases, except for an impulse. It is expected since the definition of the FS coefficients is an integral and the integrand includes the complex exponential with k, the order of the harmonic, in its exponent. Integrating a signal makes it smoother. For the square waveform, the coefficients decrease at the rate of $1/k$, where k is the order of the harmonic. The square waveform is not differentiable, as it has discontinuities. The convolution output of a square wave with itself is a triangular wave, whose first derivative has discontinuities. As convolution becomes multiplication in the frequency domain, the FS coefficients of the triangular waveform decrease at the rate of $1/k^2$, where k is the order of the harmonic. Therefore, if the nth derivative of a waveform is the first one having a discontinuity then its FS coefficients decrease at the rate of $1/k^{n+1}$. The FS representation converges uniformly at all points except at discontinuities.

The FS reconstructed waveform converges to the average of the two values of the signal at the discontinuity. Let a and b be the values at the discontinuity and c is that of the reconstructed waveform. As the reconstruction is with respect to least squares error criterion,

$$(c - a)^2 + (c - b)^2$$

must be minimum. Differentiating the expression with respect to c and equating the resulting expression to zero, we get

$$c - a + c - b = 0$$

and $c = (a + b)/2$. As the sinusoids are smooth functions, it cannot be be expected to reconstruct a discontinuity exactly. The oscillatory behavior at discontinuities is called the Gibbs phenomenon.

The expression for the FS reconstructed waveform up to the first harmonic is $x(t) = \frac{1}{2} + \frac{2}{\pi} \cos(t)$. Differentiating this expression with respect to t and equating it to zero, we get $\sin(t) = 0$. The point $t = 0$ is a solution to this equation. Substituting $t = 0$ in the expression for $x(t)$, we get the value of the peak as 1.1366, as shown in Fig. 7.6a. The maximum overshoots in other cases can be found similarly.

As the number of harmonics is increased to reconstruct the waveform, the frequency of oscillations increases and the oscillations are confined more closer to the discontinuity. However, the largest amplitude of the oscillations settles at 1.0869 for relatively small number of harmonics. Therefore, there will be deviations of 8.69% of the discontinuity for a moment using any finite number of harmonics to reconstruct a waveform, Of course, the area under the deviation tends to zero.

7.2 Properties of the Fourier Series

Properties of signals and their transforms are useful in reducing the complexity of signal analysis. For example, if the signal is even-symmetric then the determination of its cosine components alone is adequate and the limits of the integration or summation operations also get reduced. Further, with a knowledge of corresponding operations in the time domain and frequency domain, they can be implemented in the appropriate domain efficiently. For example, the implementation of the convolution operation of long and arbitrary signals is more efficient in the frequency domain.

7.2.1 Linearity

The FS coefficients of a linear combination of a set of periodic signals of the same period is equal to the same linear combination of their individual FS coefficients. That is,

if $x(t) \leftrightarrow X_{fs}(k)$ and $y(t) \leftrightarrow Y_{fs}(k)$ then $ax(t) + by(t) \leftrightarrow aX_{fs}(k) + bY_{fs}(k)$, where a and b are arbitrary constants. It is the linearity property of the Fourier analysis that makes it suitable in the analysis of linear systems. For example, the FS coefficients for $\cos(t)$ and $\sin(t)$ are $X_{fs}(\pm 1) = \frac{1}{2}$ and $X_{fs}(\pm 1) = \mp\frac{j}{2}$, respectively. The FS coefficients for $2\cos(t) - 2j\sin(t) = 2e^{-jt}$ are $X_{fs}(\pm 1) = 1 + (-j)(\mp j)$. That is, the only nonzero FS coefficient is $X_{fs}(-1) = 2$.

7.2.2 Symmetry

Taking advantage of the symmetry properties, evaluation of the FS coefficients can be made easier. The spectrum of a real-valued signal is conjugate-symmetric. That is, the real part of its spectrum is even and the imaginary part is odd. As the harmonics are orthogonal, at each frequency index k, the terms of the FS have to be of the form

$$X_{fs}(k)e^{jk\frac{2\pi}{T}t} \text{ and } X_{fs}^*(k)e^{-jk\frac{2\pi}{T}t}$$

so that they combine to produce a real sinusoid

$$X_{fs}(k)e^{jk\frac{2\pi}{T}t} + X_{fs}^*(k)e^{-jk\frac{2\pi}{T}t} = 2|X_{fs}(k)| \cos\left(k\frac{2\pi}{T}t + \angle(X_{fs}(k))\right)$$

Therefore, $X_{fs}^*(-k) = X_{fs}(k)$. Let $X(-1) = \sqrt{3} - j$, $X(1) = \sqrt{3} + j$ and the fundamental frequency be $\omega_0 = 1$. Then,

$$(\sqrt{3} - j)e^{-jt} + (\sqrt{3} + j)e^{jt} = 4\cos\left(t + \frac{\pi}{6}\right)$$

Although we use the exponential form of the Fourier analysis for mathematical convenience, the redundancy in the exponential form of the FS can be taken care of in the practical implementations of the algorithms to compute the Fourier representation of real-valued signals.

Even and Odd Symmetry

The FS coefficients are defined by an integral. The integrand is the product of the signal to be analyzed and the complex exponential. The complex exponential has cosine and sine components. Therefore, if the signal is even-symmetric, as the cosine waveform is even-symmetric and the sine waveform is odd-symmetric, the coefficients corresponding to the cosine component only are nonzero. The spectrum is real and even. Due to the symmetry, the evaluation of the defining integral over one-half of the period is sufficient.

$$X_c(0) = \frac{2}{T} \int_0^{\frac{T}{2}} x(t)\, dt,$$

$$X_c(k) = \frac{4}{T} \int_0^{\frac{T}{2}} x(t) \cos(k\omega_0 t)\, dt, \quad k = 1, 2, \ldots, \infty$$

For example, the FS coefficients of the impulse train and the square waveform in the earlier examples are even-symmetric.

If the signal is odd-symmetric, the coefficients corresponding to the sine component only are nonzero. The spectrum is imaginary and odd.

$$X_s(k) = \frac{4}{T} \int_0^{\frac{T}{2}} x(t) \sin(k\omega_0 t)\, dt, \quad k = 1, 2, \ldots, \infty$$

The square wave, in Example 7.4, pushed down by 0.5 along the vertical axis (so that there is no DC bias) and shifted by $\pi/2$ s to the right becomes odd-symmetric and its FS is

$$\frac{2}{\pi} \left(\sin(t) + \frac{1}{3} \sin(3t) + \frac{1}{5} \sin(5t) + \cdots \right)$$

Any signal $x(t)$ can be decomposed into its even and odd components, $x_e(t)$ and $x_o(t)$. The real part of the FS coefficients, $\mathrm{Re}(X_{fs}(k))$, of a real-valued signal $x(t)$ are the FS coefficients of its even component $x_e(t)$ and $j\,\mathrm{Im}(X_{fs}(k))$ are those of its odd component $x_o(t)$.

Half-Wave Symmetry

This property is defined for periodic signals. If the first- and second-halves of a waveform $x(t)$, with period T, are the same, then it is said to have even half-wave symmetry. That is $x(t \pm \frac{T}{2}) = x(t)$. Obviously, it completes two cycles of a pattern in the interval T and consists of even-indexed frequency components only. The FS coefficients can be expressed as

$$X_{fs}(k) = \frac{1}{T} \int_0^{\frac{T}{2}} \left(x(t) + (-1)^k x\left(t + \frac{T}{2}\right) \right) e^{-jk\omega_0 t}\, dt \qquad (7.15)$$

The odd-indexed FS coefficients are zero. The even-indexed FS coefficients are given by

$$X_{fs}(k) = \frac{2}{T} \int_0^{\frac{T}{2}} x(t) e^{-jk\omega_0 t}\, dt, \quad k = 0, 2, 4, \ldots$$

The full-wave rectified sine wave, presented later, is an example of this type of waveform.

If the first- and second-halves of a waveform $x(t)$, with period T, are the negatives of each other, then it is said to have odd half-wave symmetry. That is $-x(t \pm \frac{T}{2}) =$

$x(t)$. The even-indexed FS coefficients are zero. The odd-indexed FS coefficients are given by

$$X_{fs}(k) = \frac{2}{T} \int_0^{\frac{T}{2}} x(t) e^{-jk\omega_0 t}\, dt, \ k = 1, 3, 5, \ldots$$

The square wave in Example 7.4, with the DC bias subtracted, is an example of this type of waveform.

Any periodic signal $x(t)$, with period T, can be decomposed into its even and odd half-wave symmetric components $x_{eh}(t)$ and $x_{oh}(t)$, respectively. That is $x(t) = x_{eh}(t) + x_{oh}(t)$, where

$$x_{eh}(t) = \frac{1}{2}\left(x(t) + x\left(t \pm \frac{T}{2}\right)\right) \quad \text{and} \quad x_{oh}(t) = \frac{1}{2}\left(x(t) - x\left(t \pm \frac{T}{2}\right)\right)$$

This decomposition is the basis for the fast implementation of the Fourier analysis in practical applications. A waveform composed of N frequency components is decomposed into two waveforms, each of which is composed of $N/2$ frequency components. This process is recursively continued resulting in fast algorithms.

7.2.3 Time Shifting

Time shifting is an often used operation in signal analysis. The variable t in $x(t)$ is replaced by $(t - t_0)$. The origin of the signal is shifted by t_0. Shifting a signal does not change its magnitude profile. Consider a typical harmonic component

$$Ae^{j(k\omega_0 t + \theta)}$$

Shifting it by t_0 results in

$$Ae^{j(k\omega_0(t-t_0)+\theta)} = Ae^{j(k\omega_0 t + (\theta - k\omega_0 t_0))} = Ae^{-jk\omega_0 t_0}e^{j(k\omega_0 t + \theta)}$$

The phase has been changed to $(\theta - k\omega_0 t_0)$. The change in phase is proportional to the harmonic order k. Such a change of the phase of the harmonic components of a waveform $x(t)$ shifts it by t_0. Therefore, if $x(t) \leftrightarrow X_{fs}(k)$ with the fundamental frequency $\omega_0 = \frac{2\pi}{T}$, then

$$x(t \pm t_0) \leftrightarrow e^{\pm jk\omega_0 t_0} X_{fs}(k)$$

Consider the FS of the square wave of Example 7.4.

$$x(t) = \frac{1}{2} + \frac{2}{\pi}\left(\cos(t) - \frac{1}{3}\cos(3t) + \frac{1}{5}\cos(5t) - \cdots\right)$$

Replacing t by $t - \pi/2$, we get

$$x(t) = \frac{1}{2} + \frac{2}{\pi} \left(\cos \left(t - \frac{\pi}{2} \right) - \frac{1}{3} \cos \left(3 \left(t - \frac{\pi}{2} \right) \right) + \frac{1}{5} \cos \left(5 \left(t - \frac{\pi}{2} \right) \right) - \cdots \right)$$
$$= \frac{1}{2} + \frac{2}{\pi} \left(\sin(t) + \frac{1}{3} \sin(3t) + \frac{1}{5} \sin(5t) + \cdots \right)$$

7.2.4 Frequency Shifting

If the signal $x(t)$ is replaced by $x(t)e^{\pm jk_0\omega_0 t}$, where k_0 is an integer and ω_0 is the fundamental frequency, the two complex exponentials in the resulting integrand, in the definition of the FS, combine to become $e^{-j(k \mp k_0)\omega_0 t}$. The net effect is that the independent variable $k\omega_0$ is replaced by $(k \mp k_0)\omega_0$. The spectrum is shifted. Therefore, we get

$$x(t)e^{\pm jk_0\omega_0 t} \leftrightarrow X_{fs}(k \mp k_0)$$

For example, the FS coefficients for $\cos(t)$ are $X_{fs}(1) = \frac{1}{2}$ and $X_{fs}(-1) = \frac{1}{2}$. Let us find the FS coefficients for $\cos(3t)\cos(t)$. Since $\cos(3t) = \frac{1}{2}(e^{j3t} + e^{-j3t})$, the FS coefficients for $\cos(3t)\cos(t)$ is the sum of the FS coefficients for $\cos(t)$ shifted to the right and left by 3, in addition to the scale factor $\frac{1}{2}$. That is,

$$\left\{ X_{fs}(-4) = \frac{1}{4}, X_{fs}(-2) = \frac{1}{4}, X_{fs}(2) = \frac{1}{4}, X_{fs}(4) = \frac{1}{4} \right\}$$

The corresponding time-domain function is

$$\frac{1}{2}(\cos(2t) + \cos(4t)) = \cos(3t)\cos(t)$$

7.2.5 Convolution in the Time Domain

As pointed in Chap. 3, the convolution of complex exponentials with the same frequency is a complex exponential of the same frequency. However, the convolution of complex exponentials with different harmonic frequencies is zero. Therefore, the periodic convolution of $x(t)$ and $h(t)$, with a common period T, is given by

$$y(t) = \sum_{k=-\infty}^{\infty} TX_{fs}(k)H_{fs}(k)e^{jk\omega_0 t},$$

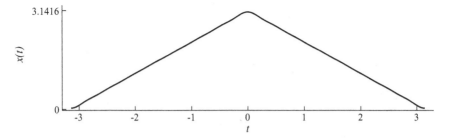

Fig. 7.7 The reconstructed triangular wave, which is the convolution of the square wave with itself

where $X_{fs}(k)$ and $H_{fs}(k)$ are, respectively, the FS coefficients for $x(t)$ and $h(t)$. That is,

$$\int_0^T x(\tau)h(t-\tau)d\tau = \sum_{k=-\infty}^{\infty} TX_{fs}(k)H_{fs}(k)e^{jk\omega_0 t} \leftrightarrow TX_{fs}(k)H_{fs}(k)$$

Consider the convolution of the square wave in Example 7.4 with itself. Its FS representation is

$$x(t) = \frac{1}{2} + \frac{2}{\pi}\left(\cos(t) - \frac{1}{3}\cos(3t) + \frac{1}{5}\cos(5t) - \cdots\right)$$

The complex FS coefficients of the square wave are squared pointwise, multiplied by the period 2π and the Eulers' formula is applied. The convolution output is given by

$$y(t) = 2\pi\left(\frac{1}{4} + \frac{2}{\pi^2}\left(\cos(t) + \frac{1}{9}\cos(3t) + \frac{1}{25}\cos(5t) + \cdots\right)\right)$$

The reconstructed waveform using up to 15 harmonics is shown in Fig. 7.7. The convolution of the square wave with itself is a triangular wave. FS for different functions may be used to find the closed-form values for large number of infinite series. Putting $t = 0$ in $y(t)$ with $y(0) = \pi$, we get

$$\frac{\pi^2}{8} = \left(1 + \frac{1}{9} + \frac{1}{25} + \cdots\right)$$

7.2.6 Convolution in the Frequency Domain

Given that $z(t) = x(t)y(t)$ with period T, the transform of $z(t)$, $Z_{fs}(k)$, is to be expressed in terms of those of the component functions, $X_{fs}(k)$ and $Y_{fs}(k)$. Let the FS representations of $x(t)$ and $y(t)$ be

$$x(t) = X_{fs}(1)e^{j\omega_0 t} + X_{fs}(2)e^{j2\omega_0 t} \quad \text{and} \quad y(t) = Y_{fs}(1)e^{j\omega_0 t} + Y_{fs}(2)e^{j2\omega_0 t}$$

with each one having just two frequency components. The product of the two functions is given by

$$z(t) = x(t)y(t)$$
$$= X_{fs}(1)Y_{fs}(1)e^{j\omega_0 t} + (X_{fs}(1)Y_{fs}(2) + X_{fs}(2)Y_{fs}(1))e^{j3\omega_0 t} + X_{fs}(2)Y_{fs}(2)e^{j2\omega_0 t}$$

This is just multiplication of the exponential polynomial representation of the two functions. The coefficients of the product polynomial are the convolution of those of the individual polynomials. Let the orders of the polynomials be p and q and $p + q = k$. Then, we get

$$z(t) = x(t)y(t) = \sum_{k=-\infty}^{\infty} \left(\sum_{p=-\infty}^{\infty} X_{fs}(p)Y_{fs}(k-p) \right) e^{jk\omega_0 t}$$

This is a FS for $z(t) = x(t)y(t)$ with coefficients

$$Z_{fs}(k) = \sum_{p=-\infty}^{\infty} X_{fs}(p)Y_{fs}(k-p)$$

$$x(t)y(t) \leftrightarrow \frac{1}{T} \int_0^T x(t)y(t)e^{-jk\omega_0 t}\, dt = \sum_{p=-\infty}^{\infty} X_{fs}(p)Y_{fs}(k-p)$$

The convolution is aperiodic, as the FS spectra are aperiodic.

Consider the multiplication of the square wave in Example 7.4 by itself, which is itself. Therefore, if we convolve its spectrum by itself and take the inverse of the resulting spectrum, we should get back the square wave. Let us approximate the square wave by convolving 13 of the FS coefficients

$$\{0, 0.0637, 0, -0.1061, 0, 0.3183, 0.5000, 0.3183, 0, -0.1061, 0, 0.0637, 0\}$$

The result of convolving this set with itself yields

$$\{0.4833, 0.3183, 0.0203, -0.1061, -0.0270, 0.0637, 0.0518\}$$

As the spectrum is symmetric, only part of the one side of the coefficients are given. The odd-indexed coefficients are quite accurate. The even-indexed (except the DC component) are close to zero, but not zero as expected. Convolving a set of 23 coefficients with itself yields

$$\{0.4916, 0.3183, 0.0092, -0.1061, -0.0102, 0.0637, 0.0116\}$$

The value of the DC component becomes still closer to the exact value and the values of the other even-indexed coefficients become more closer to zero. Remember that, the rate of convergence of the coefficients of the square wave is low due to its discontinuities.

7.2.7 Time Scaling

It is often required to change the time scale of the function $x(t)$ with period T by replacing t by at $(a \neq 0)$. The result of time scaling is that the spectrum remains the same with the fundamental frequency changed to $a\omega_0$, where $\omega_0 = 2\pi/T$. For example, let $a = 2$. The square wave in Example 7.4, with period 2π, becomes periodic with period π. Therefore, the fundamental frequency becomes $2\pi/\pi = 2$. The waveform gets compressed. The compression of a function in the time domain results in the expansion of its spectrum. The FS series representation of the square wave becomes

$$x(t) = \frac{1}{2} + \frac{2}{\pi}\left(\cos(2t) - \frac{1}{3}\cos(6t) + \frac{1}{5}\cos(10t) - \cdots \right)$$

With $a > 1$, the signal is compressed and the spectrum is expanded. With $0 < a < 1$, the signal is expanded and the spectrum is compressed. With

$$x(t) \leftrightarrow X_{fs}(k)$$

and the fundamental frequency $\omega_0 = \frac{2\pi}{T}$, we get

$$x(at) \leftrightarrow X_{fs}(k)$$

with the fundamental frequency $a\omega_0$ and $a > 0$. For negative values of a, the spectrum is also frequency-reversed with the fundamental frequency $|a|\omega_0$.

7.2.8 Time Differentiation and Integration

The FS representation can be differentiated or integrated, as long the result can be defined by a FS. Differentiating the FS representation of $x(t)$

$$x(t) = \sum_{k=-\infty}^{\infty} X_{fs}(k)e^{j\omega_0 tk},$$

with respect to t, we get

$$x^{(1)}(t) = j\omega_0 k \sum_{k=-\infty}^{\infty} X_{fs}(k) e^{j\omega_0 t k}$$

The nth derivative is given by

$$x^{(n)}(t) = (j\omega_0 k)^n \sum_{k=-\infty}^{\infty} X_{fs}(k) e^{j\omega_0 t k}$$

Therefore,

$$\frac{d^n x(t)}{dt^n} \leftrightarrow (jk\omega_0)^n X_{fs}(k)$$

Note that the DC component becomes zero in the differentiation process. Similarly,

$$\int_{-\infty}^{t} x(\tau)d\tau \leftrightarrow \frac{1}{jk\omega_0} X_{fs}(k)$$

provided the dc component of $x(t)$ is zero ($X_{fs}(0) = 0$). Successive integration of $x(t)$ can be carried out.

One period of the square wave, with $\omega_0 = 1$, in Example 7.4 can be expressed, in terms of time-shifted unit-step signals, as

$$u\left(t + \frac{\pi}{2}\right) - u\left(t - \frac{\pi}{2}\right), \quad -\pi < t < \pi$$

The derivative, with respect to t, of this signal in a period is composed of impulses

$$\delta\left(t + \frac{\pi}{2}\right) - \delta\left(t - \frac{\pi}{2}\right)$$

The FS coefficients are given by

$$\frac{1}{T}\left(e^{jk\frac{\pi}{2}} - e^{-jk\frac{\pi}{2}}\right) = \frac{1}{T}j2 \sin\left(k\frac{\pi}{2}\right)$$

Dividing by jk and with $T = 2\pi$, we get the FS coefficients of the square wave as

$$X_{fs}(k) = \frac{1}{k\pi} \sin\left(k\frac{\pi}{2}\right)$$

A detailed derivation is as follows. Using the time-shifting property and the FS coefficients of the impulse $\delta(t)$, we get

$$\delta\left(t + \frac{\pi}{2}\right) - \delta\left(t - \frac{\pi}{2}\right)$$

$$\longleftrightarrow \left\{ \cdots, \frac{1}{T}, -\frac{1}{T}, \frac{\overset{\vee}{1}}{T}, \frac{1}{T}, -\frac{1}{T}, \cdots \right\}$$

$$- \left\{ \cdots, -\frac{1}{T}, \frac{1}{T}, \frac{\overset{\vee}{1}}{T}, -\frac{1}{T}, \frac{1}{T}, \cdots \right\}$$

where $T = 2\pi$ and $\omega_0 = 1$. The resulting FS representation is

$$\frac{1}{\pi}(\cdots - e^{-j5t} + e^{-j3t} - e^{-jt} + e^{jt} - e^{j3t} + e^{j5t} + \cdots)$$

Integrating this function, we get

$$\frac{1}{\pi}\left(\cdots + \frac{1}{5}e^{-j5t} - \frac{1}{3}e^{-j3t} + e^{-jt} + e^{jt} - \frac{1}{3}e^{j3t} + \frac{1}{5}e^{j5t} + \cdots\right)$$

Adding the DC component $\frac{1}{2}$, we get the FS representation of the square

$$\frac{1}{\pi}(\cdots + \frac{1}{5}e^{-j5t} - \frac{1}{3}e^{-j3t} + e^{-jt} + \frac{1}{2} + e^{jt} - \frac{1}{3}e^{j3t} + \frac{1}{5}e^{j5t} + \cdots)$$

7.2.9 Parseval's Theorem

As Fourier analysis belongs to the class of orthogonal transforms, it has the power preserving property. That is, the average signal power remains the same in its transformed representation. The signal is expressed as a sum of complex exponentials of harmonic frequencies in its FS representation. The magnitude of each complex exponential is 1. Therefore, the total power in a period of period T is T and the average power is $T/T = 1$. For a specific harmonic component with coefficient $X_{fs}(k)$, the average power is $|X_{fs}(k)|^2$. The total power is the sum of those of all the harmonics. That is, the total average power of a signal is

$$P = \frac{1}{T}\int_0^T |x(t)|^2 dt = \sum_{k=-\infty}^{\infty} |X_{fs}(k)|^2$$

Example 7.5 Find the power of the square wave of Example 7.4 in both the time and frequency domains. Find the power upto the fifth harmonic.

Solution

From the time-domain representation, we get

$$P = \frac{1}{T} \int_0^T |x(t)|^2 dt = \frac{1}{\pi} \int_0^{\frac{\pi}{2}} dt = \frac{1}{2}$$

From the frequency-domain representation, we get

$$P = \sum_{k=-\infty}^{\infty} |X_{fs}(k)|^2 = \left(\frac{1}{2}\right)^2 + 2 \sum_{k=1,3,}^{\infty} \left(\frac{1}{k\pi}\right)^2 = \frac{1}{4} + \frac{2}{\pi^2}\frac{\pi^2}{8} = \frac{1}{2}$$

The sum of the power of the components of the signal up to the fifth harmonic is

$$\frac{1}{4} + \frac{2}{\pi^2} + \frac{2}{9\pi^2} + \frac{2}{25\pi^2} = 0.4833$$

As stated earlier, Fourier analysis approximates a signal adequately with a finite number of frequency components with respect to the power or amplitude of the signal.   ∎

7.3 Applications of the Fourier Series

Apart from signal compression applications, the two major applications of Fourier analysis are: (i) the amplitude and/or power spectrum of a signal reveals important characteristics of the signal pertinent to its use in applications and (ii) efficient implementation of system input–output models such as convolution and differential equation. In Fourier analysis, a signal can be adequately approximated, for practical purposes, by a sum of a finite number of sinusoids. Large number of practical systems can be modeled adequately as linear systems. Then, the response of a system to an arbitrary input signal can be obtained by a linear combination of the responses to the constituent sinusoids. The result is that the response of systems can be obtained faster than other methods in most cases.

The FS representation is important in linear systems analysis due to its approximation of arbitrary signals over a given interval by a sum of sinusoids. The linearity property of such systems permits the superposition of the responses to individual sinusoids. This procedure makes the simplicity of sinusoidal steady-state analysis applicable to arbitrary input signals.

Some of the important Fourier series applications include network analysis and synthesis, vibration analysis, electrical power system waveform analysis, power conversion circuit analysis, acoustics, medical signal analysis, communication circuit analysis, and control system analysis. For example, the electrocardiogram shows the cardiac cycle of a patient. It is a periodic signal with frequency about 4/3 Hz. The

FS representation of the electrocardiogram of a patient is an aid in the diagnosis of the nature and cause of an illness. In power system applications, the frequency of the waveform is 50 or 60 Hz. While its ideal form is of a sinusoid, due to various disturbances, it gets distorted. The harmonic content of the distorted waveform is important. As the maximum allowed harmonic content is specified, corrective measures have to be taken to restore it to acceptable form.

7.3.1 Analysis of Rectified Power Supply

Invariably, all electronic devices contain power conversion circuits to convert the alternating current input supply voltage to DC voltage. This requires rectifiers followed by lowpass filters. As the input voltage is periodic, FS is the appropriate tool in the analysis of the rectified power supply. Let us find the FS representation of the full-wave rectified waveform given by

$$x(t) = |\sin(t)|$$

with period 2π and the fundamental frequency $\omega_0 = 1$. The waveform is shown in Fig. 7.8a.

As the waveform is even-symmetric $x(-t) = x(t)$, it is composed of cosine components only. Further, the first-half and second-half of the waveform are identical. That is $x(t + \frac{T}{2}) = x(t)$. Therefore, the waveform is composed of even-indexed cosine components only.

$$X_c(0) = \frac{1}{\pi} \int_0^\pi \sin(t)dt = \frac{2}{\pi}$$

For even-indexed k, except $k = 0$, we get

$$
\begin{aligned}
X_c(2k) &= \frac{4}{2\pi} \int_0^\pi \sin(t) \cos(2kt)dt \\
&= \frac{1}{\pi} \int_0^\pi (\sin((1-2k)t) + \sin((1+2k)t))dt \\
&= -\frac{1}{\pi} \left(\frac{\cos((1-2k)t)}{(1-2k)} + \frac{\cos((1+2k)t)}{(1+2k)} \right) \Big|_0^\pi \\
&= \frac{4}{\pi} \left(\frac{1}{1-4k^2} \right)
\end{aligned}
$$

$$
\begin{aligned}
x(t) &= \frac{2}{\pi} + \frac{4}{\pi} \sum_{k=1}^\infty \left(\frac{1}{1-4k^2} \right) \cos(2kt) \\
&= \frac{2}{\pi} - \frac{4}{3\pi} \cos(2t) - \frac{4}{15\pi} \cos(4t) - \frac{4}{35\pi} \cos(6t) - \cdots
\end{aligned}
$$

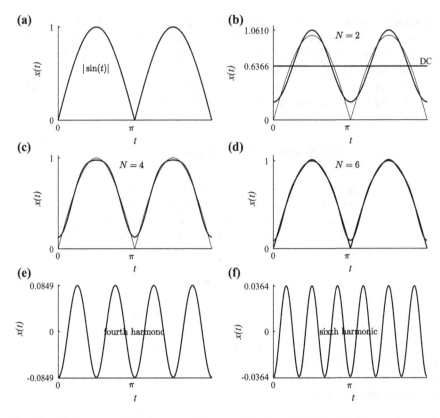

Fig. 7.8 a Full-wave rectified sine wave; **b** its reconstruction with DC and 2nd harmonic; **c** and **d** its reconstruction with up to 4th and 6th harmonics, respectively; **e** the 4th harmonic; **f** the 6th harmonic

Figure 7.8b shows its DC component and its reconstruction with the DC and the 2nd harmonic. Figure 7.8c shows its reconstruction with the DC and the 2nd and 4th harmonics. Figure 7.8d shows its reconstruction with the DC and the 2nd, 4th, and 6th harmonics. As expected, the reconstructed waveform becomes more closer to the original with the addition of more harmonic components.

While signal reconstruction is an important aspect of Fourier analysis, the magnitude of the harmonics is another important aspect. Figure 7.8e, f show, respectively, the fourth and sixth harmonics of the waveform in (a). The harmonics make 4 and 6 cycles in a period. In most applications, the harmonics are considered as unwanted components of the waveform. The harmonic magnitudes have to be smaller than a given specification. That requires some sort of filtering. In order to design a suitable filter, Fourier analysis is required, in the first place, to estimate the magnitudes of the harmonics.

7.3.2 Steady-State Response of Linear Systems

When the excitation to a linear system is a complex exponential, the form of the response at any part of the system is also the same exponential with changes only in magnitude and phase. The frequency remains unchanged. This property, along with the linearity property, makes the steady-state output of a stable system to be determined by the superposition sum of the responses due to all the constituent exponentials of an arbitrary input signal. A continuous periodic input signal can be decomposed into a sum of exponentials with harmonic frequencies by the FS.

The steady-state output of a LTI system to an input $e^{jk_0\omega_0 t}$ is the same function multiplied by the complex scale factor, $H(jk_0\omega_0)$, called the frequency response. Therefore, the output of the system is $H(jk_0\omega_0)e^{jk_0\omega_0 t}$. The function $H(jk\omega_0)$ is obtained by sampling the continuous frequency response $H(j\omega)$ of the system at the discrete frequencies $\omega = k\omega_0$.

For an arbitrary continuous periodic input signal, as

$$x(t) = \sum_{k=-\infty}^{\infty} X_{fs}(k)e^{jk\omega_0 t},$$

the system output is

$$y(t) = \sum_{k=-\infty}^{\infty} H(jk\omega_0)X_{fs}(k)e^{jk\omega_0 t}$$

Therefore, solving a differential equation or evaluating a convolution integral in the time domain is reduced to the evaluation of an algebraic operation in the frequency domain.

While an electrical circuit is characterized by a differential equation in the time domain, it is characterized by an algebraic equation in the frequency domain. Therefore, we first represent the circuit in the frequency domain. For example, the voltage across an inductor with L henries due a current $i(t)$ flowing through it is modeled, in the time domain, as

$$L\frac{di(t)}{dt}$$

With $i(t) = Ie^{j\omega t}$, the voltage across the inductor is

$$j\omega L I e^{j\omega t}$$

Therefore, the model for an inductor, in the frequency domain, is $j\omega L$.

Similarly, the voltage across a capacitor with C farads due a current $i(t)$ flowing through it is modeled, in the time domain, as

$$\frac{1}{C}\int i(t)dt$$

With $i(t) = Ie^{j\omega t}$, the voltage across the capacitor is

$$\frac{Ie^{j\omega t}}{j\omega C}$$

Therefore, the model for a capacitor, in the frequency domain, is $1/(j\omega C)$. The model for a resistor, in the frequency domain, is R itself. The value is unaffected by the frequency.

The opposition to the flow of current through an inductor or capacitor is called the reactance. The sum of resistance and reactance is called the impedance. For example, the impedance of a series connected resistor R and inductor L is $R + j\omega L$. Similarly, the impedance of a series connected resistor R and capacitor C is $R + (1/(j\omega C))$. The reactance due to an inductor is zero at $\omega = 0$ and increases with increasing frequency. The reactance due to a capacitor is infinite at $\omega = 0$ and decreases with increasing frequency. Therefore, for a given current, the voltage across an inductor increases with increasing frequency. The voltage across a capacitor decreases with increasing frequency. A resistor is required to limit the current at low and high frequencies. These characteristics enable to build filter circuits with resistors, capacitors, and inductors.

Let the input $x(t) = |\sin(t)|$ be applied to the lowpass filter circuit, shown in Fig. 7.9. It is a series resistor–capacitor network, with the resistor $10\,\Omega$ and capacitor $1/2\,\mathrm{F}$. The task is to find an expression for the output voltage across the capacitor, $y(t)$.

The FS for the full-wave rectified sine wave is

$$x(t) = \frac{2}{\pi} + \frac{4}{\pi} \sum_{k=1}^{\infty} \left(\frac{1}{1 - 4k^2}\right) \cos(2kt)$$

$$= \frac{2}{\pi} - \frac{4}{3\pi} \cos(2t) - \frac{4}{15\pi} \cos(4t) - \frac{4}{35\pi} \cos(6t) + \cdots$$

The voltage across the capacitor, by voltage division, is

$$y(t) = x(t)\frac{1/(j\omega C)}{R + 1/(j\omega C)} = x(t)\frac{1}{1 + j\omega RC}$$

Fig. 7.9 A resistor–capacitor
lowpass filter circuit

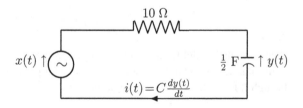

When the input is a sinusoid or a complex exponential, the Ohm's law, for circuits with resistances, is extended to circuits with energy storage elements inductor and capacitor with the impedance replacing the resistance. For DC, with the fundamental frequency $\omega_0 = 1$ and $k = 0$,

$$y(t) = \frac{2}{\pi}$$

since the capacitor is an open circuit to DC. For the 2nd harmonic, with the fundamental frequency $\omega_0 = 1$ and $k = 2$,

$$y(t) = -\frac{4}{3\pi}\frac{1}{1+j2RC}\cos(2t)$$

With $R = 10$ and $C = 0.5$, the magnitude of the output is

$$\left|\frac{4}{3\pi}\frac{1}{1+j10}\right| = \frac{4}{3\pi}0.0995$$

The magnitude of the 2nd harmonic has been reduced by a factor of 0.0995. Therefore, the output waveform will have ripples of a smaller magnitude and is more closer to the required output, which is DC. The factor in the denominator attenuates the frequency components other than DC in proportion to the frequency. As the frequency approaches infinity, the input components are almost blocked.

The total response of the circuit is the sum of the responses to all the harmonic components of the input. The RC circuit is called a lowpass filter as it passes more readily the low-frequency components of the input compared with those of the high-frequency components. In practice, the FS, approximated by the DFT, can only have a finite number of terms. Therefore, the number of terms has to be fixed depending on the accuracy required.

For the resistor–inductor series circuit shown in Fig. 7.10, the input–output relationship is given by

$$y(t) = x(t)\frac{j\omega L}{1+j\omega RL}$$

For DC, with the fundamental frequency $\omega_0 = 1$ and $k = 0$,

$$y(t) = 0$$

Fig. 7.10 A resistor–inductor highpass filter circuit

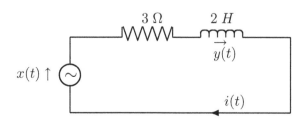

since the inductor is a short circuit to DC. For the 2nd harmonic, with the fundamental frequency $\omega_0 = 1$ and $k = 2$,

$$y(t) = -\frac{4}{3\pi} \frac{j2L}{1 + j2RL} \cos(2t)$$

With $R = 3$ and $L = 2$, the magnitude of the output is

$$\left| \frac{4}{3\pi} \frac{j4}{1 + j12} \right| = \frac{4}{3\pi} 0.3322$$

The magnitude of the 2nd harmonic has been reduced by a factor of 0.3322. As the frequency approaches infinity, the input components are more readily passed. The total response of the circuit is the sum of the responses to all the harmonic components of the input. The RL circuit is called a highpass filter as it passes more readily the high-frequency components of the input compared with those of the low-frequency components.

The frequency-domain analysis is similar in other applications. For example, in mechanical engineering, resistor, inductor, and capacitor correspond, respectively, to friction, spring, and mass.

7.4 Numerical Evaluation of the Fourier Series

In practice, the amplitude profiles of practical waveforms are arbitrary and, invariably, the other versions of the Fourier analysis have to be approximated by the DFT and the IDFT. The integral in Eq. (7.10) is numerically evaluated by using the DFT. The DFT uses the samples over one period of the periodic signal $x(t)$ with period T. The sampling interval is $T_s = \frac{T}{N}$ and the number of samples is N. The N samples of $x(t)$ over one period are

$$\{x(0), x(T_s), x(2T_s), \ldots, x((N-1)T_s)\}$$

Now, Eq. (7.10) is approximated as

$$X_{fs}(k) = \frac{1}{T} \int_0^T x(t) e^{-jk\omega_0 t} dt$$

$$= \lim_{T_s \to 0} \frac{1}{T} \sum_{n=0}^{N-1} x(nT_s) e^{-jk\omega_0 nT_s} T_s$$

$$= \lim_{T_s \to 0} \frac{1}{N} \sum_{n=0}^{N-1} x(nT_s) e^{-jk\frac{2\pi}{N}n}$$

In numerical approximation, sampling interval has to be a finite value, but not zero, to keep the number of samples finite. Let us ignore the limiting process for sufficiently small values of T_s, which will result in aliasing error. Then, the FS analysis equation is approximated by

$$X_{fs}(k) = \frac{1}{N} \sum_{n=0}^{N-1} x(nT_s) e^{-j\frac{2\pi}{N}nk}, \quad k = 0, 1, \ldots, N-1 \tag{7.16}$$

The synthesis Eq. (7.9) is approximated as

$$x(n) = \sum_{k=0}^{N-1} X_{fs}(k) e^{j\frac{2\pi}{N}nk}, \quad n = 0, 1, \ldots, N-1.$$

The numerical approximations of the FS analysis and synthesis equations are the same as the DFT and IDFT equations, with a scale factor. Remember that $X_{fs}(k)$ is aperiodic and the DFT spectrum is periodic. If $x(t)$ is band-limited and adequate number of samples is taken over a period, the assumed periodicity of the DFT is not a problem. If it is not the case, aliasing of the spectrum will arise. Then, it has to be ensured that the quality of the spectrum is adequate by taking sufficient number of samples.

For N even, comparing the coefficients of the DFT with that in Eq. (7.10), we get, for real signals,

$$X_{fs}(k) = \frac{X(k)}{N}, \quad k = 0, 1, \ldots, \frac{N}{2} - 1 \quad \text{and} \quad \mathrm{Re}\left(X_{fs}\left(\frac{N}{2}\right)\right) = \frac{X\left(\frac{N}{2}\right)}{2N} \tag{7.17}$$

For example, let us compute the FS coefficients, using the DFT, of

$$x(t) = 2 + \cos\left(\frac{2\pi}{4}t\right) + \cos\left(2\frac{2\pi}{4}t + \frac{\pi}{4}\right)$$

$$= 2 + 0.5e^{j\left(\frac{2\pi}{4}t\right)} + 0.5e^{-j\left(\frac{2\pi}{4}t\right)} + 0.5e^{j\left(\frac{\pi}{4}\right)}e^{j\left(\frac{2\pi}{4}2t\right)} + 0.5e^{-j\left(\frac{\pi}{4}\right)}e^{-j\left(\frac{2\pi}{4}2t\right)}$$

The FS coefficients $X_{fs}(k)$ are

$$\left\{2, 0.5, \frac{\sqrt{2}}{4} + j\frac{\sqrt{2}}{4}, \frac{\sqrt{2}}{4} - j\frac{\sqrt{2}}{4}, 0.5\right\}$$

With $T_s = 1$, the samples are

$$\left\{3 + \frac{1}{\sqrt{2}}, 2 - \frac{1}{\sqrt{2}}, 1 + \frac{1}{\sqrt{2}}, 2 - \frac{1}{\sqrt{2}}\right\} = \{3.7071, 1.2929, 1.7071, 1.2929\}$$

With 4 samples, the DFT coefficients are

$$\{8, 2, 2\sqrt{2}, 2\}$$

Dividing $X(2)$ by 8 and the rest by 4, we get

$$\{2, 0.5, 0.25\sqrt{2}, 0.5\}$$

The component with frequency index 2 is

$$\cos\left(2\frac{2\pi}{4}t + \frac{\pi}{4}\right) = \frac{1}{\sqrt{2}}\cos\left(2\frac{2\pi}{4}t\right) - \frac{1}{\sqrt{2}}\sin\left(2\frac{2\pi}{4}t\right)$$

With $N = 4$ samples, the samples of both the $\cos(2\frac{2\pi}{4}t)$ and the DFT basis function with index 2 are $\{1, -1, 1, -1\}$ and the sum of pointwise multiplication yields 4. By multiplying with the constant $\frac{1}{\sqrt{2}}$, we get the DFT coefficient as $2\sqrt{2}$. Dividing $2\sqrt{2}$ by $2N = 8$, we get the real part of the FS coefficient $\frac{\sqrt{2}}{4}$ according to Eq. (7.17). The samples of the sine component with frequency index 2 are all zero and get no representation.

With $T_s = 0.8$, the 5 samples are

$$\{3.7071, 1.3213, 2.0820, 0.7370, 2.1526\}$$

The DFT coefficients are

$$\{10, 2.5, 1.7678 + j1.7678, 1.7678 - j1.7678, 2.5\}$$

which are the correct FS coefficients scaled by 5.

In practice, the number of samples has to be a power of 2 to suit fast practical DFT algorithms. With $T_s = 0.5$, the 8 samples are

$$\{3.7071, 2.0000, 1.2929, 2.0000, 1.7071, 0.5858, 1.2929, 3.4142\}$$

The DFT coefficients are

$$\{16, 4, 2.8284 + j2.8284, 0, 0, 0, 2.8284 - j2.8284, 4\}$$

which are the correct FS coefficients scaled by 8.

7.4.1 Aliasing Effect

When a continuous time-domain function is sampled, its spectrum becomes periodic due to aliasing, as the frequency range gets reduced from infinite to some finite value.

The period of the spectrum is determined by the sampling interval in the time domain. The sampling process can be modeled as multiplying the continuous function by an impulse train. Multiplication in the time domain corresponds to convolution of the spectra of the impulse train, which is also an impulse train, and the continuous function. The convolution of the FS spectrum with an impulse is relocating the aperiodic spectrum at the location of the impulse. The combined spectrum of these shifted spectra is the spectrum of the sampled function. To recover the continuous signal back from the spectrum of the sampled signal to a good approximation requires that the periodic repetition of the spectrum does not overlap within each period. In practice, negligible overlapping is acceptable.

Consider the waveform $x(t)$

$$x(t) = 0.2 + \cos\left(\frac{2\pi}{16}t\right) + 0.3\cos\left(\frac{2\pi}{16}3t\right)$$

$$= 0.2 + \left(\frac{1}{2}e^{j\frac{2\pi}{16}t} + \frac{1}{2}e^{-j\frac{2\pi}{16}t}\right) + \left(\frac{0.3}{2}e^{j\frac{2\pi}{16}3t} + \frac{0.3}{2}e^{-j\frac{2\pi}{16}3t}\right)$$

The FS coefficients in the center-zero format are

$$X_{fs}(k), k = -3, -2, -1, 0, 1, 2, 3 = \{0.15, 0.00, 0.50, 0.\check{2}0, 0.50, 0.00, 0.15\}$$

If we sum the shifted copies of the spectral values placed at a distance of 4 samples, we get a corrupted periodic spectrum $\{0.\check{2}0, 0.65, 0.00, 0.65\}$.

0.15 0.00 0.50 0.$\check{2}$0, 0.50 0.00 0.15
    0.15 0.00 0.50 0.$\check{2}$0, 0.50 0.00 0.15
        0.15 0.00 0.50 0.$\check{2}$0, 0.50 0.00 0.15

$$, \ldots, 0.\check{2}0, 0.65\ 0.00\ 0.65, \ldots,$$

The 4 samples of $x(t)$ are

$$\{1.\check{5}, 0.2, -1.1, 0.2\}$$

The DFT coefficients, divided by $N = 4$, are

$$\{0.\check{2}0, 0.65, 0.00, 0.65\}$$

Equation (7.18)

$$X(k) = N \sum_{m=-\infty}^{\infty} X_{fs}(k - mN), \quad k = 0, 1, \ldots, N - 1 \tag{7.18}$$

shows how the DFT spectrum is corrupted due to aliasing.

7.5 Summary

- FS is one of the four versions of Fourier analysis that provides the representation of a continuous periodic time-domain waveform by a discrete aperiodic spectrum in the frequency domain.
- FS represents a continuous periodic waveform as a linear combination of sinusoidal or, equivalently, complex exponential basis functions of harmonically related frequencies.
- Harmonics are any of the frequency components whose frequencies are integral multiples of a fundamental. The frequency of the fundamental is the same as that of the periodic waveform being analyzed.
- The FS is the limiting case of the DFT as the sampling interval of the time-domain sequence tends to zero with the period fixed.
- While physical devices generate real sinusoidal waveforms, it is found that the analysis is mostly carried out using complex exponentials due to its compact form and ease of manipulation.
- While an infinite number of frequency components are required to represent an arbitrary waveform exactly, it is found that, in practice, a finite number of frequency components provides an adequate representation.
- The orthogonality property of the basis signals makes it easy to determine the FS coefficients.
- The representation of a signal in terms of its spectrum is just as complete and specific as its time-domain representation in every respect.
- The independent variable of the waveform may be other than time, such as distance.
- The conditions for the existence of a Fourier representation of a waveform are met by signals generated by physical devices.
- The amplitude versus frequency plot of the harmonics is called the spectrum. As the spectrum is usually complex, it is represented by two plots, either the real and imaginary parts or the magnitude and phase. While the time-domain waveform is continuous and periodic, its FS spectrum is aperiodic and discrete.
- The more smoother the waveform, the faster is the convergence of its spectrum.
- A signal can be reconstructed using its spectral components.
- The exponential and trigonometric forms of Fourier analysis are related by the Euler's formula.
- The least squares error is the criterion for the representation of a signal by the Fourier analysis. With respect to this criterion, there is no better approximation than that provided by the Fourier analysis.
- The properties of Fourier analysis help to relate the effects of characteristics of signals in one domain into the other.
- LTI system analysis is simpler with the Fourier representation of signals and systems.
- The Fourier spectrum can be adequately approximated by the DFT in practical applications.

- FS is important in the analysis of signals such as acoustical, vibration, power system, communication, electrocardiogram, and frequency response.

Exercises

7.1 Express the frequency components of the waveform $x(t)$ in terms of complex exponentials and, hence, find its complex FS coefficients $X_{fs}(k)$. Verify that the samples of the two forms are the same by finding the 7 uniformly spaced samples of the waveform over a cycle starting from $t = 0$. What is the fundamental frequency ω_0?

7.1.1

$$x(t) = -2 + 2\sin\left(\pi t + \frac{\pi}{3}\right) - 4\cos\left(3\pi t - \frac{\pi}{6}\right)$$

7.1.2

$$x(t) = 8\cos\left(2\pi t - \frac{\pi}{3}\right) - 6\sin\left(4\pi t - \frac{\pi}{4}\right) + 4\cos\left(6\pi t + \frac{\pi}{6}\right)$$

* 7.1.3

$$x(t) = 1 - 4\sin\left(2t + \frac{\pi}{6}\right) + 8\cos\left(3t - \frac{\pi}{3}\right)$$

7.2 Express the frequency components of the waveform $x(t)$ in terms of complex exponentials and, hence, find its complex FS coefficients $X_{fs}(k)$. What is the fundamental frequency ω_0?

7.2.1

$$x(t) = 3 + \cos\left(\frac{2}{3}t\right) + 2\sin\left(\frac{5}{7}t\right)$$

* 7.2.2

$$x(t) = 1 - 2\sin\left(\frac{3}{6}t\right) + \sin\left(\frac{2}{9}t\right)$$

7.2.3

$$x(t) = 4 + 3\cos\left(\frac{5}{7}t\right) + 3\cos\left(\frac{4}{6}t\right)$$

7.3
* 7.3.1 Find the FS for a periodic pulse train defined over one period as

$$x(t) = \begin{cases} 1 & \text{for } |t| < a \\ 0 & \text{for } a < |t| < \frac{T}{2} \end{cases}$$

Deduce the FS with $T = 10$ and $a = 2$.
7.3.2 Find the FS for a periodic exponential signal defined over one period as

$$x(t) = e^{at} \quad \text{for } 0 \leq t < 2\pi$$

7.3.3
Find the FS for a saw-tooth signal defined over one period as

$$x(t) = t \text{ for } 0 \leq t < 1$$

7.4
* 7.4.1 The FS representation of the full-wave rectified sine wave $x(t) = |\sin(t)|$ is

$$x(t) = \frac{2}{\pi} + \frac{4}{\pi} \sum_{k=1}^{\infty} \left(\frac{1}{1 - 4k^2} \right) \cos(2kt)$$

$$= \frac{2}{\pi} - \frac{4}{3\pi} \cos(2t) - \frac{4}{15\pi} \cos(4t) - \frac{4}{35\pi} \cos(6t) + \cdots$$

Using the time-shift property, find the FS representation of the full-wave rectified cosine wave $x(t) = |\cos(t)|$.

7.4.2 Find the convolution $y(t)$ of $x(t) = \sin(t + \frac{\pi}{3})$ and $h(t) = \sin(t - \frac{\pi}{6})$ using the FS time-domain convolution theorem. Verify the answer by directly convolving $x(t)$ and $h(t)$ in the time domain.

7.4.3 Using the time-differentiation property, find the derivative of $x(t) = 2\sin(3t)$.

7.5
Compute the FS coefficients of $x(t)$, using the DFT, with the sampling interval equal to 1 and 0.8 s. Compare the result with Eq. (7.18).

7.5.1
$$x(t) = 1 + \sin\left(\frac{2\pi}{4}t - \frac{\pi}{3}\right) + \cos\left(2\frac{2\pi}{4}t - \frac{\pi}{3}\right)$$

* 7.5.2
$$x(t) = 3 + \sin\left(\frac{2\pi}{4}t - \frac{\pi}{6}\right) + \sin\left(2\frac{2\pi}{4}t + \frac{\pi}{3}\right)$$

7.5.3
$$x(t) = 2 + \cos\left(\frac{2\pi}{4}t - \frac{\pi}{4}\right) + \cos\left(2\frac{2\pi}{4}t + \frac{\pi}{4}\right)$$

Chapter 8
The Discrete-Time Fourier Transform

The DTFT is the dual of the FS. In the FS representation of signals, the time-domain waveform is continuous and periodic and the spectrum is discrete and aperiodic. In the DTFT representation of signals, the spectrum is continuous and periodic and the time-domain waveform is discrete and aperiodic. Another difference is that the continuous signal is usually real and the discrete spectrum is complex in the FS. On the other hand, the continuous spectrum is complex and the discrete signal is usually real in the DTFT. By interchanging the roles of time-domain and frequency-domain variables, we can obtain one representation from the other. However, it is better to derive the DTFT as the limiting case of the DFT, as the period of the time-domain sequence tends to infinity with the sampling interval fixed. The effective frequency range of the spectrum is fixed by the sampling interval. In other than the DFT version of the Fourier analysis, the signal or the spectrum or both are of continuous type. In the Fourier analysis and synthesis definitions, the summation in the DFT becomes an integral for continuous type of signals and spectra. Basically, Fourier analysis, despite the differences between the versions, is always the decomposition of an arbitrary signal in terms of sinusoidal waveforms.

8.1 The DTFT

The DTFT, $X(e^{j\omega})$, of the sequence $x(n)$, with the sampling interval $T_s = 1$, is defined by the infinite summation

$$X(e^{j\omega}) = \sum_{n=-\infty}^{\infty} x(n)e^{-j\omega n} \qquad (8.1)$$

© Springer Nature Singapore Pte Ltd. 2018
D. Sundararajan, *Fourier Analysis—A Signal Processing Approach*,
https://doi.org/10.1007/978-981-13-1693-7_8

The exponential $e^{j\omega}$ in $X(e^{j\omega})$ emphasizes the fact that it is a periodic function of ω. The frequency-domain representation is a complex function of ω and is periodic in ω with period $2\pi/T_s = 2\pi$, since $e^{-j(\omega+2\pi)n} = e^{-j\omega n}$. The inverse DTFT, $x(n)$, of the spectrum $X(e^{j\omega})$ is defined by the integral over one period of $X(e^{j\omega})$ as

$$x(n) = \frac{1}{2\pi} \int_{-\pi}^{\pi} X(e^{j\omega})e^{j\omega n}d\omega, \quad n = 0, \pm 1, \pm 2, \ldots \qquad (8.2)$$

The infinite time-domain sequence $x(n)$ is the FS for the continuous periodic spectrum $X(e^{j\omega})$.

Example 8.1 Find the DTFT of $x(n)$. The nonzero samples are $\{x(-1) = 1, x(1) = 1\}$.
Solution
From the definition,

$$X(e^{j\omega}) = e^{j\omega} + e^{-j\omega} = 2\cos(\omega)$$

The inverse DTFT is

$$x(n) = \frac{1}{2\pi} \int_{-\pi}^{\pi} (e^{j\omega} + e^{-j\omega})e^{j\omega n}d\omega = 1, \quad n = -1, 1 \text{ and } 0 \text{ otherwise}$$

since

$$\int_{-\pi}^{\pi} e^{j\omega k}d\omega = \begin{cases} 1 \text{ for } k = 0 \\ 0 \text{ for } k = \pm 1, \pm 2, \ldots \end{cases}$$

One period of the magnitude spectrum $|X(e^{j\omega})|$ of $x(n)$ and the phase spectrum $\angle X(e^{j\omega})$ is shown in Fig. 8.1a, b, respectively.   ∎

The spectrum is a complex function of ω and it is periodic in ω with period 2π. For real-valued signals, the magnitude spectrum is even-symmetric and the phase spectrum is odd-symmetric. Therefore, only one half of the spectrum from $\omega = 0$ to $\omega = \pi$ is usually displayed. Frequencies near π are high frequencies and those near zero are low frequencies. High frequency signals are characterized by the high

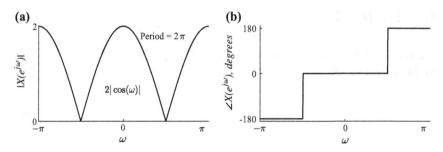

Fig. 8.1 a The magnitude spectrum $|X(e^{j\omega})|$ for $x(n)$ and **b** the phase spectrum

difference between successive sample values. Unlike those of the continuous-time signals, the spectra of discrete signals are always periodic. This implies that useful frequency range of the spectra is finite.

8.1.1 DTFT as a Limiting Case of the DFT

In this section, we derive the DTFT as a limiting case of the DFT by letting the period of the time-domain sequence tends to infinity with the sampling interval fixed. As in the case of the FS, we start with the IDFT expression and substitute the expression for $X(k)$ in that. We use the center-zero format for convenience. Let the DFT of sequence $x(n), N \le n \le N$ be $X(k), N \le k \le N$. Then, the Fourier representation of $x(n)$ is given as

$$x(n) = \frac{1}{2N+1} \sum_{k=-N}^{N} X(k) e^{j\frac{2\pi}{(2N+1)}nk}, \quad n = 0, \pm 1, \pm 2, \ldots, \pm N \qquad (8.3)$$

where

$$X(k) = \sum_{m=-N}^{N} x(m) e^{-j\frac{2\pi}{(2N+1)}mk} \qquad (8.4)$$

Substituting for $X(k)$ in Eq. (8.3), we get

$$x(n) = \frac{1}{2N+1} \sum_{k=-N}^{N} \left(\sum_{m=-N}^{N} x(m) e^{-j\frac{2\pi}{(2N+1)}mk} \right) e^{j\frac{2\pi}{(2N+1)}nk} \qquad (8.5)$$

Let the $2N + 1$ samples be obtained by sampling a periodic signal of period $2N + 1$ s over a period with a sampling interval of $T_s = 1$ s. The frequency increment of the spectrum becomes

$$\Delta\omega = \frac{2\pi}{(2N+1)}$$

radians per second.

With the same T_s, let the period $2N + 1$ of $x(n)$ be longer. Then, with $N \to \infty$, the frequency increment of the spectrum becomes

$$\frac{2\pi}{(2N+1)} \to 0$$

and the spectrum becomes continuous with the same period, as T_s is fixed. The discrete and periodic time-domain sequence becomes discrete and aperiodic and the discrete and periodic spectrum becomes continuous and periodic. Then, the frequency increment, $\frac{2\pi}{(2N+1)}$, is formally replaced by the differential $d\omega$ in the limit. The discrete frequency variable $\frac{2\pi}{(2N+1)}k$ is replaced by the continuous frequency variable ω. Eventually, the limits of the inner summation become $-\infty$ and ∞ and the DTFT $X(e^{j\omega})$ of the signal $x(n)$ is defined as

$$X(e^{j\omega}) = \sum_{n=-\infty}^{\infty} x(n)e^{-j\omega n} \qquad (8.6)$$

The outer summation becomes an integral with limits $-\pi$ and π. As the spectrum is periodic, the limits could be any continuous interval of 2π. Therefore, the synthesis expression of the DTFT becomes

$$x(n) = \frac{1}{2\pi} \int_{-\pi}^{\pi} X(e^{j\omega})e^{j\omega n} d\omega, \ \ n = 0, \pm 1, \pm 2, \ldots \qquad (8.7)$$

The analysis equation of the DTFT is a summation, since the time-domain signal is discrete. The synthesis equation is an integral, since the spectrum is continuous.

Consider the signal $x(n)$ with $N = 5$ nonzero samples and its DFT spectrum, shown in Fig. 8.2a, b, respectively. The sampling interval is $T_s = 1$ and the frequency increment is $2\pi/5$ radians per sample. With the same $T_s = 1$, consider the signal $x(n)$ with $N = 9$ and its DFT spectrum, shown in Fig. 8.2c, d, respectively. The sampling interval is $T_s = 1$ and the frequency increment is $2\pi/9$ radians per sample. The time-domain range is increased making the record length longer and the frequency increment is decreased making the spectrum denser. With 9 independent samples in the time domain, there can only a maximum of 9 independent samples in the frequency domain. Remember that complex spectrum of real-valued signals is redundant by a factor of 2. With $N = 17$ samples, the record length is longer and the spectrum is denser, as shown in Fig. 8.2e, f. As $N \rightarrow \infty$, the time-domain signal becomes aperiodic. The frequency increment tends to zero and, therefore, the corresponding spectrum becomes continuous. The DTFT spectrum is shown by a solid line.

In all the versions of the Fourier analysis, for certain values of the frequency index or the time index, the expressions for the forward and inverse transforms take a simplified form. Consequently, the values with index such as zero or π can be computed easily. These values are useful to check the correctness of the closed-form expressions for $x(n)$ or $X(e^{j\omega})$.

$$X(e^{j0}) = \sum_{n=-\infty}^{\infty} x(n), \ \ X(e^{j\pi}) = \sum_{n=-\infty}^{\infty} (-1)^n x(n), \qquad x(0) = \frac{1}{2\pi} \int_{-\pi}^{\pi} X(e^{j\omega}) d\omega,$$

The DTFT definitions for the forward and inverse transforms, in Eqs. (8.6) and (8.7), have been derived assuming that the sampling interval of the time-domain signal, T_s, is one second. With the scaling of the frequency axis, the transform values

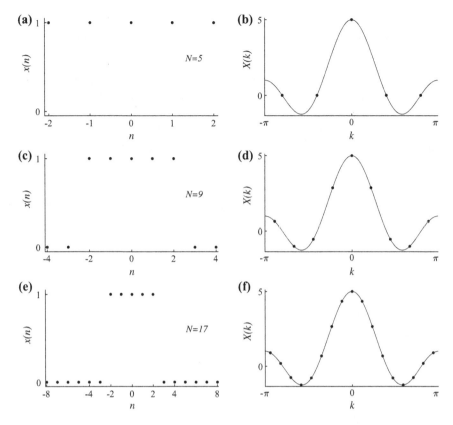

Fig. 8.2 **a** The time-domain signal $x(n)$ with $N = 5$ and **b** its DFT spectrum, $|X(k)|$; **c** the time-domain signal $x(n)$ with $N = 9$ and **d** its DFT spectrum, $|X(k)|$; **e** the time-domain signal $x(n)$ with $N = 17$ and **f** its DFT spectrum, $|X(k)|$. The DTFT spectrum is shown by a solid line

for any other values of T_s can be easily found. However, with $T_s \neq 1$, the DTFT and its inverse can also be redefined including T_s as

$$X(e^{j\omega T_s}) = \sum_{n=-\infty}^{\infty} x(nT_s)e^{-jn\omega T_s} \tag{8.8}$$

$$x(nT_s) = \frac{1}{\omega_s} \int_{-\frac{\omega_s}{2}}^{\frac{\omega_s}{2}} X(e^{j\omega T_s})e^{jn\omega T_s} \, d\omega, \ n = 0, \pm1, \pm2, \ldots, \tag{8.9}$$

where $\omega_s = \frac{2\pi}{T_s}$. The plot of the spectrum over the half period from $\omega = 0$ to $\omega = \pi/T_s$ is adequate for real-valued signals.

In the case of the DFT and the FS, the frequency of the harmonic components is discrete and, hence, the reconstruction of the waveforms is easy to visualize. However, in the case of the DTFT and FT, the synthesis process is not apparent, since all the frequency components in the range of frequencies contribute. The DTFT synthesizes a discrete aperiodic signal, $x(n)$, as integrals of a continuum of complex sinusoids $e^{j\omega n}$ (amplitude $\frac{1}{2\pi}X(e^{j\omega})d\omega$) over the finite frequency range $-\pi$ to π, which is one period of $X(e^{j\omega})$. As the amplitude of the constituent sinusoids of a signal is infinitesimal, the spectral density $X(e^{j\omega})$, which is proportional to the spectral amplitude, represents the frequency content of a signal. Hence, the DTFT of a signal is its relative amplitude spectrum.

As infinite summation is involved in finding the DTFT of a signal, its convergence has to be ensured. The convergence properties are similar to those of the FS. The summation converges uniformly, as in the case of FS, to $X(e^{j\omega})$, if $x(n)$ is absolutely summable. That is,

$$\sum_{n=-\infty}^{\infty} |x(n)| < \infty$$

The summation converges in the least squares error sense, if $x(n)$ is square summable. That is,

$$\sum_{n=-\infty}^{\infty} |x(n)|^2 < \infty$$

Since the spectrum is of finite duration, the power spectrum is integrable and the energy is finite. Gibbs phenomenon occurs in constructing the continuous spectrum, if $x(n)$ is not absolutely summable.

Example 8.2 Determine the DTFT of the unit impulse signal $x(n) = \delta(n)$.
Solution
From the DTFT definition, we get

$$X(e^{j\omega}) = \sum_{n=-\infty}^{\infty} \delta(n)e^{-j\omega n} = 1 \quad \text{and} \quad \delta(n) \leftrightarrow 1$$

Since $\delta(n)$ is zero except at $n = 0$ and the value of the exponential equals 1 at that point, the summation yields a value of unity for all values of ω. The transform of the impulse is a constant. Therefore, the unit impulse signal is composed of complex sinusoids of all the infinite frequencies from $\omega = -\pi$ to $\omega = \pi$ with equal strength.   ∎

Example 8.3 Find the DTFT spectrum of the signal

$$x(n) = u(n + 5) - u(n - 5),$$

where $u(n)$ is the unit-step function.
Solution
Signal $x(n)$ is shown in Fig. 8.3a. From the DTFT definition,

$$X(e^{j\omega}) = e^{j5\omega} + \cdots + e^{j2\omega} + e^{j\omega} + 1 + e^{-j\omega} + e^{-j2\omega} + \cdots + e^{-j5\omega}$$

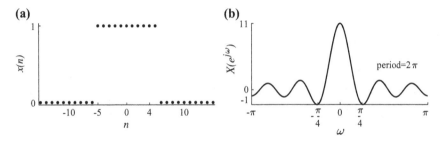

Fig. 8.3 **a** $x(n) = u(n+5) - u(n-5)$; **b** its DTFT spectrum

Using the formula for the geometric sum for complex sequences

$$\sum_{k=0}^{N} z^n = \frac{1 - z^{(N+1)}}{1 - z}, \quad z \neq 1$$

we get

$$X(e^{j\omega}) = \frac{1 - e^{-j6\omega}}{1 - e^{-j\omega}} + \frac{1 - e^{j6\omega}}{1 - e^{j\omega}} - 1$$

As the first two terms are complex conjugates, their sum is equal to twice the real part of either of the terms. Therefore,

$$X(e^{j\omega}) = \frac{2 \sin\left(\frac{6}{2}\omega\right) \cos\left(\frac{5}{2}\omega\right)}{\sin\left(\frac{\omega}{2}\right)} - 1 = \frac{\sin\left(\frac{11}{2}\omega\right) + \sin\left(\frac{\omega}{2}\right)}{\sin\left(\frac{\omega}{2}\right)} - 1 = \frac{\sin\left(\frac{11}{2}\omega\right)}{\sin\left(\frac{\omega}{2}\right)}, \quad -\pi < \omega < \pi$$

The DTFT spectrum for $x(n)$ is shown in Fig. 8.3b. In general,

$$\begin{cases} 1 \text{ for } -N \leq n \leq N \\ 0 \text{ otherwise} \end{cases} \leftrightarrow \frac{\sin\left(\omega \frac{(2N+1)}{2}\right)}{\sin\left(\frac{\omega}{2}\right)}$$

∎

The frequency-domain continuous function

$$\frac{\sin\left(\frac{11}{2}\omega\right)}{\sin\left(\frac{\omega}{2}\right)},$$

shown in Fig. 8.3b, as the DTFT of a pulse of width 11 samples, is called the sinc function, a function of great significance in signal and system analysis. The sinc function is even-symmetric. The peak value 11 occurs at $\omega = 0$, since

$$\lim_{\theta \to 0} \sin(\theta) = \theta$$

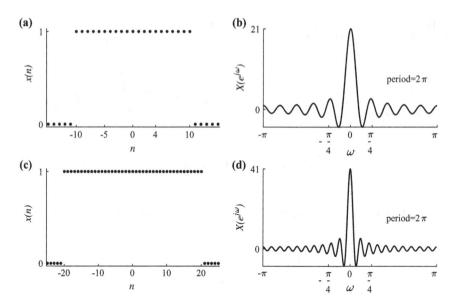

Fig. 8.4 **a** $x(n) = u(n + 10) - u(n - 10)$; **b** its DTFT spectrum; **c** $x(n) = u(n + 20) - u(n - 20)$; **d** its DTFT spectrum

Sinc$(\omega) = 0$ whenever the argument of the sine function in the numerator is $\pm\pi, \pm2\pi, \pm3\pi, \ldots$ For the example, the zeros occur at integral multiple of $2\pi/11$.

Figure 8.4b, d show the frequency-domain sinc functions corresponding to the time-domain pulses of increasing length shown in Fig. 8.4a, c, respectively. For any N, the area of the function remains the same at 2π, since

$$x(0) = 1 = \frac{1}{2\pi} \int_{-\pi}^{\pi} X(e^{j\omega}) d\omega,$$

With N increasing, the function becomes taller and slimmer and the area enclosed is concentrated around $\omega = 0$. As $N \to \infty$, the time-domain function becomes the DC function. The frequency-domain function

$$\frac{\sin\left(\frac{(2N+1)}{2}\omega\right)}{\sin\left(\frac{\omega}{2}\right)}$$

degenerates into an impulse function with strength 2π, as the zeros move to $\omega = 0$. That is, all the oscillations and the large hump coincide at $\omega = 0$ and the net area of the function occurs at that point, which characterizes an impulse. The sinc function is an energy signal, as it is square integrable. But it is not absolutely integrable. Therefore, the DTFT transform pair for the DC signal is

$$1 \leftrightarrow 2\pi\delta(\omega)$$

The DC signal is a complex exponential $x(n) = e^{j\omega n}$ with $\omega = 0$. Therefore, its nonzero spectral component occurs only at the single frequency $\omega = 0$.

Functions such as the DC, which are neither absolutely or square summable, are important in the analysis of signals and systems. Their frequency-domain representation is not defined by the Fourier analysis defining equations. Therefore, a frequency-domain representation is obtained through a limiting process, starting with a pair of defined transform pair such as the one in the last example. The impulse function is the limiting form of not only the sinc function but several other pulses including rectangular and triangular.

Example 8.4 Find the DTFT of the signal $x(n) = a^n u(n)$, $|a| < 1$.
Solution

$$X(e^{j\omega}) = \sum_{n=0}^{\infty} a^n u(n) e^{-j\omega n} = \sum_{n=0}^{\infty} (ae^{-j\omega})^n = \frac{1}{1 - ae^{-j\omega}}, \quad |a| < 1$$

The middle summation is a geometric progression with ratio $ae^{-j\omega} < 1$, since $|a| < 1$ and $|e^{-j\omega}| = 1$. Using the Euler's formula, we get

$$X(e^{j\omega}) = \frac{1}{1 - a\cos(\omega) + ja\sin(\omega)}$$

and, in terms of magnitude and phase, we get

$$|X(e^{j\omega})| = \frac{1}{\sqrt{1 + a^2 - 2a\cos(\omega)}} \quad \text{and} \quad \angle X(e^{j\omega}) = \tan^{-1}\left(\frac{a\sin(\omega)}{a\cos(\omega) - 1}\right)$$

∎

The exponential signals $0.5^n u(n)$ and $(-0.5)^n u(n)$ and their magnitude and phase spectra are shown in Fig. 8.5a–f. The magnitude spectra are even-symmetric and the phase spectra are odd-symmetric. Signal in (a) is smoother and, therefore, the magnitude of low frequency components is high. On the other hand, signal in (b) is more fluctuating and, therefore, the magnitude of high frequency components is high.

Let us derive the DTFT spectrum of *sgn* signal by considering this signal as sum of two exponentials in the limit as $a \to 1$. It is an everlasting signal. For positive values of its argument it is the same as $u(n)$, the unit-step function. Otherwise, its values are equal to -1. Figure 8.6a, b show the signal $x(n) = 0.8^n u(n) - 0.8^{-n} u(-n) + \delta(n)$ and its real and imaginary parts of the DTFT spectrum. Figure (c) and (d) show the signal $x(n) = 0.99^n u(n) - 0.99^{-n} u(-n) + \delta(n)$ and its real and imaginary parts of the DTFT spectrum. As the base of the exponentials approach 1, the sum of the exponentials approaches the *sgn* function. Consider the transform

$$X(e^{j\omega}) = \frac{1}{1 - ae^{-j\omega}} - \frac{1}{1 - ae^{j\omega}} + 1 = \frac{1 - 2ae^{j\omega} + a^2}{(1 - ae^{-j\omega})(1 - ae^{j\omega})}$$

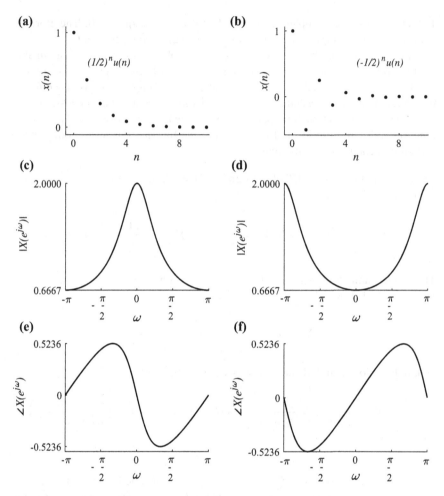

Fig. 8.5 **a** $x(n) = 0.5^n u(n)$; **b** $x(n) = (-0.5)^n u(n)$; **c** the magnitude of the DTFT spectrum of the signal in **a** and **e** its phase spectrum; **d** the magnitude of the DTFT spectrum of the signal in **b** and **f** its phase spectrum

In the limit as a tends to 1, we get

$$\lim_{a \to 1} X(e^{j\omega}) = \lim_{a \to 1} \frac{1 - 2ae^{j\omega} + a^2}{(1 - ae^{-j\omega})(1 - ae^{j\omega})} = \frac{2(1 - e^{j\omega})}{(1 - e^{-j\omega})(1 - e^{j\omega})} = \frac{2}{1 - e^{-j\omega}}$$

The inverse DTFT of this spectrum is the *sgn* function, called signum function, $sgn(n)$.

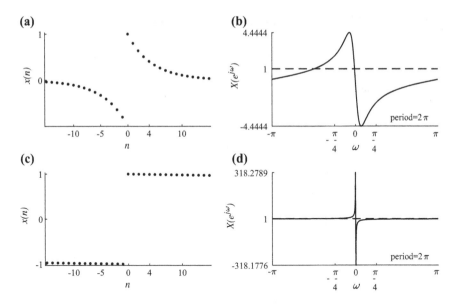

Fig. 8.6 **a** $x(n) = 0.8^n u(n) - 0.8^{-n} u(-n) + \delta(n)$ and **b** its real (dashed line) and imaginary parts of the DTFT spectrum; **c** $x(n) = 0.99^n u(n) - 0.99^{-n} u(-n) + \delta(n)$ and **d** its real and imaginary parts of the DTFT spectrum

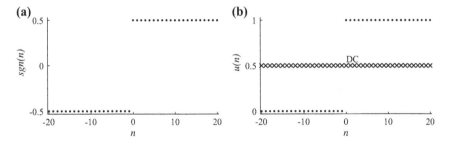

Fig. 8.7 **a** The sign function $0.5 sgn(n)$; **b** $u(n)$ and $x(n) = 0.5$

Figure 8.7a shows the *sgn* function multiplied by 0.5, $0.5 sgn(n)$. Figure 8.7b shows the unit-step function $u(n)$ and the DC function $x(n) = 0.5$. The DC function is the even component of $u(n)$ and $0.5 sgn(n)$ is the odd component of $u(n)$. Therefore,

$$u(n) = 0.5 + 0.5 sgn(n)$$

and the DTFT pair for $u(n)$, due to linearity property of the DTFT, is

$$u(n) \leftrightarrow \pi \delta(n) + \frac{1}{1 - e^{-j\omega}}$$

8.1.2   The DTFT of a Discrete Periodic Signal

The DFT of $x(n) = Ae^{k\omega_0 n}$ with N samples ($\omega_0 = \frac{2\pi}{N}$) is the discrete impulse with value $X(k) = NA$. That is, the reconstructed signal from IDFT is $\frac{X(k)}{N} e^{k\omega_0 n}$. The DTFT of the DC signal Ae^{j0n} is the continuous impulse $2A\pi\delta(\omega)$. The reconstructed signal is

$$\frac{1}{2\pi} \int_{-\pi}^{\pi} X(e^{j\omega})d\omega = \frac{1}{2\pi} \int_{-\pi}^{\pi} 2A\pi\delta(\omega)d\omega = Ae^{0n} = A$$

Therefore, the DTFT of $Ae^{j\omega_0 n}$ is $2\pi A\delta(\omega - \omega_0)$, $-\pi < \omega < \pi$. The DTFT spectrum is periodic with period 2π. Therefore, the DTFT of a periodic signal is a periodic train of impulses with strength $\frac{2\pi}{N} X(k)$ at $\frac{2\pi}{N} k$ with period 2π.

For example, consider the DFT pair

$$\sin\left(\frac{2\pi}{4}n\right) \leftrightarrow \{X(0) = 0, X(1) = -j2, X(2) = 0, X(3) = j2\}$$

with $N = 4$. The DTFT $X(e^{j\omega})$ for the same signal is

$$\left\{ X(e^{j0}) = 0, X\left(e^{j\frac{2\pi}{4}}\right) = -j\pi\delta\left(\omega - \frac{2\pi}{4}\right), \right.$$

$$\left. X\left(e^{j2\frac{2\pi}{4}}\right) = 0, X\left(e^{j3\frac{2\pi}{4}}\right) = j\pi\delta\left(\omega - 3\frac{2\pi}{4}\right) \right\}$$

8.1.3   Determination of the DFT from the DTFT

The DTFT and the DFT of a finite sequence $x(n)$ of length N are given as

$$X(e^{j\omega}) = \sum_{n=n_0}^{n_0+N-1} x(n)e^{-j\omega n} \quad \text{and} \quad X(k) = \sum_{n=n_0}^{n_0+N-1} x(n)e^{-j\frac{2\pi}{N} kn},$$

where n_0 is the starting point of $x(n)$. The DTFT is evaluated at all frequencies on the unit circle from $-\pi$ to π, whereas the DFT is evaluated at the set of discrete frequencies $2\pi k/N$. Therefore,

$$X(k) = X(e^{j\omega})|_{\omega = \frac{2\pi}{N} k} = X\left(e^{j\frac{2\pi}{N} k}\right)$$

Given the nonzero samples of a signal,

$$\{x(-2) = 1, x(-1) = -1, x(0) = 1, x(1) = -1\}$$

the DTFT of $x(n)$ is

$$X(e^{j\omega}) = e^{j2\omega} - e^{j\omega} + 1 - e^{-j\omega} = e^{j2\omega} - 2\cos(\omega) + 1$$

The set of samples of $X(e^{j\omega})$,

$$\{X(0) = 0, X(1) = 0, X(2) = 4, X(3) = 0\}$$

at $\omega = \frac{2\pi}{4}k$, $k = 0, 1, 2, 3$ is the DFT of $x(n)$.

8.2 Properties of the Discrete-Time Fourier Transform

It is often of interest to know the effect of an operation in the time domain corresponding to that in the frequency domain or vice versa. Properties help us to find the transforms with less effort. In addition, the DTFT of related signals can be found easily from those known.

8.2.1 Linearity

The DTFT of a linear combination of two or more signals is equal to the same linear combination of the DTFT of the individual signals. That is,

$$a(n) \leftrightarrow A(e^{j\omega}), \quad b(n) \leftrightarrow B(e^{j\omega}), \rightarrow pa(n) + qb(n) \leftrightarrow pA(e^{j\omega}) + qB(e^{j\omega}),$$

where p and q are arbitrary constants. As the DTFT is defined by the summation operation, which is linear, the DTFT has the linearity property. We used this property in deriving the DTFT of the unit-step signal from those of the sgn and DC signals.

8.2.2 Time Shifting

Shifting a signal $x(n)$ in the time domain by $\pm n_0$ sample intervals results in $x(n \pm n_0)$. The shifting does not affect the magnitude of the signal. But the phases of its constituent sinusoidal components are changed in proportion of their frequencies. By directly substituting $x(n \pm n_0)$ for $x(n)$ in the DTFT definition, we get

$$x(n \pm n_0) \leftrightarrow e^{\pm j\omega n_0} X(e^{j\omega})$$

Consider the transform pair

$$\cos\left(2\frac{2\pi}{8}n\right) \leftrightarrow \pi\left(\delta\left(\omega - 2\frac{2\pi}{8}\right) + \delta\left(\omega + 2\frac{2\pi}{8}\right)\right)$$

Then, due to this property, we get the transform pair

$$\cos\left(2\frac{2\pi}{8}(n-1)\right) \leftrightarrow \pi\left(e^{-j\frac{\pi}{2}}\delta\left(\omega - 2\frac{2\pi}{8}\right) + e^{j\frac{\pi}{2}}\delta\left(\omega + 2\frac{2\pi}{8}\right)\right)$$

$$= -j\pi\left(\delta\left(\omega - 2\frac{2\pi}{8}\right) - j\delta\left(\omega + 2\frac{2\pi}{8}\right)\right)$$

The shifted signal is $\sin(2\frac{2\pi}{8}n)$.

8.2.3 Frequency Shifting

Let the signal $x(n)$ with DTFT $X(e^{j\omega})$ be multiplied by the exponential $e^{\pm j\omega_0 n}$ to become $x(n)e^{\pm j\omega_0 n}$. Then, combining the two exponentials in the summand of the DTFT definition for this signal, we get $e^{-j(\omega \mp \omega_0)}$. The result is that the independent variable ω in $X(e^{j\omega})$ is replaced by $\omega \mp \omega_0$. The resulting DTFT is $X(e^{j(\omega \mp \omega_0)})$. That is,

$$x(n)e^{\pm j\omega_0 n} \leftrightarrow X(e^{j(\omega \mp \omega_0)})$$

The spectrum $X(e^{j\omega})$ gets shifted by ω_0.

The response, shown in Fig. 8.5c, is of lowpass type. The shift of this frequency response by π radians yields a highpass frequency response, shown in Fig. 8.5d. In the time domain, the corresponding signals are

$$\left(\frac{1}{2}\right)^n u(n) \quad \text{and} \quad e^{j\pi n}\left(\frac{1}{2}\right)^n u(n) = (-1)^n\left(\frac{1}{2}\right)^n u(n) = \left(-\frac{1}{2}\right)^n u(n)$$

The corresponding spectra are

$$\frac{1}{1 - (0.5)e^{-j\omega}} \quad \text{and} \quad \frac{1}{1 - (0.5)e^{-j(\omega+\pi)}} = \frac{1}{1 + (0.5)e^{-j\omega}}$$

8.2.4 Convolution in the Time Domain

$$\sum_{m=-\infty}^{\infty} x(m)h(n-m) = \frac{1}{2\pi}\int_{-\pi}^{\pi} X(e^{j\omega})H(e^{j\omega})e^{j\omega n}d\omega \leftrightarrow X(e^{j\omega})H(e^{j\omega})$$

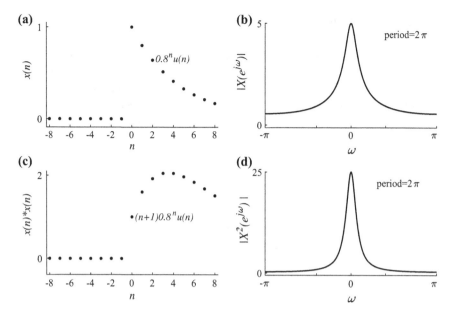

Fig. 8.8 **a** The exponential signal $0.8^n u(n)$ and **b** its spectrum. **c** The signal $0.8^n u(n) * 0.8^n u(n) = (n+1)0.8^n u(n)$ and **d** its spectrum

Consider convolution of the exponential signal $0.8^n u(n)$ with itself. The signal and its DTFT spectrum are shown in Fig. 8.8a, b, respectively. The DTFT of the signal and that of its convolution with itself are

$$\frac{1}{1 - 0.8e^{-j\omega}} \quad \text{and} \quad \frac{1}{(1 - 0.8e^{-j\omega})^2}$$

The signal $(n + 1)0.8^n u(n)$ and its spectrum are shown, respectively, in Fig. 8.8c, d.

8.2.5 Correlation

For implementation purposes, correlation of two signals is the convolution of the two signals with one of them time reversed. Consider the correlation of $\{\check{1}, 2\}$ and $\{\check{3}, 4\}$. The DTFT of the signals are $1 + 2e^{-j\omega}$ and $3 + 4e^{-j\omega}$. The conjugate of the DTFT of the second signal is $3 + 4e^{j\omega}$. The product of $3 + 4e^{j\omega}$ and $1 + 2e^{-j\omega}$ is $4e^{j\omega} + 11 + 6e^{-j\omega}$. The inverse DTFT of the product $\{4, \check{1}1, 6\}$ is the correlation of $\{\check{1}, 2\}$ and $\{\check{3}, 4\}$.

8.2.6 Convolution in the Frequency Domain

The transform of the product of two time-domain signals is the convolution of their individual transforms in the frequency domain with a scale factor. In the case of the FS, as the transform is discrete, the visualization of this operation is easy. For all purposes, the discrete spectrum is easier to interpret. When the transform is continuous, while we anticipate the result, the visualization is not so easy. Therefore, we have to use formal procedure in deriving the result.

Consider the DTFT representations of $x(n)$ and $y(n)$

$$x(n) = \frac{1}{2\pi} \int_{-\pi}^{\pi} X(e^{ju}) e^{jun} du \quad \text{and} \quad y(n) = \frac{1}{2\pi} \int_{-\pi}^{\pi} Y(e^{jv}) e^{jvn} dv$$

The DTFT of $x(n)y(n)$ is to be expressed in terms of those of $x(n)$ and $y(n)$. The DTFT representation of $x(n)y(n)$ is

$$x(n)y(n) = \frac{1}{2\pi} \int_{-\pi}^{\pi} \frac{1}{2\pi} \int_{-\pi}^{\pi} X(e^{ju}) Y(e^{jv}) e^{j(u+v)n} du \, dv$$

Letting $v = \omega - u$, we get $dv = d\omega$. Then,

$$x(n)y(n) = \frac{1}{2\pi} \int_{-\pi}^{\pi} \left(\frac{1}{2\pi} \int_{-\pi}^{\pi} X(e^{ju}) Y(e^{j(\omega-u)}) du \right) e^{j\omega n} d\omega$$

This is a DTFT representation of $x(n)y(n)$ with the DTFT

$$\left(\frac{1}{2\pi} \int_{-\pi}^{\pi} X(e^{ju}) Y(e^{j(\omega-u)}) du \right)$$

That is,

$$x(n)y(n) \leftrightarrow \sum_{n=-\infty}^{\infty} x(n)y(n) e^{-j\omega n} = \frac{1}{2\pi} \int_{-\pi}^{\pi} X(e^{ju}) Y(e^{j(\omega-u)}) du$$

As the DTFT spectrum is periodic, this convolution is periodic.

Consider the exponential signal $0.8^n u(n) 0.8^n u(n) = 0.64^n u(n)$. The signal and its DTFT spectrum are shown in Fig. 8.9a, b, respectively. The transform of the product of the signal, $0.8^n u(n)$, with itself, $x(n) = 0.8^n u(n) 0.8^n u(n) = 0.64^n u(n)$, is

$$X(e^{j\omega}) = \frac{1}{1 - 0.64 e^{-j\omega}}$$

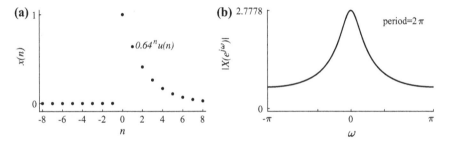

Fig. 8.9 **a** The exponential signal $0.8^n u(n)0.8^n u(n) = 0.64^n u(n)$ and **b** its spectrum

While it is easier to find the DTFT in this case, for an arbitrary signal, we have to use this property. For this example,

$$X(e^{j\omega}) = \frac{1}{2\pi} \int_0^{2\pi} \frac{1}{1 - 0.8e^{-ju}} \frac{1}{1 - 0.8e^{-j(w-u)}} du$$

The inverse DTFT of $X(e^{j\omega})$ is $x(n) = 0.64^n u(n)$. In practice, samples of the functions are obtained and the convolution is computed numerically.

8.2.7 Symmetry

There are only N independent values in a N-point real-valued sequence. Therefore, its complex-valued DTFT, with $-\pi \leq \omega < \pi$, must be redundant by a factor of 2. This happens due to the use of the complex exponential to represent a real sinusoid. The DTFT of a real-valued signal $x(n)$ is given by

$$X(e^{j\omega}) = \sum_{n=-\infty}^{\infty} x(n)e^{-j\omega n} = \sum_{n=-\infty}^{\infty} x(n)(\cos(\omega n) - j\sin(\omega n))$$

Conjugating both sides and replacing ω by $-\omega$, we get

$$X^*(e^{-j\omega}) = \sum_{n=-\infty}^{\infty} x(n)(\cos(\omega n) - j\sin(\omega n)) = X(e^{j\omega})$$

Therefore, $X^*(e^{-j\omega}) = X(e^{j\omega})$, called the conjugate symmetry. That is, if a signal is real, then the real part of its spectrum $X(e^{j\omega})$ is even-symmetric and the imaginary part is odd symmetric. Equivalently, the magnitude spectrum is even-symmetric and the phase spectrum is odd symmetric.

Given the nonzero samples of a signal,

$$\{x(-2) = 1, x(-1) = -1, x(0) = 1, x(1) = -1\}$$

the DTFT of $x(n)$ is

$$X(e^{j\omega}) = e^{j2\omega} - e^{j\omega} + 1 - e^{-j\omega} = \cos(2\omega) + j\sin(2\omega) - 2\cos(\omega) + 1$$

$$X^*(e^{j\omega}) = e^{-j2\omega} - e^{-j\omega} + 1 - e^{j\omega} = \cos(2\omega) - j\sin(2\omega) - 2\cos(\omega) + 1 = X(e^{-j\omega})$$

A real and even signal has a real and even-symmetric spectrum. Since $x(n)\cos(\omega n)$ is even and $x(n)\sin(\omega n)$ is odd, the imaginary part is zero. Therefore,

$$X(e^{j\omega}) = x(0) + 2\sum_{n=1}^{\infty} x(n)\cos(\omega n) \quad \text{and} \quad x(n) = \frac{1}{\pi}\int_0^{\pi} X(e^{j\omega})\cos(\omega n)d\omega$$

For example,

$$\{x(-1) = -1, x(0) = 1, x(1) = -1\} \leftrightarrow 1 - 2\cos(\omega)$$

A real and odd signal has an imaginary and odd symmetric spectrum. Since $x(n)\cos(\omega n)$ is odd and $x(n)\sin(\omega n)$ is even, the real part is zero. Therefore,

$$X(e^{j\omega}) = -j2\sum_{n=1}^{\infty} x(n)\sin(\omega n) \quad \text{and} \quad x(n) = \frac{j}{\pi}\int_0^{\pi} X(e^{j\omega})\sin(\omega n)d\omega$$

For example,

$$\{x(-1) = 1, x(0) = 0, x(1) = -1\} \leftrightarrow j2\sin(\omega)$$

8.2.8 Time-Reversal

Let both the variables ω and n are replaced by $-\omega$ and $-n$ in the DTFT definition

$$X(e^{j\omega}) = \sum_{n=-\infty}^{\infty} x(n)e^{-j\omega n}$$

Then, the exponent of e remains the same. But, $X(e^{j\omega})$ and $x(n)$ are replaced by $X(e^{-j\omega})$ and $x(-n)$. The change in the signs of the limits does not affect the summation. Therefore,

$$x(n) \leftrightarrow X(e^{j\omega}) \rightarrow x(-n) \leftrightarrow X(e^{-j\omega})$$

For example,

$$e^{j\omega_0 n} \leftrightarrow 2\pi\delta(\omega - \omega_0) \rightarrow e^{-j\omega_0 n} \leftrightarrow 2\pi\delta(-\omega - \omega_0) = 2\pi\delta(\omega + \omega_0)$$

8.2.9 Time-Expansion

Let

$$x(n) \leftrightarrow X(e^{j\omega})$$

Let us pad $x(n)$ with zeros to get $x_u(n)$ defined as

$$x_u(n) = x(n) \text{ if } \frac{n}{a} \text{ is an integer} \quad \text{and} \quad x_u(n) = 0 \quad \text{otherwise}$$

where $a \neq 0$ is any positive integer. Then,

$$X_u(e^{j\omega}) = \sum_{n=-\infty}^{\infty} x_u(n)e^{-j\omega n} = \sum_{n=-\infty}^{\infty} x_u(an)e^{-j\omega an} = \sum_{n=-\infty}^{\infty} x(n)e^{-j\omega an}$$

Therefore,

$$x_u(n) \leftrightarrow X(e^{ja\omega})$$

The spectrum of the expanded signal is a compressed version of that of the original. The spectral value at ω in the original spectrum occurs at ω/a in the spectrum of its expanded version. With a negative, the spectrum is also frequency-reversed, in addition.

For example, the DTFT of the signal $x(n)$ shown in Fig. 8.10a with dots, with its only nonzero values given as $x(-1) = 1$ and $x(1) = 1$, is $X(e^{j\omega}) = e^{j\omega} + e^{-j\omega} = 2\cos(\omega)$. Using the theorem, we get the DTFT of $x_u(n)$ with $a = 3$, shown in Fig. 8.10a with cross, as

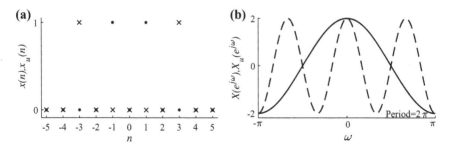

Fig. 8.10 **a** Signal $x(n)$ (dots) and its expanded version $x_u(n)$ (cross) with $a = 3$, and **b** the DTFT of $x(n)$ (solid line) and that of $x_u(n)$ (dashed line)

$$X_u(e^{j\omega}) = X(e^{j3\omega}) = 2\cos(3\omega)$$

This result is obvious from the DTFT definition. The DTFT of the signal (solid line) and that of its expanded version (dashed line) are shown in Fig. 8.10b. Since the signal is expanded by a factor of three, its spectrum is compressed by a factor of three. Since an expanded signal varies more slowly, the frequencies of its components are lowered, implying a compressed spectrum.

8.2.10   *Frequency-Differentiation*

The DTFT spectrum $X(e^{j\omega})$ of $x(n)$ can be differentiated with respect to ω, as long as the resulting functions have DTFT representations. By differentiating both sides of the DTFT defining equation, with respect to ω, we get

$$(-jn)x(n) \leftrightarrow \frac{dX(e^{j\omega})}{d\omega} \quad \text{or} \quad (n)x(n) \leftrightarrow (j)\frac{dX(e^{j\omega})}{d\omega}$$

In general, the kth derivative of $X(e^{j\omega})$ yields

$$(-jn)^k x(n) \leftrightarrow \frac{d^k X(e^{j\omega})}{d\omega^k} \quad \text{or} \quad (n)^k x(n) \leftrightarrow (j)^k \frac{d^k X(e^{j\omega})}{d\omega^k}$$

Consider the transform pair

$$\delta(n+1) + \delta(n-1) \leftrightarrow 2\cos(\omega)$$

Using the property, we get the transform pair

$$n(\delta(n+1) + \delta(n-1)) = (n\delta(n+1) + n\delta(n-1))$$
$$= (-\delta(n+1) + \delta(n-1)) \leftrightarrow -j2\sin(\omega)$$

8.2.11   *Summation*

Let

$$x(n) \leftrightarrow X(e^{j\omega})$$

Then,

$$\sum_{k=-\infty}^{n} x(k) \leftrightarrow \frac{X(e^{j\omega})}{(1 - e^{-j\omega})} + \pi X(e^{j0})\delta(\omega), \quad -\pi < \omega \leq \pi$$

As the unit-step signal is 1 for positive values of its argument and zero otherwise,

$$u(n - k) = \begin{cases} 1 \text{ for } k \le n \\ 0 \text{ for } k > n \end{cases}$$

The convolution of $x(n)$ and $u(n)$ is given by

$$x(n) * u(n) = \sum_{k=-\infty}^{\infty} x(k)u(n - k) = \sum_{k=-\infty}^{n} x(k)$$

Therefore, as convolution in the time domain becomes multiplication in the frequency domain,

$$x(n) * u(n) = \sum_{k=-\infty}^{n} x(k) \leftrightarrow \left(\frac{1}{(1 - e^{-j\omega})} + \pi\delta(\omega) \right) X(e^{j\omega}) = \frac{X(e^{j\omega})}{(1 - e^{-j\omega})} + \pi X(e^{j0})\delta(\omega)$$

The impulsive component represents the DC component of $x(n)$. It is understood that the resulting signal has a DTFT representation.

Since $u(n) \leftrightarrow 1$ and the unit-step function is a summation of the impulse, using this property, the DTFT of $u(n)$, over one period, is

$$u(n) = \sum_{k=-\infty}^{n} \delta(k) \leftrightarrow \frac{1}{(1 - e^{-j\omega})} + \pi\delta(\omega), \quad -\pi < \omega \le \pi$$

As an another example, consider the signal $x(n) = u(n) - u(n - 3)$, shown in Fig. 8.11a, and the resulting signal, shown in Fig. 8.11b, obtained by summing it. The DTFT of the given signal is, from the DTFT definition, $1 + e^{-j\omega} + e^{-j2\omega}$. Using the property, we get the DTFT of its summation as

$$\frac{1 + e^{-j\omega} + e^{-j2\omega}}{1 - e^{-j\omega}} + 3\pi\delta(\omega), \quad -\pi < \omega \le \pi$$

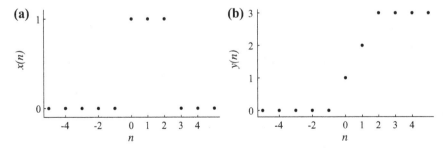

Fig. 8.11 **a** Signal $x(n) = u(n) - u(n - 3)$; **b** $y(n) = \sum_{k=-\infty}^{n} x(k)$

8.2.12   Parseval's Theorem and the Energy Transfer Function

The energy of a signal can also be expressed in terms of its spectrum, which is an equivalent representation.

$$E = \sum_{n=-\infty}^{\infty} |x(n)|^2 = \frac{1}{2\pi} \int_{0}^{2\pi} |X(e^{j\omega})|^2 d\omega$$

The energy of the signal $x(-1) = 1, x(1) = 1$ is 2. Its DTFT is $2\cos(\omega)$. From its DTFT, the energy is

$$E = \frac{1}{2\pi} \int_{0}^{2\pi} |2\cos(\omega)|^2 d\omega = \frac{1}{\pi} \int_{0}^{2\pi} (1 + \cos(2\omega)) d\omega = 2$$

In the frequency domain, the transfer function $H(e^{j\omega})$ relates the input and output of a LTI system as

$$Y(e^{j\omega}) = H(e^{j\omega})X(e^{j\omega})$$

where $X(e^{j\omega})$, $Y(e^{j\omega})$, and $H(e^{j\omega})$ are the DTFT of the input, output, and impulse response of the system. The output energy spectrum is given by

$$\begin{aligned}|Y(e^{j\omega})|^2 &= Y(e^{j\omega})Y^*(e^{j\omega}) \\ &= H(e^{j\omega})X(e^{j\omega})H^*(e^{j\omega})X^*(e^{j\omega}) = |H(e^{j\omega})|^2 |X(e^{j\omega})|^2\end{aligned}$$

As it relates the input and output energy spectral densities of the input and output of a system, $|H(e^{j\omega})|^2$ is called the energy transfer function. The quantity, such as $|X(e^{j\omega})|^2$, is the energy spectral density of the signal $x(n)$, since $\frac{1}{2\pi}|X(e^{j\omega})|^2 d\omega$ is the signal energy over the infinitesimal frequency band ω to $\omega + d\omega$.

8.3   Applications

Fourier analysis is an indispensable tool in the frequency-domain analysis of signals and systems. The analysis is carried out using sinusoidal or, equivalently, complex exponential signals. The basic principle of analysis is the same for all versions. Sinusoidal signals remain sinusoidal in any part of a linear system and this property enables fast implementation of operations. The differential equation and the convolution operation required in the time domain become algebraic operations in the frequency domain. Further, the interpretation of signals is easier using their spectra. The DTFT version of the Fourier analysis is most suitable to analyze discrete signals. Fourier analysis is applicable to all stable and linear systems to find the steady-state response.

8.3.1 Transfer Function and the System Response

The input–output relationship of a LTI system is given by the convolution operation in the time domain. It relates the input and output of a system through its impulse response. However, the convolution operation reduces to much simpler multiplication operation, when the input to a system is sinusoidal or complex exponential. That is,

$$y(n) = \sum_{m=-\infty}^{\infty} x(m)h(n-m) \leftrightarrow Y(e^{j\omega}) = X(e^{j\omega})H(e^{j\omega}),$$

where $x(n)$, $h(n)$, and $y(n)$ are, respectively, the system input, impulse response, and output, and $X(e^{j\omega})$, $H(e^{j\omega})$, and $Y(e^{j\omega})$ are their respective transforms. As multiplication of the input with $H(e^{j\omega})$ yields the output, $H(e^{j\omega})$ is called the transfer function of the system. The transfer function is the transform of the impulse response. It characterizes a system in the frequency domain just as the impulse response does in the time domain.

The spectrum of the impulse function is a constant. It is composed of complex exponentials, $e^{j\omega n}$, of all frequencies from $\omega = -\pi$ to $\omega = \pi$ with equal magnitude and zero phase. Therefore, the transform of the impulse response, the transfer function, is also called the frequency response of the system. Consequently, an exponential $Ae^{j(\omega_a n + \theta)}$ or a real sinusoidal input signal $A\cos(\omega_a n + \theta)$ is changed to, respectively, $(|H(e^{j\omega_a})|A)e^{j(\omega_a n + (\theta + \angle(H(e^{j\omega_a}))))}$ or $(|H(e^{j\omega_a})|A)\cos(\omega_a n + (\theta + \angle(H(e^{j\omega_a}))))$ at the output. The steady-state response of a stable system to the causal input $Ae^{j(\omega_a n + \theta)}u(n)$ is also the same. Since,

$$H(e^{j\omega}) = \frac{Y(e^{j\omega})}{X(e^{j\omega})},$$

the transfer function can also be described as the ratio of the transform $Y(e^{j\omega})$ of the output $y(n)$ to that of the input $x(n)$, $X(e^{j\omega})$. It is assumed that $|X(e^{j\omega})| \neq 0$ for all frequencies of interest.

Using the time-shift property of the DTFT, a difference equation with constant coefficients can be reduced to an algebraic equation that can be solved for $Y(e^{j\omega})$. Then, the inverse DTFT of $Y(e^{j\omega})$ yields the system output $y(n)$. Consider the difference equation of a causal LTI discrete system

$$y(n) + a_{K-1}y(n-1) + a_{K-2}y(n-2) + \cdots + a_0 y(n-K) =$$
$$b_M x(n) + b_{M-1}x(n-1) + \cdots + b_0 x(n-M)$$

Taking the DTFT of both sides, we get, assuming initial conditions are all zero,

$$Y(e^{j\omega})(1 + a_{K-1}e^{-j\omega} + a_{K-2}e^{-j2\omega} + \cdots + a_0 e^{-jK\omega}) =$$
$$X(e^{j\omega})(b_M + b_{M-1}e^{-j\omega} + \cdots + b_0 e^{-jM\omega})$$

Solving for $H(e^{j\omega})$, we get

$$H(e^{j\omega}) = \frac{Y(e^{j\omega})}{X(e^{j\omega})} = \frac{b_M + b_{M-1}e^{-j\omega} + \cdots + b_0 e^{-jM\omega}}{1 + a_{K-1}e^{-j\omega} + a_{K-2}e^{-j2\omega} + \cdots + a_0 e^{-jK\omega}}$$

Example 8.6 Find the response, using the DTFT, of the system governed by the difference equation

$$y(n) = x(n) + 0.8y(n-1)$$

to the input $x(n) = \cos(\frac{2\pi}{8}n + \frac{\pi}{3})$.
Solution

$$H(e^{j\omega}) = \frac{e^{j\omega}}{e^{j\omega} - 0.8}$$

Substituting $\omega = \frac{2\pi}{8}$, we get

$$H\left(e^{j\frac{2\pi}{8}}\right) = \frac{e^{j\frac{2\pi}{8}}}{e^{j\frac{2\pi}{8}} - 0.8} = 1.4022\angle(-0.9160)$$

The response of the system to the input $x(n) = \cos(\frac{2\pi}{8}n + \frac{\pi}{3})$ is $y(n) = 1.4022$ $\cos(\frac{2\pi}{8}n + \frac{\pi}{3} - 0.9160)$.   ∎

Example 8.7 Find the impulse response $h(n)$, using the DTFT, of the system governed by the difference equation

$$y(n) = x(n) - 2x(n-1) + x(n-2) + \frac{5}{6}y(n-1) - \frac{1}{6}y(n-2)$$

Solution

$$H(e^{j\omega}) = \frac{1 - 2e^{-j\omega} + e^{-j2\omega}}{\left(1 - \frac{5}{6}e^{-j\omega} + \frac{1}{6}e^{-j2\omega}\right)} = \frac{1 - 2e^{-j\omega} + e^{-j2\omega}}{\left(1 - \frac{1}{2}e^{-j\omega}\right)\left(1 - \frac{1}{3}e^{-j\omega}\right)}$$

Expanding into partial fractions, we get

$$H(e^{j\omega}) = 6 + \frac{3}{\left(1 - \frac{1}{2}e^{-j\omega}\right)} - \frac{8}{\left(1 - \frac{1}{3}e^{-j\omega}\right)}$$

Taking the inverse DTFT, we get the impulse response as

$$h(n) = 6\delta(n) + \left(3\left(\frac{1}{2}\right)^n - 8\left(\frac{1}{3}\right)^n\right)u(n)$$

The first four values of the impulse response $h(n)$ are

$$h(0) = 1, h(1) = -1.1667, \quad h(2) = -0.1389, \quad h(3) = 0.0787 \qquad \blacksquare$$

Example 8.8 Find the zero-state response, using the DTFT, of the system governed by the difference equation

$$y(n) = 2x(n) + 0.5y(n-1)$$

with the input $x(n) = 0.6^n u(n)$.
Solution

$$H(e^{j\omega}) = \frac{2}{(1 - 0.5e^{-j\omega})} \quad \text{and} \quad X(e^{j\omega}) = \frac{1}{(1 - 0.6e^{-j\omega})}$$

$$Y(e^{j\omega}) = H(e^{j\omega})X(e^{j\omega}) = \frac{2}{(1 - 0.6e^{-j\omega})(1 - 0.5e^{-j\omega})}$$

Expanding into partial fractions, we get

$$Y(e^{j\omega}) = \frac{12}{(1 - 0.6e^{-j\omega})} - \frac{10}{(1 - 0.5e^{-j\omega})}$$

Taking the inverse DTFT, we get the zero-state response.

$$y(n) = \left(12(0.6)^n - 10(0.5)^n\right) u(n)$$

The first four values of $y(n)$ are

$$y(0) = 2, \quad y(1) = 2.2, \quad y(2) = 1.82, \quad y(3) = 1.342 \qquad \blacksquare$$

8.3.2 Design of Linear-Phase FIR Digital Filters Using Windows

Filters are required in signal processing systems to suppress or modify some part of the spectrum of the input signal in a desired way. As a linear system, the filter is characterized by its impulse response in the time domain and frequency response in the transform domain. Usually, the filter is specified in terms of the required frequency response and the task is to determine the impulse response. One of the simpler methods of FIR filter design is called the window method. In this method, the impulse response of the desired filter is found by finding the inverse DTFT of the specified frequency response.

The impulse response is infinite duration and, therefore, it has to be truncated to make the impulse response stable and realizable. The truncation of the response corresponds to multiplying it by a rectangular window in the time domain. In the frequency domain, it corresponds to the convolution of their spectra. The rectangular window has discontinuities and, consequently, its frequency response has large oscillations in both the passband and the stopband. While the transition band is shorter, the oscillations are undesirable. A better frequency response can be obtained, at the cost of a longer transition band, by using tapered windows to truncate the impulse response. A variety of windows is available with different characteristics and a suitable one has to be chosen for a given requirements. Further, the truncated impulse response has to be shifted to make it causal.

8.3.3 Digital Differentiator

Digital differentiators are used in digital systems to approximate the derivative of a signal. The ideal frequency response of a differentiator must increase linearly proportional to the frequency of the component of the input signal. As a device, it is characterized by its impulse response in the time domain and its frequency response in the frequency domain. The periodic frequency response, shown in Fig. 8.12a over one period, of the ideal digital differentiator is defined as

$$H(e^{j\omega}) = j\omega, \quad -\pi < \omega < \pi$$

For example, the input and the output of the differentiator are

$$\cos(\omega_0 n) \leftrightarrow \pi(\delta(\omega + \omega_0) + \delta(\omega - \omega_0))$$
$$j\pi(-\omega_0\delta(\omega + \omega_0) + \omega_0\delta(\omega - \omega_0)) \leftrightarrow -\omega_0 \sin(\omega_0 n)$$

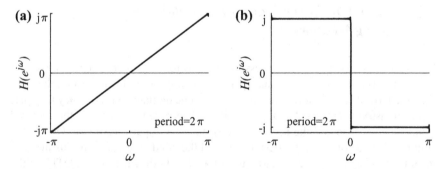

Fig. 8.12 a The frequency response of the ideal digital differentiator. **b** The frequency response of the ideal Hilbert transformer

The inverse DTFT of its frequency response is its impulse response.

$$h(n) = \frac{1}{2\pi} \int_{-\pi}^{\pi} j\omega e^{j\omega n} d\omega = \frac{\cos(\pi n)}{n} = \begin{cases} \frac{(-1)^n}{n} & \text{for } n \neq 0 \\ 0 & \text{for } n = 0 \end{cases}, \quad -\infty < n < \infty$$

As the frequency response of the differentiator is imaginary and odd-symmetric, the impulse response is real and odd-symmetric. In practice, the differentiators are designed so that their frequency responses are linear over the frequency range of interest, called the bandwidth of the differentiator.

8.3.4 Hilbert Transform

The DFT version of the Hilbert transform is presented in Chap. 5. We just present its DTFT periodic frequency response, shown in Fig. 8.12b over one period, of the ideal Hilbert transformer defined as

$$H(e^{j\omega}) = \begin{cases} -j & \text{for } 0 < \omega < \pi \\ j & \text{for } -\pi < \omega < 0 \end{cases}$$

The impulse response of the ideal Hilbert transformer is obtained by finding the inverse DTFT of its frequency response.

$$\begin{aligned} h(n) &= \frac{1}{2\pi} \int_{0}^{\pi} -je^{j\omega n} d\omega + \frac{1}{2\pi} \int_{-\pi}^{0} je^{j\omega n} d\omega \\ &= \begin{cases} \frac{2\sin^2\left(\frac{\pi n}{2}\right)}{\pi n} & \text{for } n \neq 0 \\ 0 & \text{for } n = 0 \end{cases}, \quad -\infty < n < \infty \end{aligned}$$

8.3.5 Downsampling

DTFT and DFT versions of the Fourier analysis are mostly used in multirate digital signal processing applications. Downsampling and upsampling are the two basic operations in multirate digital signal processing. Upsampling is presented in the DTFT properties section. Consider the signal $x(n)$. Retaining one out of every D samples, starting from $n = 0$, and discarding the rest is called the downsampling of a signal by a factor of D, a positive integer. The nth sample of the downsampled signal $x_d(n)$ is the (Dn)th sample of $x(n)$

$$x_d(n) = x(Dn)$$

We present the downsampling operation in the frequency domain. Let $x(n) \leftrightarrow X(e^{j\omega})$. The problem is to express the DTFT $X_d(e^{j\omega})$ of the downsampled signal $x_d(n)$ in terms of $X(e^{j\omega})$. Let $D = 2$. The approach is to find the alternating version $(-1)^n x(n)$ of $x(n)$ and express the DTFT $X_d(e^{j\omega})$ as a combination of those of $x(n)$ and $(-1)^n x(n)$. Using the frequency-shift theorem,

$$e^{-j\pi n}x(n) = (-1)^n x(n) \leftrightarrow X(e^{j(\omega+\pi)})$$

The DTFT pair for $x(n)$ with zero-valued odd-indexed samples is

$$\frac{1}{2}\left(x(n) + (-1)^n x(n)\right) \leftrightarrow \frac{1}{2}\left(X(e^{j\omega}) + X(e^{j(\omega+\pi)})\right) \qquad (8.10)$$

For example,

$$x(n) = \{x(0) = 2, x(1) = 1, x(2) = 3, x(3) = 3\}$$
$$\leftrightarrow X(e^{j\omega}) = 2 + e^{-j\omega} + 3e^{-j2\omega} + 3e^{-j3\omega}$$
$$(-1)^n x(n) = \{x(0) = 2, x(1) = -1, x(2) = 3, x(3) = -3\}$$
$$\leftrightarrow X(e^{j(\omega+\pi)}) = 2 - e^{-j\omega} + 3e^{-j2\omega} - 3e^{-j3\omega}$$

Now,

$$\frac{1}{2}\left(x(n) + (-1)^n x(n)\right) = \{x(0) = 2, x(1) = 0, x(2) = 3, x(3) = 0\}$$
$$\leftrightarrow \frac{1}{2}\left(X(e^{j\omega}) + X(e^{j(\omega+\pi)})\right) = 2 + 3e^{-j2\omega}$$

The term $3e^{-j2\omega}$ is periodic with period π $(3e^{-j2(\omega+\pi)} = 3e^{-j2\pi}e^{-j2\omega} = 3e^{-j2\omega})$ and the constant term 2 is periodic with any period. Therefore, the spectrum is periodic with period π. Now, if the take the spectral values $\{5, -1\}$ over one period and compute the IDFT, we get $\{2, 3\}$, which is the downsampled version of $x(n)$. The spectrum of $\{2, 3\}$, by definition, is $2 + 3e^{-j\omega}$, which can also be obtained from $2 + 3e^{-j2\omega}$ by replacing ω by $\omega/2$. Due to upsampling, if the DTFT of $\{a, b\}$ is $Y(e^{j\omega})$ then the DTFT of the upsampled signal $\{a, 0, b, 0\}$ is the concatenation and compression of $Y(e^{j\omega})$ and $Y(e^{j\omega})$ in the range $0 < \omega < 2\pi$, $Y(e^{j2\omega})$. The spectrum is replicated. Therefore, by replacing ω by $\omega/2$ in Eq. (8.10), we take the values of the spectrum of the upsampled signal in the frequency range from 0 to π only and expand it in forming the spectrum $X_d(e^{j\omega})$ of the downsampled signal

$$x_d(n) = x(2n) \leftrightarrow X_d(e^{j\omega}) = \frac{1}{2}\left(X\left(e^{j\frac{\omega}{2}}\right) + X\left(e^{j\left(\frac{\omega}{2}+\pi\right)}\right)\right), \quad 0 < \omega < 2\pi \quad (8.11)$$

8.4 Approximation of the Discrete-Time Fourier Transform

In learning and understanding Fourier analysis or other transforms, we find the transform of well-defined signals, such as a pulse, an exponential, a triangular etc., analytically. However, practical signals usually have arbitrary amplitude profile and they have to be approximated by a set of discrete samples enabling the use of software or digital hardware in their analysis.

Both the DFT and the DTFT spectra are periodic with period 2π. The DTFT spectrum is continuous, while the DFT spectrum is composed of uniform samples of the DTFT spectrum. Sampling the spectrum implies periodicity of the time-domain signal. That is, if all the time-domain samples of a signal are contained in a period, the samples can be obtained exactly from the samples of the DTFT spectrum by using the DFT. Otherwise, time-domain aliasing occurs.

Let

$$x(n) = \{x(0) = 2, x(1) = 1, x(2) = 3, x(3) = 4\}$$

The DTFT of $x(n)$ is

$$X(e^{j\omega}) = 2 + e^{-j\omega} + 3e^{-j2\omega} + 4e^{-j3\omega}$$

The samples of the spectrum at $\omega = 0, \pi/2, \pi, 3\pi/2$ are

$$\{10, -1 + j3, 0, -1 - j3\}$$

The 4-point DFT of $x(n)$ is the same. The IDFT of these samples yields $x(n)$. With just 2 spectral samples $\{10, 0\}$, the IDFT yields $\{5, 5\}$. Time-domain aliasing has occurred. In practice, aliasing in unavoidable due to the finite and infinite natures, respectively, of the DFT and the DTFT in the time domain. With 5 spectral samples

$$\{10, -3.3541 - j0.3633, 3.3541 - j1.5388, 3.3541 + j1.5388, -3.3541 + j0.3633\},$$

the IDFT yields

$$\{2, 1, 3, 4, 0\}$$

The conclusion is that exact time-domain samples can be obtained from the samples of the DTFT spectrum with time-limited signals. Otherwise, it has to be ensured that time-domain aliasing is negligible.

In approximating the samples of the DTFT spectrum by using the DFT, the problem is data truncation. The input time-domain samples are usually too long and truncation is required to limit the size of the data to limit the computational complexity and memory requirements of the DFT. The data has to be truncated so that sufficient energy of the signal is retained in the truncated signal. Another point is that the samples may be specified over any time-domain range, while DFT algorithms usually require them to be specified from $n = 0$ to $n = N - 1$ for a N-point signal. This can be achieved due to the assumed periodicity of the DFT. Further, zero padding may be required to make the length equal to an integral power of 2.

Let

$$x(n) = \{x(-1) = -2, x(0) = 1, x(1) = 3\}$$

The DTFT of $x(n)$ is

$$X(e^{j\omega}) = -2e^{j\omega} + 1 + 3e^{-j\omega}$$

Using the assumed periodicity of the DFT, we get

$$xp(n) = \{x(0) = 1, x(1) = 3, x(2) = -2\}$$

Let us truncate the signal $xp(n)$.

$$xp\_t(n) = \{x(0) = 1, x(1) = 3, x(2) = 0\}$$

The DFT of $xp\_t(n)$ is

$$\{4, -0.5 - j2.5981, -0.5 + j2.5981\}$$

This is the DFT of the circular convolution of a rectangular window and that of $xp(n)$ divided by 3. The DFT of $xp(n)$ is

$$\{2, 0.5 - j4.3301, 0.5 + j4.3301\}$$

The DFT of the window $w(n) = \{1, 1, 0\}$ is

$$\{2, 0.5 - j0.8660, 0.5 + j0.8660\}$$

The linear convolution of the two spectra, divided by 3, is

$$\{1.3333, 0.6667 - j3.4641, -0.5 + j2.5981, 2.6667, -1.1667 + j0.8660\}$$

We want the circular convolution output, which can be obtained by adding the last two terms with the first two terms.

8.5 Summary

- The DTFT version of the Fourier analysis is used to analyze aperiodic discrete signals. The corresponding spectrum is continuous and periodic and it is a relative amplitude spectrum.

- The DTFT is the dual of the FS with the roles of the time- and frequency domains interchanged. The aperiodic discrete sequence is the FS of the continuous periodic spectrum in the frequency domain.
- The DTFT is the limiting case of the DFT as the period of the time-domain sequence tends to infinity with the sampling interval fixed.
- As the DTFT represents sampled waveforms, its spectrum is periodic. Periodicity of the spectrum implies that the effective range of the frequencies in the spectrum is finite.
- As the DTFT is defined by an infinite sum, the infinite samples is absolutely or square summable is a sufficient condition for its existence.
- As with the other versions of the Fourier analysis, the DTFT is also effectively approximated by the DFT in practical applications.
- The DTFT is widely used in the analysis of discrete signals and stable LTI systems. The DTFT of the impulse response of a system is its frequency response.

Exercises

8.1 Find the DTFT of $\sin\left(\frac{2\pi}{8}n + \frac{\pi}{6}\right)$.

* 8.2 Find the inverse DTFT of $j2\sin(2\omega)$.

8.3 Given the nonzero samples of a signal,

$$\{x(-2) = 2, x(-1) = -1, x(0) = 3, x(1) = 4\}$$

find its DFT from the DTFT.

* 8.4 Find the convolution of $x(n) = \{3, 1, 2\}$ and $h(n) = \{1, 2, 4\}$ using the DTFT.

8.5 Find the correlation of $x(n) = \{1, 2, 3\}$ and $h(n) = \{2, -1, 4\}$ using the DTFT.

* 8.6 Find the response, using the DTFT, of the system governed by the difference equation
$$y(n) = x(n) - 0.7y(n-1)$$

to the input $x(n) = \sin(\frac{2\pi}{8}n - \frac{\pi}{6})$.

8.7 Find the impulse response $h(n)$, using the DTFT, of the system governed by the difference equation

$$y(n) = x(n) + 3x(n-1) + 2x(n-2) + y(n-1) - \frac{2}{9}y(n-2)$$

* 8.8 Find the zero-state response, using the DTFT, of the system governed by the difference equation
$$y(n) = x(n) - 0.6y(n-1)$$

with the input $x(n) = (-0.8)^n u(n)$.

8.9 Find the DTFT $X_d(e^{j\omega})$ of the downsampled version $x_d(n)$ of $x(n)$, by a factor of 2, in terms of its DTFT $X(e^{j\omega})$.

8.9.1

$$x(n) = (0.8)^n u(n)$$

8.9.2

$$x(n) = \cos\left(\frac{2\pi}{8}n\right)$$

8.9.3

$$x(n) = \begin{cases} 1 \text{ for } -3 \le n \le 3 \\ 0 \text{ otherwise} \end{cases}$$

8.10 Given 4 samples of $x(n)$, find its DTFT $X(e^{j\omega})$ analytically. Find the DFT of one period $xp(n)$, $n = 0, 1, 2, 3$ obtained by periodically extending $x(n)$. Verify that the uniform samples of $X(e^{j\omega})$ and the DFT are the same. Truncate $xp(n)$ by a window $\{w(0) = 1, w(1) = 1, w(2) = 1, w(3) = 0\}$ to get $xp\_t(n)$. Find the DFT of $xp\_t(n)$ and justify it in terms of the DFT of $w(n)$ and $xp(n)$.

* 8.10.1

$$\{x(-2) = 2, x(-1) = 1, x(0) = 3, x(1) = 4\}$$

8.10.2

$$\{x(-1) = 1, x(0) = 1, x(1) = 3, x(2) = 3\}$$

8.10.3

$$\{x(-3) = 2, x(-2) = 4, x(-1) = 3, x(0) = 1\}$$

8.11 Given 4 samples of $x(n)$, find its DTFT $X(e^{j\omega})$ analytically. Find the 4 uniform samples of $X(e^{j\omega})$ and compute the IDFT and verify that we get back $x(n)$. Find the 3 and 5 uniform samples of $X(e^{j\omega})$ and compute the respective IDFT and justify the result in terms of $x(n)$.

8.11.1

$$\{x(0) = 3, x(1) = 1, x(2) = 3, x(3) = 4\}$$

8.11.2

$$\{x(0) = 1, x(1) = 2, x(2) = 3, x(3) = 4\}$$

8.11.3

$$\{x(0) = 3, x(1) = 1, x(2) = 2, x(3) = 4\}$$

Chapter 9
The Fourier Transform

The Fourier transform (FT) is the most general version of the Fourier analysis. It is primarily used for the representation of continuous aperiodic signals with continuous aperiodic spectra. In addition, it is the tool to analyze mixed class of signals as it can represent signals represented by other versions of the Fourier analysis. It can be considered as an extension of the DTFT. As the sampling interval tends to zero, the time-domain signal becomes continuous and the continuous periodic spectrum becomes continuous and aperiodic. It can also be considered as an extension of the FS as the period of the periodic signal tends infinity. The time-domain signal corresponding to a discrete spectrum is periodic. On the other hand, the time-domain signal corresponding to a continuous spectrum is aperiodic.

9.1 The FT as a Limiting Case of the FS

Consider the FS synthesis and analysis expressions derived in Chap. 7 for a continuous periodic signal $x(t)$ with period T.

$$x(t) = \sum_{k=-\infty}^{\infty} X_{fs}(k)e^{j\omega_0 tk} \tag{9.1}$$

and

$$X_{fs}(k) = \frac{1}{T}\int_{-T/2}^{T/2} x(t)e^{-j\omega_0 tk}\,dt = \frac{\omega_0}{2\pi}\int_{-T/2}^{T/2} x(t)e^{-j\omega_0 tk}\,dt \tag{9.2}$$

© Springer Nature Singapore Pte Ltd. 2018
D. Sundararajan, *Fourier Analysis—A Signal Processing Approach*,
https://doi.org/10.1007/978-981-13-1693-7_9

As the period $T \to \infty$, the fundamental frequency $\omega_0 \to 0$ and the coefficients $X_{fs}(k) \to 0$. However, the ratio $X_{fs}(k)/\omega_0 = TX_{fs}(k)/(2\pi)$ approaches a finite limiting function. Therefore, Eqs. (9.1) and (9.2) are expressed in the form

$$x(t) = \frac{1}{2\pi} \sum_{k\omega_0=-\infty}^{\infty} \left(TX_{fs}(k)\right) e^{j\omega_0 tk} \delta(k\omega_0) \tag{9.3}$$

and

$$TX_{fs}(k) = \int_{-T/2}^{T/2} x(t) e^{-j\omega_0 tk} dt \tag{9.4}$$

The frequency increment, which is the fundamental frequency ω_0, is denoted by $\delta(k\omega_0)$. In the limit,

$$T \to \infty, \quad \omega_0 \to 0, \quad k \to \infty, \quad k\omega_0 \to \omega, \quad \delta(k\omega_0) \to d\omega, \quad TX_{fs}(k) \to X(j\omega)$$

At the end of the limit process, we get the FT and IFT expressions. The FT $X(j\omega)$ of $x(t)$ is defined as

$$X(j\omega) = \int_{-\infty}^{\infty} x(t) e^{-j\omega t} dt \tag{9.5}$$

The inverse FT $x(t)$ of $X(j\omega)$ is defined as

$$x(t) = \frac{1}{2\pi} \int_{-\infty}^{\infty} X(j\omega) e^{j\omega t} d\omega \tag{9.6}$$

Figure 9.1a, b show a square pulse with period 2π and its FS spectrum with the fundamental frequency $\omega_0 = 1$ rad. The corresponding scaled FT is also shown in a solid line. Figure 9.1c, d show the square pulse with period 4π and its scaled FS spectrum with $\omega_0 = 0.5$ rad. As the period is increased, the fundamental frequency is decreased resulting in a denser spectrum. The ratio $X_{fs}(k)/\omega_0$, in the limit, approaches a finite limiting function. The period eventually approaches ∞ and the fundamental frequency approaches zero. The result is that both the signal and its spectrum become continuous and aperiodic. Figure 9.1e, f show the square pulse with period 8π and its scaled FS spectrum with $\omega_0 = 0.25$ rad.

The limiting process can be thought of doubling and redoubling of the period T. The order k of a specific frequency component gets doubled with the doubling of the period. As the period is doubled, the frequency gets divided by 2. Therefore, $k\omega_0$ remains constant. The product $TX_{fs}(k)$ becomes a finite function $X(j\omega)$. For example,

$X_{fs}(1) = 0.3183$ with $\omega_0 = 1$, $X_{fs}(1)/1 = 1X_{fs}(1) = 0.3183$ and $k\omega_0 = \omega = 1$.
$X_{fs}(2) = 0.1592$ with $\omega_0 = 0.5$, $X_{fs}(2)/0.5 = 2X_{fs}(2) = 0.3183$ and $k\omega_0 = \omega = 1$.
$X_{fs}(4) = 0.0796$ with $\omega_0 = 0.25$, $X_{fs}(4)/0.25 = 4X_{fs}(4) = 0.3183$ and $k\omega_0 = \omega = 1$.

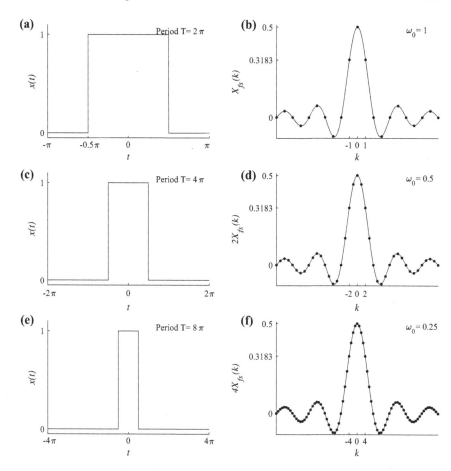

Fig. 9.1 **a** A periodic square pulse with period 2π and **b** its FS spectrum $X_{fs}(k)$ with $\omega_0 = 1$ rad; **c** the pulse with period 4π and **d** its scaled FS spectrum $2X_{fs}(k)$ with $\omega_0 = 0.5$ rad; **e** the pulse in **a** with period 8π and **f** its scaled FS spectrum $4X_{fs}(k)$ with $\omega_0 = 0.25$ rad; The corresponding scaled FT is shown in a solid line

The amplitude of each frequency component is $X(j\omega)d\omega/(2\pi)$, which is infinitesimal. However, $X(j\omega)$ is proportional to $X(j\omega)d\omega/(2\pi)$. As $X(j\omega)$ is finite, the plot of $X(j\omega)$ versus ω constitutes the FT spectrum. The FT spectrum is a relative amplitude spectrum showing the relative variations of the harmonic amplitudes versus frequency.

9.1.1   The FT Using Orthogonality

In Chap. 2, we expressed a discrete periodic signal as a sum of complex exponentials
in Eq. (2.1). Then, we used the orthogonality property of complex exponentials to
find the coefficients of the constituent frequency components of the signal. It is
straightforward to apply this procedure to derive the FS. The same procedure applies
to the DTFT, since it is the dual of the FS. It is interesting to find how the orthogonality
property can be used to derive the FT.

As both the signal and the spectrum are continuous and of infinite extent, the
signal is reconstructed as the integral of continuous-time complex exponentials of all
frequencies with their associated coefficients, rather than a sum of finite exponentials
in the case of DFT. The aperiodic continuous signal is synthesized as

$$x(t) = \frac{1}{2\pi} \int_{-\infty}^{\infty} X(j\omega) e^{j\omega t} d\omega$$

Now, the same procedure to determine $X(k)$ in DFT is used to determine $X(j\omega)$.
Multiplying both sides by $e^{-j\omega_0 t}$ and integrating over the infinite time domain, we
get

$$\int_{-\infty}^{\infty} x(t) e^{-j\omega_0 t} dt = \int_{-\infty}^{\infty} \frac{1}{2\pi} \int_{-\infty}^{\infty} X(j\omega) e^{j(\omega-\omega_0)t} d\omega \, dt$$

$$= \frac{1}{2\pi} \int_{-\infty}^{\infty} X(j\omega) d\omega \int_{-\infty}^{\infty} e^{j(\omega-\omega_0)t} dt$$

$$= \frac{1}{2\pi} \int_{-\infty}^{\infty} X(j\omega) d\omega \left(\frac{e^{j(\omega-\omega_0)t}}{j(\omega-\omega_0)} \right) \Big|_{-\infty}^{\infty}$$

Now, as sinc function degenerates into an impulse with strength 2π, irrespective of
the value T, in the limit,

$$\lim_{T\to\infty} \left(\frac{e^{j(\omega-\omega_0)t}}{j(\omega-\omega_0)} \right) \Big|_{-T}^{T} = \lim_{T\to\infty} \left(\frac{2\sin((\omega-\omega_0)T)}{(\omega-\omega_0)} \right) = 2\pi\delta(\omega-\omega_0)$$

Using this result,

$$\int_{-\infty}^{\infty} x(t) e^{-j\omega_0 t} dt = \int_{-\infty}^{\infty} \delta(\omega-\omega_0) X(j\omega) d\omega = X(j\omega_0)$$

Replacing ω_0 by ω, since the impulse is effective only at $\omega = \omega_0$, we get

$$X(j\omega) = \int_{-\infty}^{\infty} x(t) e^{-j\omega t} dt$$

Since t or ω is zero, the values $X(j0)$ and $x(0)$ can be determined easily. They can be used to check the correctness of the derived closed-form expressions for $X(j\omega)$ or $x(t)$.

$$X(j0) = \int_{-\infty}^{\infty} x(t)dt \quad \text{and} \quad x(0) = \frac{1}{2\pi} \int_{-\infty}^{\infty} X(j\omega)d\omega,$$

The FT and the IFT definitions can be expressed in terms of the the the cyclic frequency f. Since $\omega = 2\pi f$ and $d\omega = 2\pi df$, we get

$$X(j2\pi f) = \int_{-\infty}^{\infty} x(t)e^{-j2\pi ft}dt \quad \text{and} \quad x(t) = \int_{-\infty}^{\infty} X(j2\pi f)e^{j2\pi ft}df$$

Gibbs phenomenon can occur in either or both the time domain and frequency domain.

9.1.2 Existence of the FT

The sufficient conditions, called the Dirichlet conditions, for the existence of the FT are as follows. The first condition is that the signal $x(t)$ is absolutely integrable. From the definition of the FT, we get

$$|X(j\omega)| \leq \int_{-\infty}^{\infty} |x(t)e^{-j\omega t}|\,dt = \int_{-\infty}^{\infty} |x(t)||e^{-j\omega t}|\,dt = \int_{-\infty}^{\infty} |x(t)|\,dt,$$

since $|e^{-j\omega t}| = 1$. The second condition is that $x(t)$ may have only a finite number of finite maxima and minima in any finite interval. The third condition is that $x(t)$ may have only a finite number of finite discontinuities of $x(t)$ in any finite interval. All physical signals have FT representation.

As Fourier representation of a signal is equivalent with respect to the least squares error criterion,

$$\int_{-\infty}^{\infty} |x(t)|^2 dt = \frac{1}{2\pi} \int_{-\infty}^{\infty} |X(j\omega)|^2 d\omega$$

That is, a square integrable signal, $\int_{-\infty}^{\infty} |x(t)|^2 dt < \infty$, has a FT representation. A signal with finite energy satisfies the existence conditions.

Example 9.1 Find the IFT, $x(t)$, of the spectrum $X(j\omega) = \pi(u(\omega + \omega_0) - u(\omega - \omega_0))$, where $u(\omega)$ is the unit-step function.

Solution

$$x(t) = \frac{1}{2\pi} \int_{-\omega_0}^{\omega_0} \pi e^{j\omega t}d\omega = \int_{0}^{\omega_0} \cos(\omega t)d\omega = \frac{\sin(\omega_0 t)}{t}$$

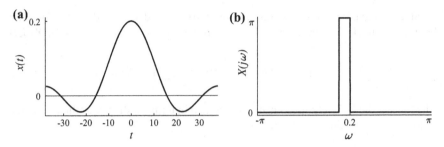

Fig. 9.2 **a** $\frac{\sin(0.2t)}{t}$ and **b** its FT spectrum, $\pi(u(\omega + 0.2) - u(\omega - 0.2))$

$$\frac{\sin(\omega_0 t)}{t} \leftrightarrow \pi(u(\omega + \omega_0) - u(\omega - \omega_0))$$

Signal $x(t)$ and its FT are shown, respectively, in Fig. 9.2a, b with $\omega_0 = 0.2$. ∎

The function of the form $\frac{\sin(at)}{t}$, a typical example is that shown in Fig. 9.2a, is called the sinc function that has important role in signal processing. As both the sine function in the numerator and the denominator function t are odd, the sinc function is even-symmetric. The value of the sinc function is zero for $at = \pm\pi, \pm2\pi, \pm3\pi, \ldots$ At $t = 0$,

$$\lim_{t \to 0} \sin \frac{at}{t} = a$$

as $\lim_{\theta \to 0} \sin(\theta) = \theta$. The oscillations of the sine function are damped by the denominator t. The period of the sine function is $2\pi/a$.

The area enclosed by the sinc function is constant, irrespective of the value of a. Finding the FT of $x(t)$ in Example 9.1 with $\omega = 0$,

$$X(j0) = \int_{-\infty}^{\infty} \frac{\sin(0.2t)}{t} dt = \pi$$

It is also known that the area enclosed by the function is equal to the area of the triangle inscribed within its main hump. The sinc function is only square integrable. As $a \to 0$, the function $\frac{\sin(at)}{t}$ is expanded and, eventually, degenerates into a DC function. The first pair of zeros at $\omega = \pm\frac{\pi}{a}$ move to infinity and the function becomes a horizontal line with amplitude a. As a becomes larger, the numerator sine function $\sin(at)$ of $\frac{\sin(at)}{t}$ alone is compressed (frequency of oscillations is increased). As a consequence, the amplitudes of all the ripples along with that of the main hump increase with fixed ratios to one another. While the ripples and the main hump become taller and narrower, the area enclosed by each and the total area enclosed by the function remains fixed. In the limit, as $a \to \infty$, the main hump and all the ripples of significant amplitude are concentrated at $t = 0$ and $\frac{\sin(at)}{t}$ degenerates into an impulse with strength π.

9.1.3 *Determination of the FS from the FT*

Let $x_p(t)$ be a periodic signal of period T. Let us define an aperiodic signal $x(t)$ that is identical with $x_p(t)$ over its one period from t_1 to $t_1 + T$ and is zero otherwise, where t_1 is arbitrary. The FT of this signal is

$$X(j\omega) = \int_{-\infty}^{\infty} x(t)e^{-j\omega t}\,dt = \int_{t_1}^{t_1+T} x_p(t)e^{-j\omega t}\,dt$$

The FS spectrum for $x_p(t)$ is

$$X_{fs}(k) = \frac{1}{T}\int_{t_1}^{t_1+T} x_p(t)e^{-jk\omega_0 t}\,dt, \quad \omega_0 = \frac{2\pi}{T}$$

Comparing the FS and FT definitions of the signals, we get

$$X_{fs}(k) = \frac{1}{T}X(j\omega)|_{\omega=k\omega_0} = \frac{1}{T}X(jk\omega_0)$$

The discrete samples of $\frac{1}{T}X(j\omega)$, at intervals of ω_0, constitute the FS spectrum for the periodic signal $x_p(t)$. While the spectral values at discrete frequencies are adequate to reconstruct one period of the periodic waveform using the inverse FS, spectral values at continuum of frequencies are required to reconstruct one period of the periodic waveform and the infinite extent zero values of the aperiodic waveform using the IFT. A similar relationship exists between the DTFT and the DFT.

Example 9.2 Find the FS spectrum for the periodic signal $x_p(t)$, one period of which is defined as

$$x_p(t) = \begin{cases} 1 & \text{for } |t| < \frac{\pi}{2} \\ 0 & \text{for } \frac{\pi}{2} < |t| < \pi \end{cases}$$

Solution

The derivative of the corresponding aperiodic pulse is composed of two shifted impulses, $\delta(t + \frac{\pi}{2})$ and $-\delta(t - \frac{\pi}{2})$. The integral of the derivative is the square pulse. Therefore, we find the FT of the impulses and, using the derivative and integral properties of the FT, we find the FT. The FT of $\delta(t)$ is 1 and that of $\delta(t - T_0)$ is $e^{-jT_0\omega}$ (using the FT shift property). It is relatively easier to find the FT using this method for this type of signals. This method is called the derivative method of obtaining the FT of $x(t)$.

Let $X(j\omega)$ be the FT of $x(t)$. The derivative of $X(j\omega)$ is $j\omega X(j\omega)$. Then, for the example,

$$j\omega X(j\omega) = e^{j\frac{\pi}{2}\omega} - e^{-j\frac{\pi}{2}\omega} \quad \text{and} \quad X(j\omega) = \frac{e^{j\frac{\pi}{2}\omega} - e^{-j\frac{\pi}{2}\omega}}{j\omega} = 2\frac{\sin\left(\frac{\pi}{2}\omega\right)}{\omega}$$

Since $X_{fs}(k) = \frac{1}{T}X(jk\omega_0)$, with $T = 2\pi$ and $\omega = k\omega_0 = k\frac{2\pi}{2\pi} = k$, we get

$$X_{fs}(k) = \frac{2}{2\pi}\frac{\sin\left(\frac{\pi}{2}k\right)}{k} = \frac{1}{\pi}\frac{\sin\left(\frac{\pi}{2}k\right)}{k}$$

∎

The waveform is even-symmetric and, therefore, the coefficients of all the sine components are zero. From another point of view, since the complex coefficients are real, the waveform is composed of cosine components only. Further, subtracting the DC bias, the first-half and second-half of $x(t)$ are antisymmetric. This is called the odd half-wave symmetry. This implies that all the even-indexed, except the DC, components are absent. The Fourier series for the square pulse is

$$x(t) = \frac{1}{2} + \frac{2}{\pi}\left(\cos(t) - \frac{1}{3}\cos(3t) + \frac{1}{5}\cos(5t) - \cdots\right), \tag{9.7}$$

as given in Chap. 7.

Example 9.3 Find the FT of the unit impulse signal $x(t) = \delta(t)$.

Solution

Using the sampling property of the impulse, we get

$$X(j\omega) = \int_{-\infty}^{\infty}\delta(t)e^{-j\omega t}dt = e^{-j\omega 0}\int_{-\infty}^{\infty}\delta(t)dt = 1 \quad \text{and} \quad \delta(t) \leftrightarrow 1$$

The unit impulse signal is composed of complex sinusoids, with zero phase shift, of all frequencies from $\omega = -\infty$ to $\omega = \infty$ in equal proportion. That is,

$$\delta(t) = \frac{1}{2\pi}\int_{-\infty}^{\infty}e^{j\omega t}d\omega = \frac{1}{2\pi}\int_{-\infty}^{\infty}\cos(\omega t)d\omega = \frac{1}{\pi}\int_{0}^{\infty}\cos(\omega t)d\omega$$

∎

Example 9.4 Find the IFT of $X(j\omega) = 2\pi\delta(\omega)$.

Solution

Using the sampling property of the impulse, we get

$$x(t) = \frac{1}{2\pi}\int_{-\infty}^{\infty}2\pi\delta(\omega)e^{j\omega t}d\omega = 1 \quad \text{and} \quad 1 \leftrightarrow 2\pi\delta(\omega)$$

The FT of the DC signal is $2\pi\delta(\omega)$. ∎

Example 9.5 Find the FT $X(j\omega)$ of the real, causal, and decaying exponential signal $x(t) = e^{-at}u(t)$, $a > 0$. Find the value of $x(0)$ from $X(j\omega)$.

Solution

$$X(j\omega) = \int_0^\infty e^{-at}e^{-j\omega t}\,dt = \int_0^\infty e^{-(a+j\omega)t}\,dt = -\left.\frac{e^{-(a+j\omega)t}}{a+j\omega}\right|_0^\infty = \frac{1}{a+j\omega}$$

$$e^{-at}u(t),\ a > 0 \leftrightarrow \frac{1}{a+j\omega}$$

$$x(0) = \frac{1}{2\pi}\int_{-\infty}^\infty \frac{1}{a+j\omega}\,d\omega = \frac{1}{2\pi}\int_{-\infty}^\infty \frac{a}{\omega^2+a^2}\,d\omega - \frac{j}{2\pi}\int_{-\infty}^\infty \frac{\omega}{\omega^2+a^2}\,d\omega$$

As the imaginary part of $X(j\omega)$ is odd, its integral evaluates to zero. Therefore,

$$x(0) = \frac{1}{2\pi}\int_{-\infty}^\infty \frac{a}{\omega^2+a^2}\,d\omega = \frac{1}{2\pi}\int_{-\infty}^\infty \frac{d\left(\frac{\omega}{a}\right)}{\left(\frac{\omega}{a}\right)^2+1} = \frac{1}{2\pi}\tan^{-1}\left(\frac{\omega}{a}\right)\Big|_{-\infty}^\infty = \frac{1}{2}$$

For any value of a, $x(t)$, at $t = 0$, is $\frac{1}{2}$. Any Fourier reconstructed waveform converges to the average of the right- and left-hand limits at a discontinuity.

Figure 9.3a, b show, respectively, the signal $e^{-2t}u(t)$ and its spectra. The real part of the spectrum (continuous line) is an even function with a peak value of $\frac{1}{2}$ at $\omega = 0$ and the imaginary part (dashed line) is an odd function with peaks of value $\pm\frac{1}{4}$ at $\omega = \mp 2$. ∎

Example 9.6 Find the FT of the *sgn* signal

$$sgn(t) = \begin{cases} 1 & \text{for } t > 0 \\ -1 & \text{for } t < 0 \end{cases}$$

Solution

The sgn function, $sgn(t)$, is the limit of a linear combination of two decaying exponential signals. That is,

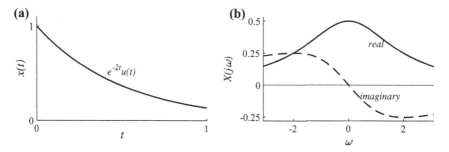

Fig. 9.3 **a** $x(t) = e^{-2t}u(t)$ and **b** its FT spectrum

$$sgn(t) = \lim_{a \to 0}(e^{-at}u(t) - e^{at}u(-t))$$

Consequently,

$$X(j\omega) = \lim_{a \to 0}\left(\frac{1}{a + j\omega} - \frac{1}{a - j\omega}\right) = \frac{2}{j\omega}$$

∎

As the unit-step signal $u(t)$ is a combination of the DC and $sgn(t)$, we get

$$u(t) = 0.5 + 0.5sgn(t) \leftrightarrow \pi\delta(\omega) + \frac{1}{j\omega}$$

Example 9.7 Find the IFT of $X(j\omega) = \delta(\omega - \omega_0)$.

Solution

$$x(t) = \frac{1}{2\pi}\int_{-\infty}^{\infty}\delta(\omega - \omega_0)e^{j\omega t}d\omega = \frac{1}{2\pi}e^{j\omega_0 t} \quad \text{and} \quad e^{j\omega_0 t} \leftrightarrow 2\pi\delta(\omega - \omega_0)$$

That is, the spectrum of the complex sinusoid $e^{j\omega_0 t}$ is an impulse at $\omega = \omega_0$ with strength 2π. ∎

It follows that

$$\cos(\omega_0 t) \leftrightarrow \pi(\delta(\omega + \omega_0) + \delta(\omega - \omega_0))$$
$$\sin(\omega_0 t) \leftrightarrow j\pi(\delta(\omega + \omega_0) - \delta(\omega - \omega_0)).$$

9.2   Properties of the Fourier Transform

Properties present the frequency-domain effect of time-domain characteristics and operations on signals and vice versa. In addition, they are used to find new transform pairs more easily from the FT of known functions and they also simplify the determination of the FT from the definition.

9.2.1   Linearity

The FT of a linear combination of a set of signals is the same linear combination of their individual FT. That is,

$$x(t) \leftrightarrow X(j\omega), \quad y(t) \leftrightarrow Y(j\omega), \quad ax(t) + by(t) \leftrightarrow aX(j\omega) + bY(j\omega),$$

where a and b are arbitrary constants. As the integral defining the FT is a linear operation, the FT is also a linear operation. Consider the FT pairs

$$\cos(\omega_0 t) \leftrightarrow \pi(\delta(\omega + \omega_0) + \delta(\omega - \omega_0))$$
$$\sin(\omega_0 t) \leftrightarrow j\pi(\delta(\omega + \omega_0) - \delta(\omega - \omega_0))$$

Using the linearity property,

$$\cos(\omega_0 t) + j\sin(\omega_0 t) = e^{j\omega_0 t} \leftrightarrow 2\pi\delta(\omega - \omega_0)$$

9.2.2 Duality

The definitions of FT and IFT are almost identical in form. The three differences between the definitions are the change of the signs of the exponents of the complex exponentials, the interchange of the variables t and ω, and the constant appearing in the definition of the IFT. Therefore, $x(t)$ and $X(j\omega)$ are each other's transform, with some minor changes. That is, the variables t and ω can be interchanged with some modifications.

The IFT is defined as

$$x(t) = \frac{1}{2\pi}\int_{-\infty}^{\infty} X(j\omega)e^{j\omega t}d\omega$$

By replacing t by $-t$, we get

$$x(-t) = \frac{1}{2\pi}\int_{-\infty}^{\infty} X(j\omega)e^{-j\omega t}d\omega \quad \text{and} \quad 2\pi x(-t) = \int_{-\infty}^{\infty} X(j\omega)e^{-j\omega t}d\omega$$

That is,

$$2\pi x(-t) \leftrightarrow X(j\omega)$$

This is a forward transform with $2\pi x(-t)$ being the FT of $X(j\omega)$. That is, we get $2\pi x(-t)$ by taking the FT of $x(t)$ twice in succession, $2\pi x(-t) = \text{FT}(\text{FT}(x(t)))$.

$$x(t) \leftrightarrow X(j\omega) \rightarrow X(\pm t) \leftrightarrow 2\pi x(\mp(j\omega))$$

For an even $x(t)$, as $X(j\omega)$ is also even, the sign change of either t or ω may be omitted. If the FT and IFT definitions using the cyclic frequency f are used, the 2π factor gets eliminated leaving only the sign change.

For example, consider the FT pair

$$e^{-at}u(t), \ a > 0 \leftrightarrow \frac{1}{a + j\omega}$$

Using the property, we get the transform pair

$$2\pi e^{a\omega} u(-\omega), \ a > 0 \leftrightarrow \frac{1}{a + jt}$$

As an another example, consider the FT pair

$$\sin(t) \leftrightarrow j\pi(\delta(\omega + 1) - \delta(\omega - 1))$$

Considering ω as the time-domain variable, we get

$$j\pi(\delta(\omega + 1) - \delta(\omega - 1)) \leftrightarrow j\pi(e^{jt} - e^{-jt}) = -2\pi \sin(t) = 2\pi \sin(-t)$$

9.2.3 Symmetry

The FT of $x(t)$ is given by

$$X(j\omega) = \int_{-\infty}^{\infty} x(t)e^{-j\omega t}dt = \int_{-\infty}^{\infty} x(t)(\cos(\omega t) - j\sin(\omega t))dt \qquad (9.8)$$

Let $x(t)$ be real and even. Since $x(t)\cos(\omega t)$ is even and $x(t)\sin(\omega t)$ is odd, the second term in the integrand in Eq. (9.8) contributes the value zero to the value of this integral and we get

$$X(j\omega) = 2\int_{0}^{\infty} x(t)\cos(\omega t)dt \quad \text{and} \quad x(t) = \frac{1}{\pi}\int_{0}^{\infty} X(j\omega)\cos(\omega t)d\omega$$

If a signal $x(t)$ is real and even, then its spectrum is also real and even. The FT 1 of $\delta(t)$ is an example of the FT of an even function.

Let $x(t)$ be real and odd. Since $x(t)\sin(\omega t)$ is even and $x(t)\cos(\omega t)$ is odd, the first term in the integrand in Eq. (9.8) contributes the value zero to the value of this integral and we get

$$X(j\omega) = -j2\int_{0}^{\infty} x(t)\sin(\omega t)dt \quad \text{and} \quad x(t) = \frac{j}{\pi}\int_{0}^{\infty} X(j\omega)\sin(\omega t)d\omega$$

If a signal $x(t)$ is real and odd, then its spectrum is imaginary and odd.

$$\begin{cases} 1 \text{ for } t > 0 \\ -1 \text{ for } t < 0 \end{cases} \leftrightarrow \frac{2}{j\omega}$$

is an example of the FT of an odd function.

An arbitrary real signal $x(t)$ can be decomposed into even and odd components, $x(t) = x_e(t) + x_o(t)$, as presented in Chap. 1. Therefore, the FT of $x(t)$, which is neither even nor odd, is $X(j\omega) = X_e(j\omega) + X_o(j\omega)$. If a signal $x(t)$ is real, then the real part of its spectrum $X(j\omega)$ is even and the imaginary part is odd, called the conjugate symmetry. That is, $X(-j\omega) = X^*(j\omega)$. An example is

$$x(t) = \delta(t - t_0) \leftrightarrow X(j\omega) = e^{-j\omega t_0} = \cos(\omega t_0) - j\sin(\omega t_0)$$

9.2.4 Time Shifting

The shifting of a time-domain signal merely adds phase angles to its constituent sinusoidal components, which are linearly proportional to their frequencies. If we replace $e^{\pm j\omega t_0} X(j\omega)$ in the IFT definition and combine the exponentials, we get

$$x(t \pm t_0) \leftrightarrow e^{\pm j\omega t_0} X(j\omega)$$

Consider the FT pair of the even-symmetric square pulse with width $2a$

$$u(t + a) - u(t - a) \leftrightarrow 2\frac{\sin(a\omega)}{\omega}$$

Using the theorem,

$$u(t + a - b) - u(t - a - b) \leftrightarrow 2\frac{\sin(a\omega)}{\omega}e^{-jb\omega}$$

9.2.5 Frequency Shifting

Let $x(t)$ is replaced by $x(t)e^{\pm j\omega_0 t}$ in the FT definition. Then, the two exponentials in the resulting integrand can be combined to get

$$x(t)e^{\pm j\omega_0 t} \leftrightarrow X(j(\omega \mp \omega_0))$$

The duality properties hold to both transform pairs and properties. The time-shifting property is the dual of the frequency-shifting property.

Consider the FT pair

$$e^{-3t}u(t) \leftrightarrow \frac{1}{3 + j\omega}$$

Then,

$$e^{-3t}\sin(2t)u(t) = e^{-3t}\frac{(e^{j2t} - e^{-j2t})}{j2}u(t)$$

$$\leftrightarrow \frac{1}{j2}\left(\frac{1}{3+j(\omega - 2)} - \frac{1}{3+j(\omega + 2)}\right) = \frac{2}{(3+j\omega)^2 + 4}$$

For easier visualization, consider the FT pair

$$e^{-3t}e^{j2t}u(t) \leftrightarrow \frac{1}{3+j(\omega - 2)}$$

$$\frac{1}{3+j\omega}\bigg|_{\omega=0} = \frac{1}{3} \quad \text{and} \quad \frac{1}{3+j(\omega - 2)}\bigg|_{\omega=2} = \frac{1}{3}$$

9.2.6 Convolution in the Time Domain

Fourier analysis is considered as the decomposition of a signal into its harmonic components, even for continuous aperiodic signals and their spectra. Therefore, as pointed out in earlier chapters, the convolution of two complex exponentials of the same frequency is a complex exponential of the same frequency with its coefficient being the product of those of the two exponentials. This property holds only for complex exponentials and their linear combination, the real sinusoid. As the Fourier spectrum is a representation of a signal in terms of its harmonic components, the product of the two spectra of the two signals to be convolved is the spectrum of the convolution of the two signals in the time domain.

$$\int_{-\infty}^{\infty} x(\tau)h(t - \tau)d\tau = \frac{1}{2\pi}\int_{-\infty}^{\infty} X(j\omega)H(j\omega)e^{j\omega t}d\omega \leftrightarrow X(j\omega)H(j\omega)$$

That is, the convolution operation in the time domain corresponds to much simpler multiplication operation in the frequency domain. This result, due to the representation of the signals in terms of complex exponentials, is a major reason for the dominant role of the frequency-domain analysis in the study of signals and systems.

Consider the convolution of $x(t) = e^{-4t}u(t)$ and $h(t) = e^{-2t}u(t)$. In the time domain, we get

$$y(t) = x(t) * h(t) = \int_{0}^{t} e^{-2\tau}e^{-4(t-\tau)}d\tau = e^{-4t}\int_{0}^{t} e^{2\tau}d\tau = 0.5(e^{-2t} - e^{-4t})u(t)$$

In the frequency domain, the convolution output is the product of their individual FT. That is,

$$Y(j\omega) = \frac{1}{2+j\omega}\frac{1}{4+j\omega} = \frac{0.5}{2+j\omega} - \frac{0.5}{4+j\omega}$$

The IFT of $Y(j\omega)$ is $y(t)$.

9.2.7 Convolution in the Frequency Domain

The problem is to find the FT $Z(j\omega)$ of the product $z(t) = x(t)y(t)$ of two signals $x(t)$ and $y(t)$ in the time domain, in terms of their individual FT. This is the dual of the time-domain convolution property.

$$z(t) = x(t)y(t) \leftrightarrow Z(j\omega) = \int_{-\infty}^{\infty} x(t)y(t)e^{-j\omega t}dt$$

$$= \frac{1}{2\pi}\int_{-\infty}^{\infty} X(jv)Y(j(\omega-v))dv = \frac{1}{2\pi}X(j\omega) * Y(j\omega)$$

The FT of

$$\frac{1}{2+jt}\frac{1}{2+jt} = \frac{1}{(2+jt)^2}$$

is the convolution of the FT of $\frac{1}{2+jt}$ with itself divided by 2π. That is,

$$Z(j\omega) = (2\pi)e^{2\omega}u(-\omega) * e^{2\omega}u(-\omega) = (2\pi)\int_0^{-\omega} e^{2\tau}e^{2(\omega-\tau)}d\tau$$

$$= (2\pi)e^{2\omega}\int_0^{-\omega} d\tau = -(2\pi)\omega e^{2\omega}u(-\omega)$$

9.2.8 Conjugation

Let $x(t) \leftrightarrow X(j\omega)$. Then, $x^*(\pm t) \leftrightarrow X^*(\mp j\omega)$. If we conjugate both sides of the FT and IFT definitions, we get

$$X^*(j\omega) = \int_{-\infty}^{\infty} x^*(t)e^{j\omega t}dt, \qquad x^*(t) = \frac{1}{2\pi}\int_{-\infty}^{\infty} X^*(j\omega)e^{-j\omega t}d\omega$$

Replacing t by $-t$, we get

$$X^*(j\omega) = \int_{-\infty}^{\infty} x^*(-t)e^{-j\omega t}dt, \qquad x^*(-t) = \frac{1}{2\pi}\int_{-\infty}^{\infty} X^*(j\omega)e^{j\omega t}d\omega$$

Replacing ω by $-\omega$ after conjugating, we get

$$X^*(-j\omega) = \int_{-\infty}^{\infty} x^*(t)e^{-j\omega t}dt, \quad x^*(t) = \frac{1}{2\pi}\int_{-\infty}^{\infty} X^*(-j\omega)e^{j\omega t}d\omega$$

For example,

$$e^{-t}u(t) \leftrightarrow \frac{1}{1+j\omega}$$

Conjugating both sides, we get

$$e^{-t}u(t) \leftrightarrow \frac{1}{1-j\omega}$$

Changing t by $-t$, we get

$$e^{t}u(-t) \leftrightarrow \frac{1}{1-j\omega}$$

Changing ω by $-\omega$ after conjugating, we get

$$e^{-t}u(t) \leftrightarrow \frac{1}{1+j\omega}$$

9.2.9   Cross-Correlation

The cross-correlation of two signals $x(t)$ and $y(t)$ is defined as

$$r_{xy}(\tau) = \int_{-\infty}^{\infty} x(t)y^*(t-\tau)dt$$

The correlation operation is the same as convolution operation without time-reversal. Therefore,

$$\int_{-\infty}^{\infty} x(t)y^*(t-\tau)dt = \frac{1}{2\pi}\int_{-\infty}^{\infty} X(j\omega)Y^*(j\omega)e^{j\omega t}d\omega \leftrightarrow X(j\omega)Y^*(j\omega)$$

Let $x(t) = \cos(t)$ and $y(t) = \sin(t)$. The approach is to find the correlation of $x(t)$ and $y(t)$ using the FS first.

$$\cos(t) \leftrightarrow 0.5(\delta(k+1) + \delta(k-1)) \quad \text{and} \quad \sin(t) \leftrightarrow j0.5(\delta(k+1) - \delta(k-1))$$

Then, the correlation of $x(t)$ and $y(t)$ in the frequency domain is the multiplication of their FS with the second FS conjugated and the period 2π. The result is

$$-j0.5\pi(\delta(k+1) - \delta(k-1))$$

The corresponding FT is obtained by multiplying by 2π and replacing the discrete impulses by continuous impulses. That is,

$$\pi(-j\pi(\delta(\omega+1) - \delta(\omega-1)))$$

The IFT of this is $-\pi \sin(t)$, the correlation of $x(t)$ and $y(t)$.
    In the time domain,

$$\int_0^{2\pi} \cos(\tau) \sin(\tau - t) d\tau$$

$$= \int_0^{2\pi} \cos(\tau)(\sin(\tau)\cos(t) - \cos(\tau)\sin(t))d\tau$$

$$= -\sin(t) \int_0^{2\pi} \cos^2(\tau)d\tau = -\pi \sin(t)$$

9.2.10 Time-Reversal

Let $x(t) \leftrightarrow X(j\omega)$. Then, $x(-t) \leftrightarrow X(-j\omega)$. If we replace t by $-t$ and ω by $-\omega$ in the FT definition, we get

$$X(-j\omega) = \int_{-\infty}^{\infty} x(-t)e^{-j\omega t}dt$$

The time-reversal of a signal results in its spectrum also frequency-reversed.
    For example,
$$\sin(\omega_0 t) \leftrightarrow j\pi(\delta(\omega + \omega_0) - \delta(\omega - \omega_0))$$

Replacing t by $-t$ and ω by $-\omega$, we get

$$\sin(\omega_0(-t)) = -\sin(\omega_0 t) \leftrightarrow j\pi(\delta(-\omega + \omega_0) - \delta(-\omega - \omega_0))$$
$$= j\pi(\delta(\omega - \omega_0) - \delta(\omega + \omega_0)) = -j\pi(\delta(\omega + \omega_0) - \delta(\omega - \omega_0))$$

9.2.11 Time Scaling

Sometimes, it is necessary to change the timescale of a signal $x(t)$ and it is of interest to express the FT of the time-scaled function in terms of $X(j\omega)$. For example,

$$\cos(t) \leftrightarrow \pi(\delta(\omega + 1) + \delta(\omega - 1))$$

Replacing t by $2t$, we get

$$\cos(2t) \leftrightarrow \pi(\delta(\omega+2) + \delta(\omega-2))$$

The frequency of the signal is increased and its period is decreased. The signal is compressed and its spectrum gets expanded. Replacing t by $0.5t$, we get

$$\cos(0.5t) \leftrightarrow \pi(\delta(\omega+0.5) + \delta(\omega-0.5))$$

The frequency of the signal is decreased and its period is increased. The signal is expanded and its spectrum gets compressed. If the scaling factor is negative, both the signal and its spectrum get reversed.

Let $x(t) \leftrightarrow X(j\omega)$. By replacing at by τ, t by $\frac{\tau}{a}$ and dt by $\frac{d\tau}{a}$, with $a > 0$, in the FT definition of $x(at)$, we get

$$\int_{-\infty}^{\infty} x(at)e^{-j\omega t}dt = \frac{1}{a}\int_{-\infty}^{\infty} x(\tau)e^{-j\omega\frac{\tau}{a}}d\tau = \frac{1}{a}X\left(j\left(\frac{\omega}{a}\right)\right)$$

For $a < 0$, the FT is

$$\frac{-1}{a}X\left(j\left(\frac{\omega}{a}\right)\right)$$

By combining both the results, we get

$$x(at) \leftrightarrow \frac{1}{|a|}X\left(j\left(\frac{\omega}{a}\right)\right), \qquad a \neq 0$$

The signal energy is changed by the scaling operation. The factor $\frac{1}{|a|}$ scales the energy suitably.

Consider the transform pair

$$u(t+1) - u(t-1) \leftrightarrow 2\frac{\sin(\omega)}{\omega}$$

Let $a = 0.5$. Using the property, the transform of

$$u(t+2) - u(t-2) \leftrightarrow 2\frac{\sin(2\omega)}{\omega}$$

Figure 9.4a, c show pulses with width 2 and its expanded version by a factor of 2, respectively. Figure 9.4b, d show the respective FT spectra.

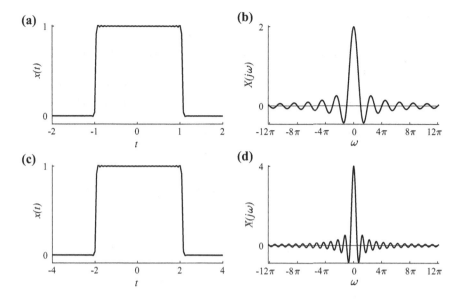

Fig. 9.4 **a** A pulse with width 2; **b** its spectrum; **c** a pulse with width 4; **d** its spectrum

9.2.12 Time Differentiation

Since the derivative of a spectral component $X(j\omega_0)e^{j\omega_0 t}$ with respect to t is $j\omega_0 X(j\omega_0)e^{j\omega_0 t}$ and an arbitrary signal $x(t)$ can be expressed in terms of its spectrum $X(j\omega)$, we get

$$\frac{dx(t)}{dt} \leftrightarrow j\omega X(j\omega)$$

In general,

$$\frac{d^n x(t)}{dt^n} \leftrightarrow (j\omega)^n X(j\omega)$$

It is assumed that the derivative of the signal is Fourier transformable. As the factor $j\omega$ appears in the spectrum of the differentiated signal, the magnitude of the high frequency components is increased proportional to the frequency. This amounts to highpass filtering of the signal. In common with other properties, this property can also be used to find the FT of related signals.

For example,

$$\cos(\omega_0 t) \leftrightarrow \pi(\delta(\omega + \omega_0) + \delta(\omega - \omega_0))$$

$$\frac{d(\cos(\omega_0 t))}{dt} = -\omega_0 \sin(\omega_0 t) \leftrightarrow (j\omega_0)(\pi(-\delta(\omega + \omega_0) + \delta(\omega - \omega_0))$$

$$= -j\omega_0 \pi(\delta(\omega + \omega_0) - \delta(\omega - \omega_0))$$

9.2.13   Time Integration

Convolution is the integral of product of two signals, with one of them time reversed. Consider the time-reversed and shifted version $u(t - \tau)$ of the unit-step signal $u(\tau)$. If we multiply a signal by $u(t - \tau)$, the product is the same signal from $-\infty$ to t and the rest from t to ∞ zero. Therefore, the integral of the product is the integral of the signal from $-\infty$ to t.

The definite integral, $y(t)$, of a time-domain signal, $x(t)$, can be expressed as the convolution of $x(t)$ and the unit-step signal, $u(t)$, as

$$y(t) = \int_{-\infty}^{t} x(\tau)d\tau = \int_{-\infty}^{\infty} x(\tau)u(t - \tau)d\tau = x(t) * u(t)$$

Using the time-domain convolution theorem, with $x(t) \leftrightarrow X(j\omega)$, and $u(t) \leftrightarrow \frac{1}{j\omega} + \pi\delta(\omega)$, we get

$$\int_{-\infty}^{t} x(\tau)d\tau \leftrightarrow X(j\omega)\left(\frac{1}{j\omega} + \pi\delta(\omega)\right) = \frac{X(j\omega)}{j\omega} + \pi X(j0)\delta(\omega)$$

It is assumed that $y(t)$ is Fourier transformable. The factor ω in the denominator of the FT of $y(t)$ indicates that, as the frequency is increased, the amplitude of the high frequency components of the signal is more and more attenuated. That is lowpass filtering. Differentiation corresponds to highpass filtering and integration corresponds to lowpass filtering.

Consider the signal $x(t) = u(t) - u(t - 1)$, shown in Fig. 9.5a with the FT $X(j\omega) = \frac{1}{j\omega}(1 - e^{-j\omega})$ and $X(j0) = 1$. Using the property,

$$y(t) = \int_{-\infty}^{t} x(\tau)d\tau \leftrightarrow Y(j\omega) = \frac{X(j\omega)}{j\omega} + \pi\delta(\omega) = \pi\delta(\omega) + \frac{(e^{-j\omega} - 1)}{\omega^2}$$

Figure 9.5b shows the integral of $x(t)$.

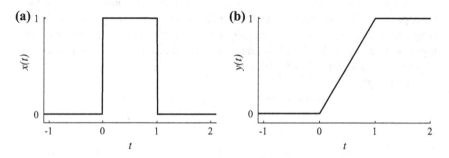

Fig. 9.5   a Signal $x(t)$. **b** The integral of $x(t)$, $y(t)$

To appreciate the use of theorems, let us do this example using the FT definition. Assume that we have obtained $y(t)$ in the time domain. First, we have to split the function $y(t)$ into the ramp section and the delayed unit-step signal. This is the use of linearity theorem.

The FT of the delayed unit-step signal is

$$\pi\delta(\omega) + \frac{e^{-j\omega}}{j\omega}$$

The FT of the time-limited ramp function is found as follows.

$$\int_0^1 xe^{-j\omega t}dx = \frac{e^{-j\omega x}}{(j\omega)^2}(-j\omega t - 1)\Big|_0^1 = -\frac{e^{-j\omega}}{j\omega} + \frac{e^{-j\omega}}{\omega^2} - \frac{1}{\omega^2}$$

Combining the two partial FT, we get the same FT as that obtained using the theorem. Properties are so convenient for theoretical analysis. Of course, the numerical approximation of the Fourier spectrum with the DFT along with fast algorithms makes Fourier analysis an indispensable tool in the applications of science and engineering.

As an another example,

$$\cos(\omega_0 t) \leftrightarrow \pi(\delta(\omega + \omega_0) + \delta(\omega - \omega_0))$$

The integral of $\cos(\omega_0 t)$ is

$$-j\pi\frac{1}{\omega_0}(-\delta(\omega + \omega_0) + \delta(\omega - \omega_0)) = \frac{\sin(\omega_0 t)}{\omega_0}$$

9.2.14 Frequency-Differentiation

As the FT spectrum is also a continuous function of ω, the spectrum $X(j\omega)$ is also differentiable. It is assumed that the resulting function is Fourier transformable. Differentiating both sides of the FT definition with respect to ω, we get

$$(-jt)x(t) \leftrightarrow \frac{dX(j\omega)}{d\omega} \quad \text{or} \quad tx(t) \leftrightarrow j\frac{dX(j\omega)}{d\omega}$$

For the nth derivative, we get,

$$(-jt)^n x(t) \leftrightarrow \frac{d^n X(j\omega)}{d\omega^n} \quad \text{or} \quad (t)^n x(t) \leftrightarrow (j)^n\frac{d^n X(j\omega)}{d\omega^n}$$

For example,

$$e^{-3t}u(t) \leftrightarrow \frac{1}{j\omega + 3} \quad \text{and} \quad te^{-3t}u(t) \leftrightarrow \frac{1}{(j\omega + 3)^2}$$

$$\delta(t+3) + \delta(t-3) \leftrightarrow 2\cos(3\omega) \quad \text{and} \quad t(\delta(t+3) + \delta(t-3)) \leftrightarrow -j(2)(3)\sin(3\omega)$$

9.2.15 Parseval's Theorem and the Energy Transfer Function

Orthogonal transforms have the energy preservation property. That is, the energy of a signal can be derived either from its time-domain or frequency-domain representations. The energy of a signal is its autocorrelation with lag zero. That is,

$$\int_{-\infty}^{\infty} x(t)x^*(t)dt = \frac{1}{2\pi} \int_{-\infty}^{\infty} X(j\omega)X^*(j\omega)d\omega$$

$$E = \int_{-\infty}^{\infty} |x(t)|^2 dt = \frac{1}{2\pi} \int_{-\infty}^{\infty} |X(j\omega)|^2 d\omega$$

This equivalent energy representation is called the Parseval's theorem. For real signals, the autocorrelation is even and we get

$$E = \int_{-\infty}^{\infty} |x(t)|^2 dt = \frac{1}{\pi} \int_{0}^{\infty} |X(j\omega)|^2 d\omega$$

The quantity $|X(j\omega)|^2$ is called the energy spectral density of the signal, since $\frac{1}{2\pi}|X(j\omega)|^2 d\omega$ is the signal energy over the infinitesimal frequency band ω to $\omega + d\omega$.

Example 9.8 Find the energy of the signal $x(t) = e^{2t}u(-t)$. Find the value of T such that 95% of the signal energy lies in the range $0 \le t \le T$. What is the corresponding signal bandwidth B, where B is such that 95% of the spectral energy lies in the range $0 \le \omega \le B$.

Solution

$$e^{2t}u(-t) \leftrightarrow \frac{1}{2 - j\omega}$$

The energy E of the signal is

$$E = \int_{-\infty}^{\infty} |x(t)|^2 dt = \int_{-\infty}^{0} e^{4t} dt = \frac{1}{4}$$

By changing the lower limit to $-T$, we get

$$\int_{-T}^{0} e^{4t}dt = \frac{1}{4}(1 - e^{-4T}) = \frac{0.95}{4} = 0.2375$$

Solving for T, we get $T = 0.7489\,s$. This value is useful in truncating the signal for numerical analysis.

From the spectrum,

$$\frac{1}{\pi}\int_{0}^{B} \frac{d\omega}{2^2 + \omega^2} = \frac{1}{2\pi}\tan^{-1}\left(\frac{B}{2}\right) = 0.2375 \quad \text{or} \quad B = 2\tan(0.2375(2\pi)) = 25.4124 \text{ rad/s}$$

The sampling interval required to sample this signal can be determined using this value. The sampling frequency must be greater than $(2)(25.4124)\,$rad/s. Therefore, the sampling interval must be smaller than $\frac{2\pi}{(2)(25.4124)} = 0.1236\,s$. For practical signals, this type of analysis can be carried out using numerical methods. ∎

The input and output of a LTI system, in the frequency domain, is related by the transfer function $H(j\omega)$ as $Y(j\omega) = H(j\omega)X(j\omega)$, where $X(j\omega)$, $Y(j\omega)$, and $H(j\omega)$ are the FT of the input, output, and impulse response of the system. An ideal filter will pass the amplitude of the frequency components of a deterministic signal, unattenuated, which lie in its passband.

The output energy spectrum is given by

$$|Y(j\omega)|^2 = Y(j\omega)Y^*(j\omega)$$
$$= H(j\omega)X(j\omega)H^*(j\omega)X^*(j\omega) = |H(j\omega)|^2|X(j\omega)|^2$$

The quantity $|H(j\omega)|^2$ is called the energy transfer function, as it relates the input and output energy spectral densities of a system. An ideal filter will pass the average power of the frequency components of a random signal, unattenuated, which lie in its passband.

9.3 Fourier Transform of Mixed Class of Signals

Fourier transform version of the Fourier analysis is its most general version. It is capable of representing all types of signals. Therefore, it is the only version that can be used for the analysis of mixed class of signals. In both the time and frequency domains, the FT is continuous and aperiodic. Using scaled and shifted continuous impulses, the FT represents other classes of signals. If its spectrum is sampled, then we get a periodic version of the time-domain signal corresponding to the FS representation of signals. If the time-domain signal is sampled, then we get a periodic version of the spectrum corresponding to the DTFT representation of signals. The sampling interval in one domain determines the period in the other domain. By carrying out sampling in both the domains, we get the DFT version of signal representation.

9.3.1 The FT of a Continuous Periodic Signal

The FS representation of a periodic signal $x(t)$ in terms of its FS coefficients $X_{fs}(k)$ is

$$x(t) = \sum_{k=-\infty}^{\infty} X_{fs}(k)e^{jk\omega_0 t},$$

where ω_0 is the fundamental frequency.

In the FT representation of a continuous periodic signal, the discrete impulses in the FS spectrum are changed to continuous impulses, scaled by the FS coefficients $X_{fs}(k)$ and shifted by $k\omega_0$, the frequency of the signal. The FT of $e^{jk\omega_0 t}$ is $2\pi\delta(\omega - k\omega_0)$. Then, from the linearity property of the FT, we get

$$x(t) = \sum_{k=-\infty}^{\infty} X_{fs}(k)e^{jk\omega_0 t} \leftrightarrow X(j\omega) = 2\pi \sum_{k=-\infty}^{\infty} X_{fs}(k)\delta(\omega - k\omega_0) \qquad (9.9)$$

Therefore, the FT of a periodic signal is a sum of continuous impulses of strength $2\pi X_{fs}(k)$ occurring periodically with period ω_0. Then, the FT representation of $x(t)$ becomes

$$x(t) = \frac{1}{2\pi} \int_{-\infty}^{\infty} \left(2\pi \sum_{k=-\infty}^{\infty} X_{fs}(k)\delta(\omega - k\omega_0) \right) e^{j\omega t} d\omega$$

Example 9.9 Find the FT of the signal $x(t) = \sin(2t)$.

Solution

The FS spectrum for $\sin(2t)$ is

$$\frac{-j}{2}(\delta(k - 1) - \delta(k + 1)), \quad \omega_0 = 2$$

Multiplying this result by 2π and replacing the discrete impulses by the corresponding continuous impulses, we get the FT pair

$$\sin(2t) \leftrightarrow -j\pi(\delta(\omega - 2) - \delta(\omega + 2))$$

In general,

$$\cos(\omega_0 t + \theta) \leftrightarrow \pi(e^{-j\theta}\delta(\omega + \omega_0) + e^{j\theta}\delta(\omega - \omega_0))$$

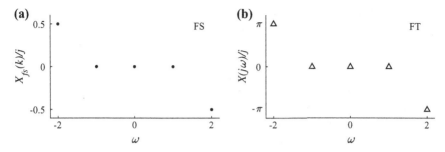

Fig. 9.6 **a** The FS spectrum, $X_{fs}(k)$, of $\sin(2t)$; **b** the FT, $X(j\omega)$, of $\sin(2t)$

The FS and FT spectra of $\sin(2t)$ are shown in Fig. 9.6a, b, respectively. Finding the IFT, we get

$$x(t) = (-j\pi)\frac{1}{2\pi}\int_{-\infty}^{\infty}(\delta(\omega - 2) - \delta(\omega + 2))e^{j\omega t}d\omega = \frac{1}{j2}(e^{j2t} - e^{-j2t}) = \sin(2t)$$

The spectra in Fig. 9.6a, b are equivalent representations of the same waveform $\sin(2t)$ by the FS and the FT. ∎

9.3.2 The FT of a Sampled Signal and the Aliasing Effect

While most naturally occurring signals are continuous, as digital processing is advantageous, the signal has be digitized invariably. The sampling interval is an important parameter in this process. In this section, we study the relation between the sampling interval and the aliasing effect. In the last section, we sampled the FT spectrum to determine the FT of a periodic signal. Now, we are going to sample the time-domain waveform to study the characteristics of the resulting periodic spectrum. The period of this spectrum is $\omega_s = 2\pi/T_s$, where T_s is the sampling interval in the time domain.

Two completely equivalent expressions for the spectrum of a sampled signal are derived. Once we represent a continuous signal by samples at discrete intervals, the effective frequency range is reduced from infinity to a finite value and its spectrum becomes periodic. A periodic signal has a FS representation. That is, the periodic spectrum has a time-domain representation composed of discrete samples. This is the same as DTFT representation with the time-domain and frequency-domain roles interchanged in the FS representation. The resulting expression for the spectrum of a sampled signal is useful for computational purposes. In a later section, we take the samples of a continuous signal and approximate its spectrum using the DFT.

The other point of view of a sampled signal is to consider it is as a sequence of continuous impulses. The continuous signal is multiplied by a sampling train to get the sequence of continuous impulses. Multiplication in the time domain becomes convolution of their spectra in the frequency domain. The FT of the impulse train is also an impulse train in the frequency domain. Convolution of a signal with an impulse is relocating it at the location of the impulse. Therefore, using this approach, the spectrum of the sampled signal is represented as a periodic repetition of that of the continuous signal being sampled. This representation enables us to visualize the spectrum from a knowledge of that of the continuous signal. Further, it clearly demonstrates the aliasing effect. It helps to truncate the bandwidth of an infinite extent spectrum to a finite extent so that a sufficiently short sampling interval can be selected to sample a continuous signal.

A continuous periodic unit-impulse train is given by

$$s(t) = \sum_{n=-\infty}^{\infty} \delta(t - nT_s),$$

where T_s is the period and n is an integer. The FS representation of the impulse train, from Chap. 7, is given as

$$s(t) = \frac{1}{T_s} \sum_{k=-\infty}^{\infty} e^{jk\omega_s t}, \quad \omega_s = \frac{2\pi}{T_s}$$

Let $x(t) \leftrightarrow X(j\omega)$. A sampled function is a sum of impulses. By multiplying $x(t)$ with an impulse train, we get its sampled version, $x_s(t)$. Therefore,

$$x_s(t) = x(t)s(t) = x(t) \sum_{n=-\infty}^{\infty} \delta(t - nT_s) = \sum_{n=-\infty}^{\infty} x(nT_s)\delta(t - nT_s)$$

The sampled signal $x_s(t)$, which is the product of $x(t)$ and $s(t)$, is also given by, using the FS representation of the impulse train,

$$x_s(t) = \frac{1}{T_s} \sum_{k=-\infty}^{\infty} x(t)e^{jk\omega_s t} = \frac{1}{T_s}(\cdots + x(t)e^{-j\omega_s t} + x(t) + x(t)e^{j\omega_s t} + \cdots)$$

Let $x_s(t) \leftrightarrow X_s(j\omega)$. Then, from the linearity and frequency-shift properties of the FT, we get

$$X_s(j\omega) = \frac{1}{T_s}(\cdots + X(j(\omega + \omega_s)) + X(j\omega) + X(j(\omega - \omega_s)) + \cdots)$$

$$= \frac{1}{T_s} \sum_{k=-\infty}^{\infty} X(j(\omega - k\omega_s)) \tag{9.10}$$

Equation (9.10) expresses the FT spectrum of the sampled signal $X_s(j\omega)$ as a scaled periodic repetition of $X(j\omega)$. The factor $\frac{1}{T_s}$ arises from the fact that

$$x(t) = \int_{-\infty}^{\infty} x(\tau)\delta(t-\tau)d\tau = \lim_{T_s \to 0} \sum_{n=-\infty}^{\infty} x(nT_s)T_s\delta(t-nT_s) = \lim_{T_s \to 0} T_s x_s(t)$$

The spectrum becomes periodic, since sampling reduces the infinite frequency range to a finite one. The longer is the sampling interval, the shorter is the unique frequency range available for the representation of the signal.

This form of representation of the FT spectrum of a sampled signal makes it easy to visualize the spectrum from a knowledge of $X(j\omega)$. If the spectrum $X(j\omega)$ is band-limited and the period of $X_s(j\omega)$ is longer than the bandwidth of $X(j\omega)$, then there will be no overlap between the copies of $X(j\omega)$ in $X_s(j\omega)$. If this condition is fulfilled, by lowpass filtering, we can recover $X(j\omega)$ from $X_s(j\omega)$ and $x(t)$ can be reconstructed exactly from its samples. This is the ideal situation. In practice, $X(j\omega)$ usually has a long tail so that there will be some overlap between the copies of $X(j\omega)$ in $X_s(j\omega)$, resulting in aliasing in the frequency domain. It has to be ensured that the sampling interval in the time domain is sufficiently short to keep the distortion of the reconstructed signal, due to aliasing, negligible. If the spectrum is sampled, then aliasing occurs in the time domain.

Figure 9.7a, b show, respectively, the continuous sinc function and its aperiodic FT spectrum.

$$x(t) = \frac{\sin\left(\frac{\pi}{2}t\right)}{\pi t} \leftrightarrow X(j\omega) = \left(u\left(\omega + \frac{\pi}{2}\right) - u\left(\omega - \frac{\pi}{2}\right)\right)$$

The signal is continuous and its spectrum is aperiodic, with value 1 between $\omega = -\frac{\pi}{2}$ to $\omega = \frac{\pi}{2}$. The peak value of the sinc function is 0.5.

Figure 9.7c, d show the sampled sinc function, with $T_s = 0.5$, and its periodic FT spectrum with period $\frac{2\pi}{0.5} = 4\pi$ rad and amplitude $\frac{1}{0.5} = 2$.

$$x_s(t) = \sum_{n=-\infty}^{\infty} \frac{\sin\left(\frac{\pi}{2}(0.5n)\right)}{\pi(0.5n)}\delta(t - 0.5n) \leftrightarrow$$

$$X_s(j\omega) = \sum_{k=-\infty}^{\infty} 2\left(u\left(\omega + \frac{\pi}{2} - 4k\pi\right) - u\left(\omega - \frac{\pi}{2} - 4k\pi\right)\right)$$

At any discontinuity of the time-domain function, the strength of the sample should be equal to the average value of the right- and left-hand limits.

The sampled signal is a sum of impulses and

$$\delta(t - nT_s) \leftrightarrow e^{-jn\omega T_s}$$

Therefore, due to the linearity property of the FT,

$$x_s(t) = \sum_{n=-\infty}^{\infty} x(nT_s)\delta(t - nT_s) \leftrightarrow X_s(j\omega) = \sum_{n=-\infty}^{\infty} x(nT_s)e^{-jn\omega T_s} \qquad (9.11)$$

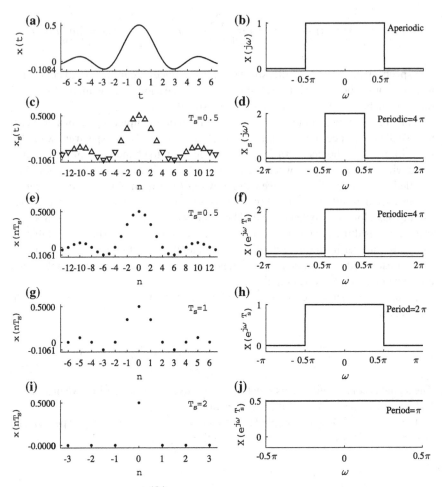

Fig. 9.7 **a** The sinc function $\frac{\sin(\frac{\pi}{2}t)}{\pi t}$ and **b** its FT spectrum; **c** samples of **a** with $T_s = 0.5$ s and **d** its periodic FT spectrum; **e** discrete samples of **a** with $T_s = 0.5$ s and **f** its DTFT spectrum with period 4π rad, which is the same as in **d**; **g** the same samples as in **e** with $T_s = 1$ and **h** its DTFT spectrum with period 2π rad; **i** samples of **a** with $T_s = 2$ s and **j** its periodic DTFT spectrum with period π radians

This form for $X_s(j\omega)$ implies that it is similar to the FS with the roles of the time and frequency domains interchanged and corresponds to the DTFT. The time-domain samples $x(nT_s)$ are the FS coefficients of the corresponding continuous periodic spectrum $X_s(j\omega)$. Equation (9.11) is, of course, completely equivalent to that in Eq. (9.10) and they express the FT of a sampled signal in different forms.

9.3.3 The FT and the DTFT of Sampled Aperiodic Signals

Let $x(nT_s)$ be the discrete sequence constructed from the sampled signal $x(nT_s)\delta(t - nT_s)$ with the amplitude of $x(nT_s)$ being the same as the strength of the corresponding continuous impulse. The DTFT of $x(nT_s)$ is defined as

$$X(e^{j\omega T_s}) = \sum_{n=-\infty}^{\infty} x(nT_s)e^{-jn\omega T_s} \tag{9.12}$$

The expressions in Eqs. (9.10) and (9.11), defining the FT of the sampled signal and the DTFT, Eq. (9.12), of the corresponding discrete sequence yield the same spectrum. Figure 9.7e, f show, respectively, the discrete samples of the sinc function $x(0.5n) = \frac{\sin(\frac{\pi}{2}(0.5n))}{\pi(0.5n)}$ with $T_s = 0.5\,\text{s}$ and its DTFT spectrum with period 4π rad, which is the same as in (d).

As scaling the frequency axis is easier than computing the DTFT with the sampling interval in the defining equation, the DTFT is usually computed assuming that $T_s = 1\,\text{s}$. Then, the frequency axis is rescaled. Figure 9.7g, h show, respectively, the samples as in (e) with $T_s = 1\,\text{s}$ and its DTFT spectrum with period 2π rad. The FT of the corresponding sampled continuous signal $x_s(t)$ is obtained by scaling the frequency axis of this DTFT spectrum so that the period of the spectrum becomes $\frac{2\pi}{T_s}$, as can be seen from Figs 9.7g, h, c, and d.

If the sampling frequency is not greater than two times that of the frequency of highest frequency component of the signal, we get a corrupted version of its FT spectrum, as shown in Fig. 9.7i, j, since the periodic repetition of $X(j\omega)$ results in the overlapping of its nonzero portions. Further, the sine component at half the sampling frequency cannot be properly represented. With $T_s = 2$, except at $t = 0$, we get zero sample values of the time-domain signal.

$$x(nT_s) = \sum_{n=-\infty}^{\infty} \frac{\sin\left(\frac{\pi}{2}(2n)\right)}{\pi(2n)} \leftrightarrow$$

$$X(e^{j\omega T_s}) = \sum_{k=-\infty}^{\infty} 0.5\left(u\left(\omega + \frac{\pi}{2} - k\pi\right) - u\left(\omega - \frac{\pi}{2} - k\pi\right)\right)$$

9.3.4 The FT and the DFT of Sampled Periodic Signals

With continuous signals, both the time-domain form and its spectrum cannot be periodic. Only in the case of the DFT, both the time- and frequency-domain forms can be periodic with the signal expressed as a linear combination of discrete harmonic components.

Assuming that a periodic signal $x(t)$ is band-limited, its FT from Eq. (9.9) is

$$x(t) = \sum_{k=-N}^{N} X_{fs}(k)e^{jk\omega_0 t} \leftrightarrow X(j\omega) = 2\pi \sum_{k=-N}^{N} X_{fs}(k)\delta(\omega - k\omega_0),$$

where $\omega_0 = \frac{2\pi}{T}$, the fundamental frequency of $x(t)$. The FT pair for the impulse train is

$$s(t) = \sum_{n=-\infty}^{\infty} \delta(t - nT_s) \leftrightarrow S(j\omega) = \frac{2\pi}{T_s} \sum_{m=-\infty}^{\infty} \delta(\omega - m\omega_s)$$

with the period in the time domain being $T_s = \frac{2\pi}{\omega_s}$.

Let us sample the periodic signal, $x(t)$. The FT $X_s(j\omega)$ of the sampled signal $x_s(t) = x(t)s(t)$ is the convolution $\frac{1}{2\pi}X(j\omega) * S(j\omega)$. As convolution of a signal with an impulse is the relocation of the origin of the signal at the location of the impulse, the FT is

$$X_s(j\omega) = \frac{2\pi}{T_s} \sum_{m=-\infty}^{\infty} \sum_{k=-N}^{N} X_{fs}(k)\delta(\omega - k\omega_0 - m\omega_s)$$

As $X(k) = (2N+1)X_{fs}(k)$, where $X(k)$ is the DFT of the $2N+1$ discrete samples of $x(t)$ over one period, we get

$$x_s(t) = \sum_{n=-\infty}^{\infty} x(nT_s)\delta(t - nT_s) \leftrightarrow X_s(j\omega)$$

$$= \frac{2\pi}{(2N+1)T_s} \sum_{m=-\infty}^{\infty} \sum_{k=-N}^{N} X(k)\delta(\omega - k\omega_0 - m\omega_s)$$

The period of the time-domain signal $x(n)$ of the DFT is $2N+1$ samples and that of corresponding sampled continuous signal $x_s(t)$ is $(2N+1)T_s = T$ s. The period of the FT spectrum is $\omega_s = \frac{2\pi}{T_s}$ rad and the spectral samples are placed at intervals of $\omega_0 = \frac{2\pi}{(2N+1)T_s} = \frac{2\pi}{T}$ rad.

The discrete samples of the continuous cosine wave $\cos(\frac{2\pi}{8}t)$ with sampling interval $T_s = 0.5$ s and its DFT spectrum are shown, respectively, in Fig. 9.8a, b. The waveform and its spectrum are periodic with period $N = 16$ samples. The sampled continuous version of the cosine wave is shown in Fig. 9.8c, which is periodic with period $NT_s = T = 8$ s. The corresponding FT spectrum is shown in Fig. 9.8d, which is periodic with period $\frac{2\pi}{T_s} = 4\pi$ rad. The sampling interval of the spectrum is $\frac{2\pi}{NT_s} = \frac{2\pi}{8} = \frac{\pi}{4}$ rad. The FT pair for the waveform is

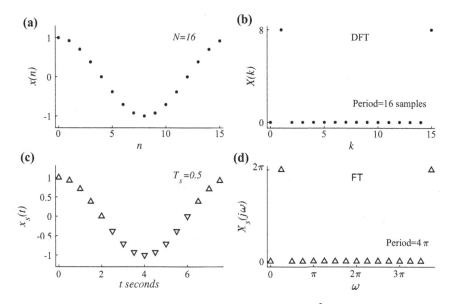

Fig. 9.8 **a** The discrete samples of the continuous cosine wave $\cos(\frac{2\pi}{8}t)$ with sampling interval $T_s = 0.5$ s and **b** its DFT spectrum; **c** the sampled version of the cosine wave $\cos(\frac{2\pi}{8}t)$ and **d** its periodic FT spectrum

$$x_s(t) = \sum_{n=-\infty}^{\infty} \cos\left(\frac{2\pi}{8}(0.5n)\right) \delta(t - 0.5n) \leftrightarrow$$

$$X_s(j\omega) = \frac{2\pi}{(16)(0.5)} \sum_{m=-\infty}^{\infty} \left(8\delta\left(\omega - \frac{2\pi}{8} - 4m\pi\right) + 8\delta\left(\omega + \frac{2\pi}{8} - 4m\pi\right)\right)$$

Both the DFT and FT spectra are equivalent representations of the same waveform. The 16-point DFT spectrum, with amplitude 8 at $k = 1$ and $k = 15$, indicates a cosine waveform $\cos(\frac{2\pi}{16}n)$. Given that the sampling interval is 0.5 s, the corresponding continuous waveform is $\cos(\frac{2\pi}{8}t)$.

The term $4m\pi$ in the FT representation indicates that the spectrum is periodic with period 4π rad and, hence, the sampling interval in the time domain is 0.5 s. The strength 8/16 of the 2 impulses is $X_{fs}(\pm 1) = 0.5$. The product $(2\pi)(0.5) = \pi$ indicates that the amplitude of the cosine waveform is 1 with frequency $\frac{2\pi}{8}$ rad. The value 0.5 in the denominator also indicates that the sampling interval is 0.5 s.

9.3.5   Reconstruction of the Continuous Signal from Its Sampled Version

A continuous signal is usually converted to digital form, processed, and reconstructed from its sampled version. In theory, the signal can be reconstructed exactly, if it is properly sampled. The major step in this process is passing the sampled signal through a lowpass filter. Ideal reconstruction requires ideal filter.

A commonly used filter for practical reconstruction is the zero-order hold filter. This filter is the simplest to approximate a continuous signal $x(t)$ from its sampled version $x_s(t)$. When an impulse is applied, its response is a rectangular pulse of unit height and width T_s, $h(t) = u(t) - u(t - T_s)$, where T_s is the sampling interval of $x_s(t)$. When $x_s(t)$ is passed through this filter, it holds the current sample value until the next sample arrives. The output signal, which is a staircase approximation of $x(t)$, is the convolution of $x_s(t)$ and $h(t)$. Convolution of a signal with an impulse is relocating it at the location of the impulse. Therefore, the convolution of $x_s(t)$, which is a sum of impulses, with $h(t)$ results in replacing each impulse of $x_s(t)$ by a pulse of width T_s and height equal to its strength.

9.4   Applications of the Fourier Transform

9.4.1   Transfer Function and the System Response

The input–output relationship of a LTI system is usually based on the impulse response in the time domain or the transfer function (frequency response) in the frequency domain. The convolution operation, in the time domain, is based on decomposing the input signal into scaled and shifted impulses. The transfer function model is based on the frequency response of the system. As convolution operation corresponds to multiplication in the frequency domain, we get

$$y(t) = \int_{-\infty}^{\infty} x(\tau)h(t - \tau)d\tau \leftrightarrow Y(j\omega) = X(j\omega)H(j\omega),$$

where $x(t)$, $h(t)$, and $y(t)$ are, respectively, the system input, impulse response, and output, and $X(j\omega)$, $H(j\omega)$, and $Y(j\omega)$ are their respective transforms. The entity in the frequency domain, $H(j\omega)$, is called the transfer function, since multiplying it with the input yields the output. The transfer function is the transform of the impulse response and the system representations in the respective domains are equivalent.

An impulse in the time domain has a uniform transform in the frequency domain, with each frequency component having the same magnitude and zero phase. Therefore, the output of a system in the frequency domain for an impulse input is called its frequency response, in addition to transfer function.

The frequency of a sinusoidal or complex exponential input signal remains the same in any part of a LTI system, with only changes in magnitude and phase. Therefore, an exponential $Ae^{j(\omega_a t+\theta)}$ or a real sinusoidal input signal $A\cos(\omega_a t + \theta)$ is changed to $(|H(j\omega_a)|A)e^{j(\omega_a t+(\theta+\angle(H(j\omega_a))))}$ and $(|H(j\omega_a)|A)\cos(\omega_a t + (\theta + \angle(H(j\omega_a))))$, respectively, at the output. The steady-state response of a stable system to the causal input $Ae^{j(\omega_a t+\theta)}u(t)$ is also the same.

The transfer function can also be characterized as the ratio of the transforms of the output and input of the system, $H(j\omega) = \frac{Y(j\omega)}{X(j\omega)}$. It is assumed that the system is initially relaxed (zero input conditions) and $X(j\omega)$ is nonzero over the range of frequencies of interest.

The differential equation characterizing a system in the time domain becomes an algebraic equation in the frequency domain, as the transform of the derivative of a signal is the transform of the signal multiplied by a factor. Consider the second-order differential equation of a stable and initially relaxed LTI continuous system.

$$\frac{d^2y(t)}{dt^2} + a_1\frac{dy(t)}{dt} + a_0y(t) = b_2\frac{d^2x(t)}{dt^2} + b_1\frac{dx(t)}{dt} + b_0x(t)$$

Taking the FT of both sides, we get

$$(j\omega)^2Y(j\omega) + a_1(j\omega)Y(j\omega) + a_0Y(j\omega) =$$
$$(j\omega)^2b_2X(j\omega) + b_1(j\omega)X(j\omega) + b_0X(j\omega)$$

The transfer function $H(j\omega)$ is obtained as

$$H(j\omega) = \frac{Y(j\omega)}{X(j\omega)} = \frac{(j\omega)^2b_2 + (j\omega)b_1 + b_0}{(j\omega)^2 + a_1(j\omega) + a_0}$$

Example 9.10 Find the steady-state current $i(t)$, using the FT, in the series circuit, consisting of a resistance R and an inductance L, excited by the input voltage $Ee^{j(\omega t+\theta)}$. Given that $R = 10\,\Omega$, $L = 0.01\,\text{H}$, $E = 100$, $\theta = \frac{\pi}{3}$, and $\omega = 2\pi60\,\text{rad/s}$. Deduce the steady-state current $i(t)$, if the excitation is $3\cos(4\pi60t + \frac{\pi}{6})$

Solution

The circuit is governed by the differential equation

$$L\frac{di(t)}{dt} + Ri(t) = Ee^{j(\omega t+\theta)}$$

In the frequency domain, we get

$$(R + j\omega L)I(j\omega) = Ee^{j\theta} \quad \text{and} \quad I(j\omega) = \frac{Ee^{j\theta}}{R + j\omega L}$$

In the time domain, we get

$$i(t) = \frac{E}{\sqrt{R^2 + \omega^2 L^2}} e^{j(\omega t + \theta - \tan^{-1}(\frac{\omega L}{R}))}$$

Substituting the values, we get

$$i(t) = 9.3572 e^{j(2\pi 60t + \frac{\pi}{3} - 0.3605))}$$

The response of the system to the input $3\cos(4\pi 60t + \frac{\pi}{6})$ is $y(t) = (3)(0.0798)$ $\cos(4\pi 60t + \frac{\pi}{6} - 0.6460)$. ∎

Example 9.11 Find the impulse response, using the FT, of the system governed by the differential equation

$$\frac{dy(t)}{dt} + 2y(t) = 4\frac{dx(t)}{dt} + 3x(t)$$

Solution

$$H(j\omega) = \frac{3 + 4j\omega}{2 + j\omega} = 4 - \frac{5}{2 + j\omega}$$

The impulse response of the system, which is the IFT of $H(j\omega)$, is $h(t) = 4\delta(t) - 5e^{-2t}u(t)$. ∎

Example 9.12 Find the zero-state response of the system governed by the differential equation

$$\frac{d^2 y(t)}{dt^2} + 5\frac{dy(t)}{dt} + 6y(t) = \frac{d^2 x(t)}{dt^2} + \frac{dx(t)}{dt} + x(t)$$

with the input $x(t) = e^{-t}u(t)$.

Solution

$$H(j\omega) = \frac{(j\omega)^2 + (j\omega) + 1}{(j\omega)^2 + 5(j\omega) + 6}$$

With $X(j\omega) = \frac{1}{1+j\omega}$,

$$Y(j\omega) = H(j\omega)X(j\omega) = \frac{(j\omega)^2 + (j\omega) + 1}{((j\omega)^2 + 5(j\omega) + 6)(1 + j\omega)}$$

Expanding into partial fractions, we get

$$Y(j\omega) = \frac{0.5}{1 + j\omega} - \frac{3}{j\omega + 2} + \frac{3.5}{(j\omega + 3)}$$

Taking the IFT, we get the zero-state response

$$y(t) = (0.5e^{-t} - 3e^{-2t} + 3.5e^{-3t})u(t)$$

∎

For all applications where Fourier analysis is suitable, it is preferred since it can be implemented using fast algorithms. The generalized version of the Fourier analysis, the Laplace transform, is suitable for stable and unstable system analysis and design with or without initial conditions.

9.4.2 Ideal Filters and Their Unrealizability

Filters are important components in signal and system analysis. The major types of filters, such as lowpass or highpass, are presented using ideal filter models. As always, in designing engineering devices, both the theoretical and practical aspects must be considered. In essence, the ideal characteristics of devices designed using the theory can be achieved only with acceptable limitations in practice. We consider the limitations in the realization of practical filters.

The ideal frequency response of a lowpass filter is shown in Fig. 9.9. As the response is even-symmetric, the specification of the response over the interval from $\omega = 0$ to $\omega = \infty$, shown in thick lines, characterizes a filter.

$$|H(j\omega)| = \begin{cases} 1 \text{ for } 0 \le \omega < \omega_c \\ 0 \text{ for } \omega > \omega_c \end{cases}$$

In the passband, from $\omega = 0$ to $\omega = \omega_c$, the filter, with gain 1, passes frequency components of a signal unattenuated and rejects the other frequency components, as they are in the stopband with gain zero. This behavior is evident from the input–output relationship in the frequency domain, $Y(j\omega) = H(j\omega)X(j\omega)$. This ideal filter model is practically unrealizable since its impulse response (IFT of $H(j\omega)$) extends from $t = -\infty$ to $t = \infty$, which requires a noncausal system. Practical filters approximate the ideal filter model to meet a given set of specifications.

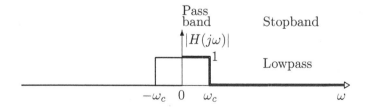

Fig. 9.9 The ideal frequency response of a lowpass filter

The impulse response of physically realizable systems must be causal. The even and odd components, for $t > 0$, of a causal time function $x(t)$ are given as

$$x_e(t) = \frac{x(t) + x(-t)}{2} = \frac{x(t)}{2} \quad \text{and} \quad x_o(t) = \frac{x(t) - x(-t)}{2} = \frac{x(t)}{2}$$

That is,

$$x(t) = 2x_e(t) = 2x_0(t), \ t > 0 \quad \text{and} \quad x_e(t) = -x_0(t), \ t < 0$$

The FT of an even real signal is real and even and that of an odd signal is imaginary and odd. Therefore, $x(t)$ can be obtained by finding the inverse FT of either the real part or the imaginary part of its spectrum $X(j\omega)$. That is,

$$x(t) = \frac{2}{\pi} \int_0^\infty \text{Re}(X(j\omega)) \cos(\omega t) d\omega = -\frac{2}{\pi} \int_0^\infty \text{Im}(X(j\omega)) \sin(\omega t) d\omega, \ t > 0$$

For example, the unit-step signal $u(t)$ can be expressed as

$$u(t) = 0.5 + 0.5 sgn(t) \leftrightarrow \pi\delta(\omega) + \frac{1}{j\omega}$$

Now,

$$u(t) = 2u_e(t) = 2u_0(t), \ t > 0 \quad \text{and} \quad u_e(t) = -u_0(t), \ t < 0$$

The point is that the real and imaginary parts or, equivalently, the magnitude and the phase of the FT of a causal signal are related. This implies that there are constraints, for the realizability, on the magnitude of the frequency response, $H(j\omega)$, of a practical filter. These constraints are given by the Paley–Wiener criterion as

$$\int_{-\infty}^\infty \frac{|\log_e |H(j\omega)||}{1 + \omega^2} d\omega < \infty$$

To satisfy this criterion, the magnitude of the frequency response $|H(j\omega)|$ can be zero at discrete points but not over any continuous band of frequencies. If $H(j\omega)$ is zero over a band of frequencies, $|\log_e |H(j\omega)|| = \infty$ and the condition is violated. However, despite zeros of $H(j\omega)$ at a finite set of discrete frequencies, the value of the integral may still be finite, although the integrand is infinite at these frequencies. In addition, any transition of this function cannot vary more rapidly than by exponential order. Obviously, the frequency response of the ideal filters fails to meet the Paley–Wiener criterion. Further, an infinite order filter is required to have a flat passband. Therefore, neither the flatness of the bands nor the sharpness of the transition between the bands of ideal filters is realizable by practical filters.

9.5 Approximation of the Fourier Transform

As the FT is continuous and aperiodic in both the domains and the DFT is discrete and periodic, both the sampling interval and record length have to be fixed in approximating the samples of the FT by the DFT. The criteria in selecting these parameters are that both the reconstructed signal and its spectrum must be adequate representations of the signal. The defining integrals of the FT and the IFT are approximated using the rectangular rule of numerical integration. As periodicity is assumed in DFT and IDFT computation, the sampling can start at $t = 0$ or $\omega = 0$. The given truncated signal or spectrum, whatever range it is defined, can be periodically extended for this purpose.

The problem of numerical integration is the numerical evaluation of a definite integral. That is, to find the area under the given function between the given limits. Using the rectangular rule of integration, we divide the period T into N intervals of width $T_s = \frac{T}{N}$ and represent the signal at N points as

$$\left\{ x(0), x\left(\frac{T}{N}\right), x\left(2\frac{T}{N}\right), \ldots, x\left((N-1)\frac{T}{N}\right) \right\}$$

Then, the total area is the sum of those of the N rectangles. The sampling interval in the time domain is T_s s and that in the frequency domain is $\frac{2\pi}{NT_s} = \frac{2\pi}{T}$ rad/s.

The FT definition is approximated as

$$X\left(j\frac{2\pi k}{NT_s}\right) = T_s \sum_{n=0}^{N-1} x(nT_s) e^{-j\frac{2\pi}{N}nk}, \quad k = 0, 1, \ldots, N-1 \quad (9.13)$$

The IFT definition is approximated as

$$x(nT_s) = \frac{1}{NT_s} \sum_{k=0}^{N-1} X\left(j\frac{2\pi k}{NT_s}\right) e^{j\frac{2\pi}{N}nk}, \quad n = 0, 1, \ldots, N-1 \quad (9.14)$$

The approximate samples of the FT spectrum are obtained by multiplying the DFT coefficients of the samples of the input signal by the sampling interval T_s. By dividing the IDFT values of the samples of the input FT spectrum by T_s, we get the approximate samples of the time-domain signal.

Example 9.13 Find the approximate samples of the FT magnitude spectrum of the signal $x(t) = e^{-2t}u(t)$ using the DFT.

Solution

$$X(j\omega) = \frac{1}{2 + j\omega} \quad \text{and} \quad |X(j\omega)| = \frac{1}{\sqrt{4 + \omega^2}} \approx \frac{1}{\omega} \quad \text{for} \quad \omega \gg 2$$

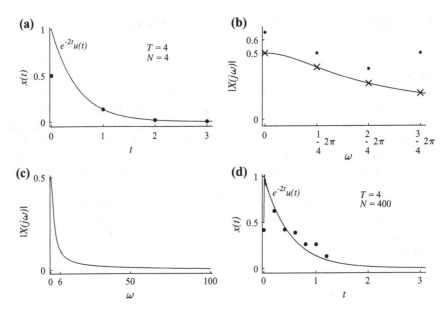

Fig. 9.10 **a** The exponential waveform $x(t) = e^{-2t}u(t)$, with 4 samples over the range $0 \leq t < 4$; **b** the magnitude of the FT (solid line) and the samples of the FT obtained through the DFT with $N = 4$ (dots) and $N = 1024$ (crosses) samples; **c** the magnitude of the FT obtained through the DFT with $N = 1024$ samples; **d** the partially and fully reconstructed waveform

Figure 9.10a shows the exponential signal $e^{-2t}u(t)$ with 4 samples over a record length of $T = 4$ s. The record length has to be sufficiently long so that the amplitude or power of the signal becomes negligible beyond the record length chosen. With $T = 4$, $e^{-8} = 0.00033546$, which is very small compared with the peak value of the signal 1. Figure 9.10b shows the magnitude of the FT and the samples of the FT obtained through the DFT with $N = 4$ and $N = 1024$ samples. The 4 sample values of the signal in Fig. 9.10a are

$$\{0.5000, 0.1353, 0.0183, 0.0025\}$$

It should be remembered that the number of samples has to be quite large for adequate representation. To make the calculation simple for illustration purpose, we use 4 samples. As the first sample value occurs at a discontinuity, it is assigned the average of the left- and right-hand limits of the discontinuity. The bandwidth of the signal is also infinite. As with typical practical signals, the spectral values become negligible beyond some finite range. The magnitudes of the DFT of these values, after scaling by $T_s = 1$ s, are

$$\{0.6561, 0.4997, 0.3805, 0.4997\}$$

The corresponding complex values are

$$\{0.6561, 0.4817 - j0.1329, 0.3805, 0.4817 + j0.1329\}$$

For example, the DC coefficient 0.6561 is the sum of the input samples. As the second half of the DFT spectrum is redundant, only the first three values are useful. The magnitudes of the first 5 samples of the FT, $X(j\omega)$, are

$$\{0.5000, 0.3932, 0.2685, 0.1953, 0.1517\}$$

The corresponding complex values are

$$\{0.5, 0.3092 - j0.2429, 0.1442 - j0.2265, 0.0763 - j0.1798, 0.0460 - j0.1445\}$$

With $\omega = 0$, we get $X(j\omega) = 0.5$. Since the sampling interval $T_s = 1$ s is too long, due to aliasing, the spectral samples obtained by the DFT are very inaccurate. For example, with $N = 4$ samples, the DC coefficient is aliased with every fourth spectral sample. Adding all the aliased components, the spectral value becomes 0.6561. In a similar manner, all the coefficients get aliased. The magnitudes of the first few of the DFT values, with $T_s = 0.5$ s, are

$$\{0.5407, 0.4100, 0.2499, 0.1523\}$$

The magnitudes of the first few of the DFT values, with $T_s = 0.25$ s, are

$$\{0.5102, 0.3965, 0.2610, 0.1779, 0.1249, 0.0877, 0.0598\}$$

The aliasing gets reduced and the accuracy improves with more samples, as expected. The magnitudes of the first five samples of the FT obtained through the DFT with $N = 1024$ are
$$\{0.4998, 0.3931, 0.2684, 0.1953, 0.1516\}$$

The DFT values are quite close to the theoretical values. For an arbitrary waveform, with no FT expression available, one has to resort to trial and error method in approximating the FT spectrum. In general, the DFT values will never be exactly equal to the analytical values, but they can be made adequate, for practical purposes, by selecting appropriate sampling interval and record length. If the number of spectral samples are insufficient to represent the spectrum properly (picket-fence effect), for a given record length, the signal may be zero padded to increase the number of spectral samples.   ∎

A similar procedure for the approximation of the IFT is required. Now, we have to fix the record length of the spectrum. The spectrum, as shown in Fig. 9.10c for positive values of ω (the spectrum is conjugate-symmetric), is slowly decaying and is of infinite extent. The cutoff frequency has to be selected to suit the accuracy requirements. The magnitude of the spectrum is approximately $1/\omega$ for large values of ω. The peak value of the spectrum is 0.5. If we want to discard the values of the spectrum those are less than one-hundredth of the peak, then $\omega = 200\,\text{rad}$ is the cutoff frequency. Of course, the cutoff frequency can also be fixed based on signal energy.

In contrast to most signals in the time domain, the spectra of signals are almost always two-sided. For real signals, the spectrum is conjugate-symmetric and, usually, the positive frequency side is shown, as can be seen in Fig. 9.10b, c. For the example waveform, the record length T in the time domain is 4 s and ω_s, the sampling frequency of the spectrum is $\pi/2\,\text{rad}$.

Let the sampling interval of the reconstructed signal be T_s s. Then, the record length of the reconstructed signal is NT_s s, where N is the number of samples. Let the number of samples N be 7 and the sampling interval T_s be 0.2 s. Then, the record length T in the time domain is 1.4 s. Let the frequency increment be 2.5π. Then, the 4 samples of the FT spectrum at

$$\omega = \{0, 7.8540, 15.7080, 23.5619\}$$

are

$$\{0.5, 0.0304 - j0.1196, 0.0080 - j0.0626, 0.0036 - j0.0421\}$$

Conjugating the last 3 spectral samples and concatenating in the reverse order, we get

$$\{0.5, 0.0304 - j0.1196, 0.008 - j0.0626, 0.0036 - j0.0421, 0.0036$$
$$+ j0.0421, 0.008 + j0.0626, 0.0304 + 0.1196\}$$

The IDFT of these values divided by $T_s = 0.2$ are

$$\{0.4171, 0.6240, 0.4210, 0.3868, 0.2611, 0.2597, 0.1302, \}$$

shown in Fig. 9.10d. The actual samples of e^{-2t} at

$$t = \{0, 0.2, 0.4, 0.6, 0.8, 1.0, 1.2\}$$

are

$$\{1, 0.6703, 0.4493, 0.3012, 0.2019, 0.1353, 0.0907\}$$

The fully reconstructed waveform, with $N = 400$ and $T = 4$, is also shown in Fig. 9.10d.

The appropriate sampling interval and record length, to approximate the FT of an arbitrary signal, are fixed by a trial and error procedure. First, we select an initial record length based on the signal magnitude or energy. Then, the DFT of the samples of the signal is computed with an initial sampling frequency. Then, we keep on doubling the frequency so that the spectral values at the middle of the spectrum (around $X(N/2)$) become negligible for real signals. For complex signals, the spectral values should become negligible at the end of the spectrum. That frequency is appropriate for sampling the signal. Once the sampling frequency is fixed, we keep increasing the record length and compute the DFT of the samples. When the DFT values are almost the same for two consecutive record lengths, the record length is fixed. Further, zero pad the signal sufficiently and compute the DFT so that the there is no picket-fence effect.

9.6 Summary

- The FT is the most general version of the Fourier analysis. It can represent all types of signals and, therefore, most useful in the analysis of mixed class of signals in addition to the analysis of continuous aperiodic signals.
- The FT is primarily used in the analysis of continuous aperiodic signals. The spectrum is also continuous and aperiodic.
- The FT can be considered as the limiting case of the FS with the period of the waveform tending to infinity. The FT can also be considered as the limiting case of the DTFT with the sampling interval of the time-domain sequence tending to zero.
- The FT spectrum is a relative amplitude spectrum.
- The FT is defined by an infinite integral. Therefore, a sufficient condition for its existence is that the time-domain signal is absolutely integrable.
- The FT can be effectively approximated by the DFT in practical applications.

Exercises

9.1 Find the FT of the rectangular pulse $x(t) = u(t+3) - u(t-3)$ using the defining integral. Find $X(j0)$ and $X(j0.5\pi)$.

* 9.2 Find the FT of the signal $x(t) = e^{-2t}\cos(3t)u(t)$, $a > 0$ using the defining integral. Find $X(j0)$ and $X(j\pi)$.

9.3 Find the FT of the signal $x(t) = e^{-2|t|}$ using the defining integral. Find $X(j0)$ and $X(j2)$.

9.4 Find the FT of the signal

$$x(t) = \begin{cases} \cos(2t) & \text{for } |t| < 2 \\ 0 & \text{for } |t| > 2 \end{cases}$$

using the defining integral. Find $X(j0)$ and $X(j\pi)$.

9.5 Find the FS spectrum, from the FT of its one period, of the periodic signal $x(t)$, one period of which is defined.

* 9.5.1

$$x(t) = t, \quad \text{for } 0 < t < 1$$

9.5.2

$$x(t) = \sin(t), \quad \text{for } 0 < t < \pi$$

9.6 Using the duality property, find the FT of the signal $x(t)$.

9.6.1. $x(t) = \frac{1}{2\pi(3+jt)}$.

* 9.6.2. $x(t) = \frac{1}{2\pi}\left(\pi\delta(t) + \frac{1}{jt}\right)$.

9.6.3. $x(t) = \frac{4}{\pi(16+t^2)}$.

9.7 Using the linearity and frequency-shifting properties, find the FT of $x(t)$.

9.7.1 $x(t) = 2\sin(3t)\cos(3t)$.

9.7.2

$$x(t) = \cos(3t)\cos(2t) - \sin(3t)\sin(2t)$$

* 9.7.3

$$x(t) = \begin{cases} \sin(3t) & \text{for } |t| < 2 \\ 0 & \text{for } |t| > 2 \end{cases}$$

9.8 Find the convolution of $x(t)$ and $h(t)$ using their FT.

9.8.1

$$x(t) = \begin{cases} 1 & \text{for } -0.5 < t < 0.5 \\ 0 & \text{otherwise} \end{cases}$$

and $h(t) = x(t)$.

* 9.8.2 $x(t) = e^{-5t}u(t)$ and $h(t) = e^{-4t}u(t)$.

9.8.3 $x(t) = e^{-2t}u(t)$ and $h(t) = \begin{cases} 1 & \text{for } -0.5 < t < 0.5 \\ 0 & \text{otherwise} \end{cases}$

9.9 Find the FT of the product of $x(t)$ and $h(t)$ using the frequency-domain convolution property.

* 9.9.1 $x(t) = \cos(3t)$ and $h(t) = u(t)$.

9.9.2 $x(t) = \cos(2t)$ and $h(t) = \begin{cases} 1 & \text{for } |t| < 3 \\ 0 & \text{otherwise} \end{cases}$

9.9.3 $x(t) = e^{-t}$ and $h(t) = x(t)$

9.10 Find the FT of the signal $x(at)$ in terms of that of $x(t)$, using the time-scaling property.

9.10.1 $x(t) = \sin(2t)$ and $a = -3$.

9.10.2 $x(t) = e^{-6t}u(t)$ and $a = \frac{1}{3}$.

* 9.10.3 $x(t) = u(t - 6)$ and $a = 3$.

9.11 Find the FT of the derivative of the signal $x(t)$ in terms of that of $x(t)$, using the time-differentiation property.

9.11.1 $x(t) = \cos(3t)$.
9.11.2 $x(t) = u(t)$.
* 9.11.3

$$x(t) = \begin{cases} t & \text{for } 0 < t < 2 \\ 0 & \text{otherwise} \end{cases}$$

9.12 Find the FT of $y(t)$, where

$$y(t) = \int_{-\infty}^{t} x(\tau)d\tau$$

in terms of the FT of $x(t)$, using the time-integration property.
9.12.1 $x(t) = \delta(t + 2)$.
9.12.2 $x(t) = \sin(2t)$.
* 9.12.3 $x(t) = e^{-2t}u(t)$

9.13 Find the energy of the signal $x(t) = e^{-t}u(t)$ in the time domain. Find the value of T such that 90% of the signal energy lies in the range $0 \le t \le T$. What is the corresponding signal bandwidth B.

9.14 Find the FT of the signal $x(t)$.
* 9.14.1

$$x(t) = \sin\left(\frac{2\pi}{8}t + \frac{\pi}{6}\right).$$

9.14.2

$$x(t) = \cos\left(\frac{2\pi}{8}t + \frac{\pi}{4}\right).$$

9.15 Find the FT of $x(t)$. Find the FT of the sampled version $x_s(t)$ of $x(t)$ in 2 different forms with sampling interval $T_s = 1$ and $T_s = 0.5$ s. List the sample values of the 3 versions of the spectra at $\omega = \{0, 0.25\pi, 0.5\pi, 0.75\pi, \pi\}$ and compare the results.
* 9.15.1 $x(t) = u(t + 5) - u(t - 5)$.
9.15.2 $x(t) = (t + 1)u(t + 1) - 2tu(t) + (t - 1)u(t - 1)$.

9.16 Find the DFT of $x(t)$ with sampling interval $T_s = 0.5$ s. Express the FT representation of the sampled $x_s(t)$ in terms of the DFT coefficients.
9.16.1

$$x(t) = 3\cos\left(2\frac{2\pi}{8}t\right)$$

9.16.2

$$x(t) = \sin\left(\frac{2\pi}{8}t\right)$$

* 9.16.3

$$x(t) = 2\cos\left(3\frac{2\pi}{8}t + \frac{\pi}{3}\right)$$

9.17 Find the steady-state current $i(t)$, using the FT, in the series circuit, consisting of a resistance R and an inductance L, excited by the input voltage $Ee^{j(\omega t+\theta)}$. Given that $R = 20\,\Omega$, $L = 0.02\,\text{H}$, $E = 100$, $\theta = \frac{\pi}{6}$, and $\omega = 2\pi 60\,\text{rad/s}$.

9.18 Find the impulse response, using the FT, of the system governed by the differential equation

$$\frac{dy(t)}{dt} + y(t) = 2\frac{dx(t)}{dt} + 3x(t)$$

* 9.19 Find the zero-state response of the system governed by the differential equation

$$\frac{d^2y(t)}{dt^2} + 4\frac{dy(t)}{dt} + 3y(t) = \frac{d^2x(t)}{dt^2} + \frac{dx(t)}{dt} + x(t)$$

with the input $x(t) = e^{-3t}u(t)$.

9.20 Derive the FT $X(j\omega)$ of the signal $x(t)$. Let the record length be $T = 5\,\text{s}$. Take 4 uniform samples of $x(t)$ and compute the DFT. Compare the scaled DFT values with the corresponding samples of $X(j\omega)$.
* 9.20.1 $x(t) = e^{-t}u(t)$.
9.20.2 $x(t) = te^{-t}u(t)$.
9.20.3 $x(t) = e^{-t}\cos(0.2\pi t)u(t)$.

9.21 Derive the FT $X(j\omega)$ of the signal $x(t)$. Let the frequency increment be $\omega = 2.5\pi\,\text{rad}$ and time-domain sampling interval be $0.2\,\text{s}$. Take 7 uniform samples of $X(j\omega)$ and compute the IDFT. Compare the scaled IDFT values with the corresponding samples of $x(t)$.
* 9.21.1 $x(t) = e^{-\frac{3}{2}t}u(t)$.
9.21.2 $x(t) = te^{-2t}u(t)$.
9.21.3 $x(t) = e^{-2t}\sin(0.2\pi t)u(t)$.

Chapter 10
Fast Computation of the DFT

It is the availability of fast algorithms to compute the DFT that makes Fourier analysis indispensable in practical applications, in addition to its theoretical importance from its invention. In turn, the other versions of the Fourier analysis can be approximated adequately by the DFT. Although the algorithm was invented by Gauss in 1805, it is the widespread use of the digital systems that has given its importance in practical applications. The algorithm is based on the classical divide-and-conquer strategy of developing fast algorithms. A problem is divided into two smaller problems of half the size. Each smaller problem is solved separately and the solution to the original problem is found by appropriately combining the solutions of the smaller problems. This process is continued recursively until the smaller problems reduce to trivial cases. Therefore, the DFT is never computed using its definition. While there are many variations of the algorithm, the order of computational complexity of all of them is the same, $O(N \log_2 N)$ to compute a N-point DFT against $O(N^2)$ from its definition. It is this reduction in computational complexity by an order that has resulted in the widespread use of the Fourier analysis in practical applications of science and engineering. In this chapter, a particular variation of the algorithm, called the PM DFT algorithm, is presented. The algorithm is developed using the half-wave symmetry of periodic waveforms. This approach gives a better physical explanation of the algorithm than other approaches such as matrix factorization. The DFT is defined for any length. However, the practically most useful DFT algorithms are of length that is an integral power of 2. That is $N = 2^M$, where M is a positive integer. If necessary, zero padding can be employed to the sequences so that they satisfy this constraint.

© Springer Nature Singapore Pte Ltd. 2018
D. Sundararajan, *Fourier Analysis—A Signal Processing Approach*,
https://doi.org/10.1007/978-981-13-1693-7_10

10.1   Half-Wave Symmetry of Periodic Waveforms

A physical understanding of the basics of the DFT algorithms can be obtained using
the half-wave symmetry of periodic waveforms. If a given periodic function $x(n)$
with period N satisfies the condition

$$x\left(n \pm \frac{N}{2}\right) = x(n),$$

then it is said to be even half-wave symmetric.

The samples of the function over any half period are the same as those in the
succeeding or preceding half period. In effect, the period is $N/2$. If the DFT is
computed over the period N, then the odd-indexed DFT coefficients will be zero.
That is, the function is composed of even-indexed frequency components only.

If a given periodic function $x(n)$ with period N satisfies the condition

$$x\left(n \pm \frac{N}{2}\right) = -x(n),$$

then it is said to be odd half-wave symmetric. The samples of the function over any
half period are the negatives of those in the succeeding or preceding half period. If
the DFT is computed over the period N, then the even-indexed DFT coefficients will
be zero. That is, the function is composed of odd-indexed frequency components
only. It is due to the fact that any periodic function can be uniquely decomposed
into even half-wave and odd half-wave symmetric components. If the even half-
wave symmetric component is composed of the even-indexed frequency components,
then the odd half-wave symmetric component must be composed of the odd-indexed
frequency components. Therefore, if an arbitrary function is decomposed into its
even half-wave and odd half-wave symmetric components, then we have divided the
original problem of finding the N frequency coefficients into two problems, each of
them being the determination of $N/2$ frequency coefficients. First, let us go through
an example of decomposing a periodic function into its even half-wave and odd
half-wave symmetric components.

An arbitrary periodic sequence $x(n)$ of period N can be expressed as the sum of
its even and odd half-wave symmetric components $x_{eh}(n)$ and $x_{oh}(n)$, respectively, as

$$x(n) = x_{eh}(n) + x_{oh}(n) \tag{10.1}$$

in which

$$x_{eh}\left(n \pm \frac{N}{2}\right) = x_{eh}(n) \quad \text{and} \quad x_{oh}\left(n \pm \frac{N}{2}\right) = -x_{oh}(n)$$

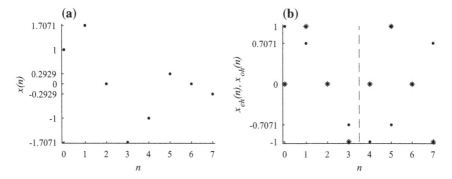

Fig. 10.1 **a** $x(n) = \cos(\frac{2\pi}{8}n) + \sin(2\frac{2\pi}{8}n)$; **b** the odd-indexed frequency component (dots) $\cos(\frac{2\pi}{8}n)$ and the even-indexed frequency component $\sin(2\frac{2\pi}{8}n)$

Adding the two components, we get

$$x\left(n \pm \frac{N}{2}\right) = x_{eh}(n) - x_{oh}(n) \tag{10.2}$$

Solving for the components using Eqs. (10.1) and (10.2), we get

$$x_{eh}(n) = \frac{1}{2}\left(x(n) + x\left(n \pm \frac{N}{2}\right)\right) \quad \text{and} \quad x_{oh}(n) = \frac{1}{2}\left(x(n) - x\left(n \pm \frac{N}{2}\right)\right)$$

Figure 10.1a shows the waveform

$$x(n) = \cos\left(\frac{2\pi}{8}n\right) + \sin\left(2\frac{2\pi}{8}n\right)$$

The first is the odd-indexed frequency component and the second is the even-indexed. Figure 10.1b shows the two components of the waveform $x(n)$. The odd-indexed component is shown by dots and the even-indexed component is shown by the symbol $*$. The samples of $x(n)$ are

$$\left\{1, 1 + \frac{\sqrt{2}}{2}, 0, -\left(1 + \frac{\sqrt{2}}{2}\right), \quad -1, 1 - \frac{\sqrt{2}}{2}, 0, \left(-1 + \frac{\sqrt{2}}{2}\right)\right\}$$

The samples of $x_{eh}(n)$ are

$$\{0, 1, 0, -1, \quad 0, 1, 0, -1\}$$

The samples of $x_{oh}(n)$ are

$$\left\{1, \frac{\sqrt{2}}{2}, 0, -\frac{\sqrt{2}}{2}, -1, -\frac{\sqrt{2}}{2}, 0, \frac{\sqrt{2}}{2}\right\}$$

It can be verified that $x(n) = x_{eh}(n) + x_{oh}(n)$. The DFT $X(k)$ of $x(n)$ is

$$\{0, 4, -j4, 0, 0, 0, j4, 4\}$$

The DFT $X_{eh}(k)$ of $x_{eh}(n)$ is

$$\{0, 0, -j4, 0, 0, 0, j4, 0\}$$

The DFT $X_{oh}(k)$ of $x_{oh}(n)$ is

$$\{0, 4, 0, 0, 0, 0, 0, 4\}$$

It can be verified that $X(k) = X_{eh}(k) + X_{oh}(k)$. Due to the reduction in the number of frequency components, $x_{eh}(n)$ and $x_{oh}(n)$ can be represented by 4 samples. The dashed line in the middle of Fig. 10.1b indicates the redundancy of data values of $x_{eh}(n)$ and $x_{oh}(n)$. The problem of computing the 8-point DFT of $x(n)$ has been reduced to two problems of 4-point DFTs of $x_{eh}(n)$ and $x_{oh}(n)$. This is the essence of all the DFT algorithms for sequence lengths those are integral powers of 2. While the sinusoidal waveforms are easy to visualize, as usual, the DFT problem using the equivalent complex exponential form is required for further description of the algorithms.

Recursively carrying out the even and odd half-wave symmetric components decomposition, along with frequency shifting and decimation of the frequency components, the frequency components are eventually isolated yielding their coefficients. In carrying out these operations, the coefficients are scaled and scrambled but remains unchanged otherwise. The decomposition of a waveform into its symmetric components is the principal operation in the algorithm, requiring repeated execution of add–subtract (plus–minus) operations. Since the frequency components are decomposed into smaller groups, this type of algorithms is named as the PM DIF DFT algorithms, where PM stands for (plus–minus) and DIF stands for (decimation-in-frequency).

10.2   The PM DIF DFT Algorithm

A given waveform $x(n)$ can be expressed as the sum of N frequency components (e.g., with $N = 8$),

$$x(n) = X(0)e^{j0\frac{2\pi}{8}n} + X(1)e^{j1\frac{2\pi}{8}n} + X(2)e^{j2\frac{2\pi}{8}n} + X(3)e^{j3\frac{2\pi}{8}n}$$
$$+ X(4)e^{j4\frac{2\pi}{8}n} + X(5)e^{j5\frac{2\pi}{8}n} + X(6)e^{j6\frac{2\pi}{8}n} + X(7)e^{j7\frac{2\pi}{8}n}, \quad n = 0, 1, \ldots, 7$$

For the most compact algorithms, the input sequence $x(n)$ has to be expressed as 2-element vectors

$$a^0(n) = \{a_0^0(n), a_1^0(n)\} = 2\left\{x_{eh}(n), x_{oh}(n), \ n = 0, 1, \ldots, \frac{N}{2} - 1\right\}$$

The first and second elements of the vectors are, respectively, the scaled even and odd half-wave symmetric components $x_{eh}(n)$ and $x_{oh}(n)$ of $x(n)$. That is, the DFT expression is reformulated as

$$X(k) = \sum_{n=0}^{N-1} x(n)e^{-j\frac{2\pi}{N}kn}, \quad k = 0, 1, \ldots, N - 1$$

$$= \begin{cases} \displaystyle\sum_{n=0}^{(N/2)-1}\left(x(n) + x\left(n + \frac{N}{2}\right)\right)e^{-j\frac{2\pi}{N}kn} = \sum_{n=0}^{(N/2)-1} a_0^0(n)e^{-j\frac{2\pi}{N}kn}, & k \text{ even} \\[4mm] \displaystyle\sum_{n=0}^{(N/2)-1}\left(x(n) - x\left(n + \frac{N}{2}\right)\right)e^{-j\frac{2\pi}{N}kn} = \sum_{n=0}^{(N/2)-1} a_1^0(n)e^{-j\frac{2\pi}{N}kn}, & k \text{ odd} \end{cases}$$

The division by 2 in finding the symmetric components is deferred. For the example shown in Fig. 10.1,

$$2x_{eh}(n) = a_0^0(n) = \{0, 2, 0, -2\}$$

$$X(k) = \sum_{n=0}^{3} a_0^0(n)e^{-j\frac{2\pi}{8}kn}, \ k = 0, 2, 4, 6$$

$$= \sum_{n=0}^{3} a_0^0(n)e^{-j\frac{2\pi}{4}kn}, \ k = 0, 1, 2, 3 = \{0, -j4, 0, j4\}$$

For the example shown in Figure 10.1,

$$2x_{oh}(n) = a_1^0(n) = \{2, \sqrt{2}, 0, -\sqrt{2}\}$$

$$X(k) = \sum_{n=0}^{3} a_1^0(n)e^{-j\frac{2\pi}{8}kn}, \ k = 1, 3, 5, 7$$

$$= \sum_{n=0}^{3} (e^{-j\frac{2\pi}{8}n}a_1^0(n))e^{-j\frac{2\pi}{4}kn}, \ k = 0, 1, 2, 3 = \{4, 0, 0, 4\}$$

In order to reduce this computation to a 4-point DFT, we have to multiply the samples of $a_1^0(n)$ by the twiddle factor samples $e^{-j\frac{2\pi}{8}n}$. This step results in frequency shifting of the DFT spectrum and the odd-indexed coefficients become even-indexed. Therefore, the 8-point DFT becomes a 4-point DFT. The pointwise product of $e^{-j\frac{2\pi}{8}n}$ and $a_1^0(n)$ is

$$(e^{-j\frac{2\pi}{8}n}a_1^0(n)) = \left(1, \frac{\sqrt{2}}{2} - j\frac{\sqrt{2}}{2}, -j, -\frac{\sqrt{2}}{2} - j\frac{\sqrt{2}}{2}\right)(2, \sqrt{2}, 0, -\sqrt{2})$$

$$= (2, 1 - j1, 0, 1 + j1)$$

Then,

$$\{2, 1 - j1, 0, 1 + j1\} \leftrightarrow \{4, 0, 0, 4\}$$

The recursive decomposition is continued until all the individual DFT coefficients are extracted, as shown in Fig. 10.3.

In terms of the frequency components,

$$a^0(n) = \{a_0^0(n), a_1^0(n)\} = 2\{x_{eh}(n), x_{oh}(n)\}$$

$$= 2\{X(0)e^{j0\frac{2\pi}{8}n} + X(2)e^{j2\frac{2\pi}{8}n} + X(4)e^{j4\frac{2\pi}{8}n} + X(6)e^{j6\frac{2\pi}{8}n},$$

$$X(1)e^{j1\frac{2\pi}{8}n} + X(3)e^{j3\frac{2\pi}{8}n} + X(5)e^{j5\frac{2\pi}{8}n} + X(7)e^{j7\frac{2\pi}{8}n}\}, \quad n = 0, 1, 2, 3$$

The array of vectors $a^0(n)$ is stored in the nodes at the beginning of the signal-flow graph of the algorithm shown in Fig. 10.2. Although a DFT algorithm can be expressed in other forms, the signal-flow graph is the most suitable form for its description. The repetitive nature of the basic computation is evident. The nodes, shown by discs, store a vector $a^r(n)$. Arrows indicate the signal-flow path. The first elements of vectors stored in a pair of source nodes produce the vector for the sink node connected by a upward-pointing arrow by add–subtract operation. Any integer value n near an arrow indicates that the value from a source node has to be multiplied by $e^{-j\frac{2\pi}{N}n}$ before the add–subtract operation. The second elements of vectors stored in a pair of source nodes produce the vector for the sink node connected by a downward-pointing arrow. There are $\log_2 N - 1$ stages of the algorithm. The input nodes are source nodes and the output nodes are sink nodes. Rest of the nodes serve both as source and sink nodes.

Each of the two symmetric components has only four frequency components and four samples are adequate to represent them. The even half-wave symmetric component is

$$2(X(0)e^{j0\frac{2\pi}{8}n} + X(2)e^{j2\frac{2\pi}{8}n} + X(4)e^{j4\frac{2\pi}{8}n} + X(6)e^{j6\frac{2\pi}{8}n})$$

$$= 2(X(0)e^{j0\frac{2\pi}{4}n} + X(2)e^{j1\frac{2\pi}{4}n} + X(4)e^{j2\frac{2\pi}{4}n} + X(6)e^{j3\frac{2\pi}{4}n}), \quad n = 0, 1, 2, 3$$

The set of frequency coefficients of this component is

$$2\{X(0), X(2), X(4), X(6)\}$$

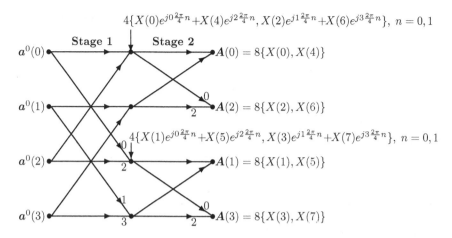

$$4\{X(0)e^{j0\frac{2\pi}{4}n}+X(4)e^{j2\frac{2\pi}{4}n}, X(2)e^{j1\frac{2\pi}{4}n}+X(6)e^{j3\frac{2\pi}{4}n}\},\ n=0,1$$

Stage 1   Stage 2

$a^0(0)$ $A(0) = 8\{X(0), X(4)\}$

$a^0(1)$ $A(2) = 8\{X(2), X(6)\}$

$$4\{X(1)e^{j0\frac{2\pi}{4}n}+X(5)e^{j2\frac{2\pi}{4}n}, X(3)e^{j1\frac{2\pi}{4}n}+X(7)e^{j3\frac{2\pi}{4}n}\},\ n=0,1$$

$a^0(2)$ $A(1) = 8\{X(1), X(5)\}$

$a^0(3)$ $A(3) = 8\{X(3), X(7)\}$

Fig. 10.2 Signal-flow graph of the PM DIF DFT algorithm, with $N = 8$. A twiddle factor $W_8^n = e^{-j\frac{2\pi}{8}n}$ is indicated only by its variable part of the exponent, n

The set of frequency coefficients of the odd half-wave symmetric component is

$$2\{X(1), X(3), X(5), X(7)\}$$

To reformulate this component as a 4-point DFT, it has to be multiplied by the exponential $e^{-j\frac{2\pi}{8}n}$ to get

$$2(X(1)e^{j1\frac{2\pi}{8}n} + X(3)e^{j3\frac{2\pi}{8}n} + X(5)e^{j5\frac{2\pi}{8}n} + X(7)e^{j7\frac{2\pi}{8}n})e^{-j\frac{2\pi}{8}n}$$
$$= 2(X(1)e^{j0\frac{2\pi}{8}n} + X(3)e^{j2\frac{2\pi}{8}n} + X(5)e^{j4\frac{2\pi}{8}n} + X(7)e^{j6\frac{2\pi}{8}n})$$
$$= 2(X(1)e^{j0\frac{2\pi}{4}n} + X(3)e^{j1\frac{2\pi}{4}n} + X(5)e^{j2\frac{2\pi}{4}n} + X(7)e^{j3\frac{2\pi}{4}n}),\ n=0,1,2,3$$

The multiplication by $e^{-j\frac{2\pi}{8}n}$ (called twiddle factor) is the frequency shifting operation necessary to shift the spectrum to the left by one sample interval. The twiddle factor is also written as

$$e^{-j\frac{2\pi}{8}n} = W_8^n,\quad W_8 = e^{-j\frac{2\pi}{8}}$$

Now, the same process is recursively continued. Decomposing the two 4-point waveforms into their even and odd half-wave symmetric components, we get

$$4\{X(0)e^{j0\frac{2\pi}{4}n} + X(4)e^{j2\frac{2\pi}{4}n}, X(2)e^{j1\frac{2\pi}{4}n} + X(6)e^{j3\frac{2\pi}{4}n}\},\ n=0,1 \qquad (10.3)$$

$$4\{X(1)e^{j0\frac{2\pi}{4}n} + X(5)e^{j2\frac{2\pi}{4}n}, X(3)e^{j1\frac{2\pi}{4}n} + X(7)e^{j3\frac{2\pi}{4}n}\},\ n=0,1 \qquad (10.4)$$

These vector arrays are stored in the middle nodes in the figure.

The even half-wave symmetric component of the waveform given by Eq. (10.3) can be expressed as a 2-point DFT

$$4(X(0)e^{j0\frac{2\pi}{4}n} + X(4)e^{j2\frac{2\pi}{4}n})$$
$$= 4(X(0)e^{j0\frac{2\pi}{2}n} + X(4)e^{j1\frac{2\pi}{2}n}), \quad n = 0, 1$$

with frequency coefficients $4\{X(0), X(4)\}$. The coefficients

$$A(0) = \{A_0(0), A_1(0)\} = 8\{X(0), X(4)\}$$

are obtained by simply adding and subtracting the two sample values. These coefficients are stored in the top node at the end of the signal-flow graph in the figure.

The odd half-wave symmetric component of the waveform given by Eq. (10.3) is multiplied by the exponential $e^{-j\frac{2\pi}{8}(2n)} = e^{-j\frac{2\pi}{4}n}$ to get

$$4(X(2)e^{j1\frac{2\pi}{4}n} + X(6)e^{j3\frac{2\pi}{4}n})e^{-j\frac{2\pi}{4}n}$$
$$= 4(X(2)e^{j0\frac{2\pi}{4}n} + X(6)e^{j2\frac{2\pi}{4}n})$$
$$= 4(X(2)e^{j0\frac{2\pi}{2}n} + X(6)e^{j1\frac{2\pi}{2}n}), \quad n = 0, 1$$

This is a 2-point DFT with frequency coefficients $4\{X(2), X(6)\}$. The coefficients

$$A(2) = \{A_0(2), A_1(2)\} = 8\{X(2), X(6)\}$$

are obtained by simply adding and subtracting the two sample values. These coefficients are stored in the second node from top at the end of the signal-flow graph in the figure.

The even half-wave symmetric component of the waveform defined by Eq. (10.4) can be expressed as

$$4(X(1)e^{j0\frac{2\pi}{4}n} + X(5)e^{j2\frac{2\pi}{4}n})$$
$$= 4(X(1)e^{j0\frac{2\pi}{2}n} + X(5)e^{j1\frac{2\pi}{2}n}), \quad n = 0, 1$$

This is a 2-point DFT with frequency coefficients $4\{X(1), X(5)\}$. The coefficients

$$A(1) = \{A_0(1), A_1(1)\} = 8\{X(1), X(5)\}$$

are obtained by simply adding and subtracting the two sample values. These coefficients are stored in the third node from top at the end of the signal-flow graph shown in the figure.

The odd half-wave symmetric component of the waveform defined by Eq. (10.4) is multiplied by the exponential $e^{-j\frac{2\pi}{4}n}$ to get

$$4(X(3)e^{j1\frac{2\pi}{4}n} + X(7)e^{j3\frac{2\pi}{4}n})e^{-j\frac{2\pi}{4}(n)}$$
$$= 4(X(3)e^{j0\frac{2\pi}{4}n} + X(7)e^{j2\frac{2\pi}{4}n})$$
$$= 4(X(3)e^{j0\frac{2\pi}{2}n} + X(7)e^{j1\frac{2\pi}{2}n}), \quad n = 0, 1$$

This is a 2-point DFT with frequency coefficients $4\{X(3), X(7)\}$. The coefficients

$$A(3) = \{A_0(3), A_1(3)\} = 8\{X(3), X(7)\}$$

are obtained by simply adding and subtracting the two sample values. These coefficients are stored in the fourth node from top at the end of the signal-flow graph shown in the figure.

The output vectors $\{A(0), A(1), A(2), A(3)\}$ are placed in the bit-reversed order. This order is obtained by reversing the order of bits of the binary number representation of the frequency indices. $\{0, 1, 2, 3\}$ is $\{00, 01, 10, 11\}$ in binary form. The bit-reversed $\{00, 10, 01, 11\}$ in binary form is $\{0, 2, 1, 3\}$ in decimal form. The bit-reversed order occurs at the output because of the repeated splitting of the frequency components into odd- and even-indexed frequency indices groups over the stages of the algorithm. Efficient algorithms are available to restore the natural order of the coefficients. In digital signal processing microprocessors, specialized instructions are available to carry out this task. The extraction of the coefficients, multiplied by 8, of $x(n) = \cos(\frac{2\pi}{8}n) + \sin(2\frac{2\pi}{8}n)$ is shown in Fig. 10.3.

The number of complex multiplications and additions required for each stage are, respectively, $N/2$ and N, where N is the sequence length that is a power of 2. With $(\log_2 N) - 1$ stages and the initial vector formation requiring N complex additions, the computational complexity of the algorithm is $O(N \log_2 N)$ compared with that of $O(N^2)$ required for the direct computation from the DFT definition. Multiplication by twiddle factors of the forms $-j$ and $(1 - j1)/\sqrt{2}$ can be handled separately reducing the number of operations. If further speed up is required, two adjacent stages of the algorithm can be implemented together. This reduces the number of data transfers between the processor registers and the memory yielding significant reduction in the execution time of the algorithm. In case the number of stages is odd, one stage can be implemented separately.

The algorithm is so regular that one can easily get the signal-flow graph for any value of N that is an integral power of 2. The signal-flow graph of the algorithm is basically an interconnection of butterflies (a computational structure), shown in Fig. 10.4. The defining equations of a butterfly at the rth stage are given by

$$a_0^{(r+1)}(h) = a_0^{(r)}(h) + a_0^{(r)}(l)$$
$$a_1^{(r+1)}(h) = a_0^{(r)}(h) - a_0^{(r)}(l)$$
$$a_0^{(r+1)}(l) = W_N^n a_1^{(r)}(h) + W_N^{n+\frac{N}{4}} a_1^{(r)}(l)$$
$$a_1^{(r+1)}(l) = W_N^n a_1^{(r)}(h) - W_N^{n+\frac{N}{4}} a_1^{(r)}(l),$$

Input values stored in vector locations	Vector formation	Stage 1 output	Stage 2 output
$x(0) = 1$ $x(4) = -1$	$a_0(0) = 0$ $a_1(0) = 2$	0 0	$X(0) = A_0(0) = 0$ $X(4) = A_1(0) = 0$
$x(1) = 1 + \frac{\sqrt{2}}{2}$ $x(5) = 1 - \frac{\sqrt{2}}{2}$	$a_0(2) = 2$ $a_1(2) = \sqrt{2}$	0 4	$X(2) = A_0(2) = -j4$ $X(6) = A_1(2) = j4$
$x(2) = 0$ $x(6) = 0$	$a_0(1) = 0$ $a_1(1) = 0$	2 2	$X(1) = A_0(1) = 4$ $X(5) = A_1(1) = 0$
$x(3) = -1 - \frac{\sqrt{2}}{2}$ $x(7) = -1 + \frac{\sqrt{2}}{2}$	$a_0(3) = -2$ $a_1(3) = -\sqrt{2}$	2 $-j2$	$X(3) = A_0(3) = 0$ $X(7) = A_1(3) = 4$

Fig. 10.3 Trace of the PM DIF DFT algorithm, with $N = 8$, in extracting the coefficients, scaled by 8, of $x(n) = \cos(\frac{2\pi}{8} n) + \sin(2\frac{2\pi}{8} n)$

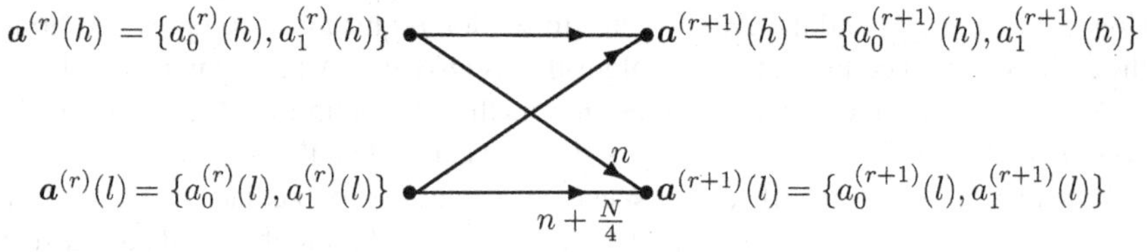

Fig. 10.4 Signal-flow graph of the butterfly of the PM DIF DFT algorithm, where $0 \le n < \frac{N}{4}$. A twiddle factor $W_N^n = e^{-j\frac{2\pi}{N}n}$ is indicated only by its variable part of the exponent, n

where $W_N^n = e^{-j\frac{2\pi}{N}n}$. There are $(\log_2 N) - 1$ stages, each with $N/4$ butterflies. With $N = 8$, therefore, we see four butterflies in Fig. 10.2.

The computation of a 4-point DFT by the PM DIF DFT algorithm is shown in Fig. 10.5. The input is

$$x(n) = \{3, 2, 1, 4\}$$

The input vectors at the 2 nodes are, respectively,

$$3 \pm 1 = \{4, 2\} \quad \text{and} \quad 2 \pm 4 = \{6, -2\}$$

$$\{x(0) = 3, x(1) = 2, x(2) = 1, x(3) = 4\}$$

$$\{X(0) = 10, X(1) = 2 + j2, X(2) = -2, X(3) = 2 - j2\}$$

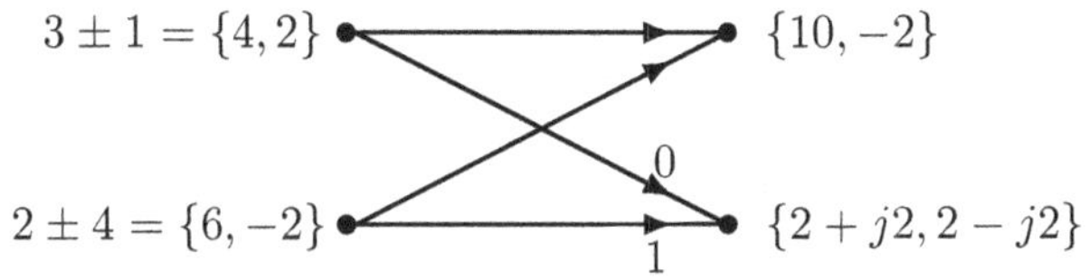

$3 \pm 1 = \{4, 2\}$ • ⟶ • $\{10, -2\}$

$2 \pm 4 = \{6, -2\}$ • ⟶ • $\{2 + j2, 2 - j2\}$

Fig. 10.5 Computation of a 4-point DFT by the PM DIF DFT algorithm. The twiddle factor $W_4^1 = e^{-j\frac{2\pi}{4}1}$ is indicated only by its exponent, 1

The output vectors at the 2 nodes are, respectively,

$$4 \pm 6 = \{10, -2\} \quad \text{and} \quad 2 \pm (-j)(-2) = \{2 + j2, 2 - j2\}$$

The computation of the IDFT can be carried out by a similar algorithm, with the twiddle factors conjugated. Further, division by N is required. However, the DFT algorithm itself can be used to carry out the IDFT computation with the interchange of the real and imaginary parts of the input and output, as shown in Chap. 4. At the end, division by N is required. Another method is to conjugate the input, compute its DFT, and conjugate the resulting output.

The computation of a 4-point IDFT by the PM DIF DFT algorithm is as follows. The input $X(k)$, from the last example, and its conjugate are

$$X(k) = \{10, 2 + j2, -2, 2 - j2\} \quad \text{and} \quad X^*(k) = \{10, 2 - j2, -2, 2 + j2\}$$

The input vectors at the 2 nodes of the PM DIF DFT algorithm are, respectively,

$$10 \pm (-2) = \{8, 12\} \quad \text{and} \quad (2 - j2) \pm (2 + j2) = \{4, -j4\}$$

The output vectors at the 2 nodes are, respectively,

$$8 \pm 4 = \{12, 4\} \quad \text{and} \quad 12 \pm (-j)(-j4) = \{8, 16\}$$

Dividing by 4, we get
$$x(n) = \{3, 2, 1, 4\}$$

By interchanging the real and imaginary parts, we get

$$X(k) = \{j10, 2 + j2, -j2, -2 + j2\}$$

The input vectors at the 2 nodes of the PM DIF DFT algorithm are, respectively,

$$j10 \pm (-j2) = \{j8, j12\} \quad \text{and} \quad (2+j2) \pm (-2+j2) = \{j4, 4\}$$

The output vectors at the 2 nodes are, respectively,

$$j8 \pm j4 = \{j12, j4\} \quad \text{and} \quad j12 \pm (-j)(4) = \{j8, j16\}$$

By interchanging the real and imaginary parts and dividing by 4, we get

$$x(n) = \{3, 2, 1, 4\}$$

The extraction of the coefficients, multiplied by 8, of $x(n) = \cos(\frac{2\pi}{8}n)$ is shown in Fig. 10.6. In the first-stage output, it is found out that there are no even-indexed frequency components in the input. In the second stage, the odd-indexed frequency components are assigned the appropriate coefficients. The operation is similar to a binary search algorithm.

The extraction of the coefficients, multiplied by 8, of $x(n) = \sin(\frac{2\pi}{8}n)$ is shown in Fig. 10.7.

Input values	Vector formation	Stage 1 output	Stage 2 output
$x(0)=1$ $x(4)=-1$	$a_0(0)=0$ $a_1(0)=2$	0 0	$X(0)=A_0(0)=0$ $X(4)=A_1(0)=0$
$x(1)=\frac{\sqrt{2}}{2}$ $x(5)=-\frac{\sqrt{2}}{2}$	$a_0(1)=0$ $a_1(1)=\sqrt{2}$	0 0	$X(2)=A_0(2)=0$ $X(6)=A_1(2)=0$
$x(2)=0$ $x(6)=0$	$a_0(2)=0$ $a_1(2)=0$	2 2	$X(1)=A_0(1)=4$ $X(5)=A_1(1)=0$
$x(3)=-\frac{\sqrt{2}}{2}$ $x(7)=\frac{\sqrt{2}}{2}$	$a_0(3)=0$ $a_1(3)=-\sqrt{2}$	2 $-j2$	$X(3)=A_0(3)=0$ $X(7)=A_1(3)=4$

Fig. 10.6 Trace of the PM DIF DFT algorithm, with $N = 8$, in extracting the coefficients, scaled by 8, of $x(n) = \cos(\frac{2\pi}{8}n)$

Input values stored in vector locations	Vector formation	Stage 1 output	Stage 2 output
$x(0) = 0$ $x(4) = 0$	$a_0(0) = 0$ $a_1(0) = 0$	0 0	$X(0) = A_0(0) = 0$ $X(4) = A_1(0) = 0$
$x(1) = \frac{\sqrt{2}}{2}$ $x(5) = -\frac{\sqrt{2}}{2}$	$a_0(2) = 0$ $a_1(2) = \sqrt{2}$	0 0	$X(2) = A_0(2) = 0$ $X(6) = A_1(2) = 0$
$x(2) = 1$ $x(6) = -1$	$a_0(1) - 0$ $a_1(1) = 2$	$-j2$ $j2$	$X(1) = A_0(1) = -j4$ $X(5) = A_1(1) = 0$
$x(3) = \frac{\sqrt{2}}{2}$ $x(7) = -\frac{\sqrt{2}}{2}$	$a_0(3) = 0$ $a_1(3) = \sqrt{2}$	$-j2$ 2	$X(3) = A_0(3) = 0$ $X(7) = A_1(3) = j4$

Fig. 10.7 Trace of the PM DIF DFT algorithm, with $N = 8$, in extracting the coefficients, scaled by 8, of $x(n) = \sin(\frac{2\pi}{8}n)$

10.3 The PM DIT DFT Algorithm

We have given the physical explanation of the decomposition of waveforms in the DIF DFT algorithm. In a decimation-in-frequency (DIF) algorithm, the transform sequence, $X(k)$, is successively divided into smaller subsequences. For example, in the beginning of the first stage, the computation of a N-point DFT is decomposed into two problems: (i) computing the $(N/2)$ even-indexed $X(k)$ and (ii) computing the $(N/2)$ odd-indexed $X(k)$. In a decimation-in-time (DIT) algorithm, the data sequence, $x(n)$, is successively divided into smaller subsequences. For example, in the beginning of the last stage, the computation of a N-point DFT is decomposed into two problems: (i) computing the $(N/2)$-point DFT of even-indexed $x(n)$ and (ii) computing the $(N/2)$-point DFT of odd-indexed $x(n)$. The DIT DFT algorithm is based on zero-padding, time-shifting, and spectral redundancy. For understanding, the DIF DFT algorithms are easier. However, the DIT algorithms are used more often, as taking care of the data scrambling problem occurring at the beginning of the algorithm is relatively easier to deal with. The DIT DFT algorithms can be considered as the algorithms obtained by transposing the signal-flow graph of the corresponding DIF algorithms, that is by reversing the direction of (signal flow) all the arrows and interchanging the input and the output.

10.3.1 Basics of the PM DIT DFT Algorithm

The DIT algorithm is based on decomposing the data sequence recursively into smaller sequences. Consider the exponential $x(n) = e^{j\frac{2\pi}{8}n}$. The sample values over one period are

$$\left\{ 1, \frac{1}{\sqrt{2}} + j\frac{1}{\sqrt{2}}, j1, -\frac{1}{\sqrt{2}} + j\frac{1}{\sqrt{2}}, -1, -\frac{1}{\sqrt{2}} - j\frac{1}{\sqrt{2}}, -j1, \frac{1}{\sqrt{2}} - j\frac{1}{\sqrt{2}} \right\}$$

The samples of $x(n)$ can be expressed as the sum of the upsampled, by a factor of 2, even-indexed and odd-indexed components

$$\{1, 0, j1, 0, -1, 0, -j1, 0\}$$

$$+ \left\{ 0, \frac{1}{\sqrt{2}} + j\frac{1}{\sqrt{2}}, 0, -\frac{1}{\sqrt{2}} + j\frac{1}{\sqrt{2}}, 0, -\frac{1}{\sqrt{2}} - j\frac{1}{\sqrt{2}}, 0, \frac{1}{\sqrt{2}} - j\frac{1}{\sqrt{2}} \right\}$$

The DFT of the even-indexed elements of $x(n)$ is

$$x_e(n) = \{1, j1, -1, -j1\} \leftrightarrow X_e(k) = \{0, 4, 0, 0\}$$

The DFT of the odd-indexed elements of $x(n)$ is

$$x_o(n) = \left\{ \frac{1}{\sqrt{2}} + j\frac{1}{\sqrt{2}}, -\frac{1}{\sqrt{2}} + j\frac{1}{\sqrt{2}}, -\frac{1}{\sqrt{2}} - j\frac{1}{\sqrt{2}}, \frac{1}{\sqrt{2}} - j\frac{1}{\sqrt{2}} \right\}$$
$$\leftrightarrow X_o(k) = \{0, 2\sqrt{2} + j2\sqrt{2}, 0, 0\}$$

Due to the upsampling theorem (Chap. 3), the DFTs of the upsampled sequences are the twofold repetition of $X_e(k)$ and $X_o(k)$ and we get

$$\{1, 0, j1, 0, -1, 0, -j1, 0\} \leftrightarrow \{0, 4, 0, 0, 0, 4, 0, 0\}$$

$$\left\{ \frac{1}{\sqrt{2}} + j\frac{1}{\sqrt{2}}, 0. - \frac{1}{\sqrt{2}} + j\frac{1}{\sqrt{2}}, 0, -\frac{1}{\sqrt{2}} - j\frac{1}{\sqrt{2}}, 0, \frac{1}{\sqrt{2}} - j\frac{1}{\sqrt{2}}, 0 \right\}$$
$$\leftrightarrow \{0, 2\sqrt{2} + j2\sqrt{2}, 0, 0, 0, 2\sqrt{2} + j2\sqrt{2}, 0, 0\}$$

Using the time-shift theorem, we get the DFT of upsampled and shifted $x_o(n)$ as

$$\left\{ 0, \frac{1}{\sqrt{2}} + j\frac{1}{\sqrt{2}}, 0, -\frac{1}{\sqrt{2}} + j\frac{1}{\sqrt{2}}, 0, -\frac{1}{\sqrt{2}} - j\frac{1}{\sqrt{2}}, 0, \frac{1}{\sqrt{2}} - j\frac{1}{\sqrt{2}} \right\} \leftrightarrow$$
$$\{0, 2\sqrt{2} + j2\sqrt{2}, 0, 0, 0, 2\sqrt{2} + j2\sqrt{2}, 0, 0\} e^{-j\frac{2\pi}{8}k}$$
$$= \{0, 4, 0, 0, 0, -4, 0, 0\}$$

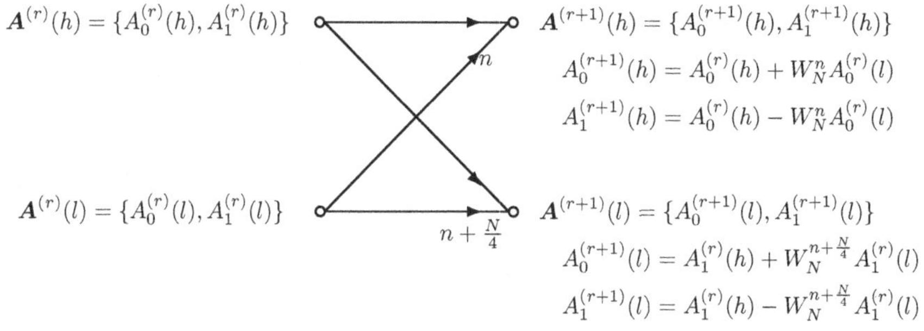

$A^{(r)}(h) = \{A_0^{(r)}(h), A_1^{(r)}(h)\}$ $A^{(r+1)}(h) = \{A_0^{(r+1)}(h), A_1^{(r+1)}(h)\}$

$$A_0^{(r+1)}(h) = A_0^{(r)}(h) + W_N^n A_0^{(r)}(l)$$

$$A_1^{(r+1)}(h) = A_0^{(r)}(h) - W_N^n A_0^{(r)}(l)$$

$A^{(r)}(l) = \{A_0^{(r)}(l), A_1^{(r)}(l)\}$ $n + \frac{N}{4}$ $A^{(r+1)}(l) = \{A_0^{(r+1)}(l), A_1^{(r+1)}(l)\}$

$$A_0^{(r+1)}(l) = A_1^{(r)}(h) + W_N^{n+\frac{N}{4}} A_1^{(r)}(l)$$

$$A_1^{(r+1)}(l) = A_1^{(r)}(h) - W_N^{n+\frac{N}{4}} A_1^{(r)}(l)$$

Fig. 10.8 Butterfly of the PM DIT DFT algorithm. A twiddle factor $W_N^n = e^{-j\frac{2\pi}{N}n}$ is indicated only by its variable part of the exponent, n

Adding the two partial DFTs, we get the DFT of $x(n)$ as

$$X(k) = \{0, 4, 0, 0, 0, 4, 0, 0\} + \{0, 4, 0, 0, 0, -4, 0, 0\} = \{0, 8, 0, 0, 0, 0, 0, 0\}$$

The decomposition continues until the sequence lengths become 1, and the DFT of the data is itself. There are $\log_2 N$ stages for a sequence of length N. The computational complexity of each stage is of the order $O(N)$. Therefore, the computational complexity of computing a N-point DFT becomes $O(N \log_2 N)$.

The butterfly and the flow graph of the PM DIT DFT algorithm are the transpose of those of the corresponding DIF algorithms. The PM DIT DFT butterfly is shown in Fig. 10.8. The butterfly input–output relations at the rth stage are

$$A_0^{(r+1)}(h) = A_0^{(r)}(h) + W_N^n A_0^{(r)}(l)$$
$$A_1^{(r+1)}(h) = A_0^{(r)}(h) - W_N^n A_0^{(r)}(l)$$
$$A_0^{(r+1)}(l) = A_1^{(r)}(h) + W_N^{n+\frac{N}{4}} A_1^{(r)}(l)$$
$$A_1^{(r+1)}(l) = A_1^{(r)}(h) - W_N^{n+\frac{N}{4}} A_1^{(r)}(l),$$

where n is an integer whose value depends on the stage of computation r and the index h. The letter A is used to differentiate this butterfly from that of the DIF algorithm.

The flow graph of the PM DIT DFT algorithm with $N = 16$ is shown in Fig. 10.9. The trace of the PM DIT DFT algorithm in extracting the coefficient scaled by 8 of $x(n) = \sin(\frac{2\pi}{8}n)$ is shown in Fig. 10.10. While it is possible to draw the flow graph in alternate ways, the usual ones are for the DIT DFT algorithm to place the input in the bit-reversed order with the output in the normal order and vice versa for the DIF DFT algorithm.

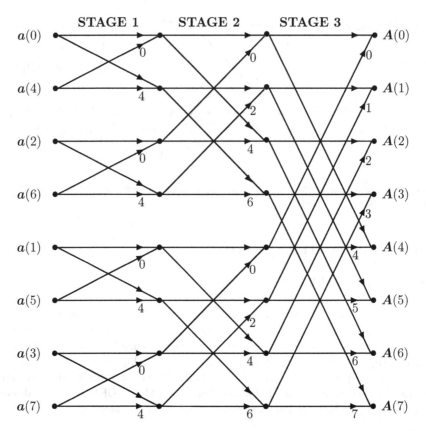

Fig. 10.9 Flow graph of the PM DIT DFT algorithm with $N = 16$. A twiddle factor $W_{16}^n = e^{-j\frac{2\pi}{16}n}$ is indicated only by its variable part of the exponent, n

10.4   Efficient Computation of the DFT of Real Data

It is the reduction of the computational complexity from $O(N^2)$ to $O(N \log_2 N)$ that is most important in most of the applications. However, if it is essential, the computational complexity and the storage requirements can be further reduced by a factor of about 2 for computing the DFT of real data. Two algorithms are commonly used to compute the DFT of real-valued data. One algorithm computes the DFT of a single real-valued data. Another algorithm computes the DFTs of two real-valued data sets simultaneously.

10.4.1   Two DFTs of Real Data Simultaneously

In this approach, we pack two real data sets into one complex data set of the same length. Let $a(n)$ and $b(n)$ be the two real-valued data sets each of length N. Let the

Input values stored in vector locations	Vector formation and swapping	Stage 1 output	Stage 2 output
$x(0) = 0$ $x(4) = 0$	$a_0(0) = 0$ $a_1(0) = 0$	0 0	$X(0) = A_0(0) = 0$ $X(4) = A_1(0) = 0$
$x(1) = \frac{\sqrt{2}}{2}$ $x(5) = -\frac{\sqrt{2}}{2}$	$a_0(2) = 0$ $a_1(2) = 2$	$-j2$ $j2$	$X(1) = A_0(1) = -j4$ $X(5) = A_1(1) = 0$
$x(2) = 1$ $x(6) = -1$	$a_0(1) = 0$ $a_1(1) = \sqrt{2}$	0 0	$X(2) = A_0(2) = 0$ $X(6) = A_1(2) = 0$
$x(3) = \frac{\sqrt{2}}{2}$ $x(7) = -\frac{\sqrt{2}}{2}$	$a_0(3) = 0$ $a_1(3) = \sqrt{2}$	$\sqrt{2}(1 - j1)$ $\sqrt{2}(1 + j1)$	$X(3) = A_0(3) = 0$ $X(7) = A_1(3) = j4$

Fig. 10.10 Trace of the PM DIT DFT algorithm, with $N = 8$ in extracting the coefficient scaled by 8 of $x(n) = \sin(\frac{2\pi}{8}n)$

respective DFTs be $A(k)$ and $B(k)$. We form the complex data $c(n)$ such that its real and imaginary parts are, respectively, $a(n)$ and $b(n)$. Let the DFT of $c(n)$ is $C(k)$. Then, using the linearity property of the DFT,

$$c(n) = a(n) + jb(n) \leftrightarrow C(k) = A(k) + jB(k)$$

Since the DFT of real-valued data is conjugate-symmetric, $A(N - k) = A^*(k)$ and $B(N - k) = B^*(k)$. Then,

$$C(N - k) = A^*(k) + jB^*(k) \quad \text{and} \quad C^*(N - k) = A(k) - jB(k)$$

Solving for $A(k)$ and $B(k)$ from the last two equations, we get

$$A(k) = \frac{C(k) + C^*(N - k)}{2} \quad \text{and} \quad B(k) = \frac{C(k) - C^*(N - k)}{j2} \tag{10.5}$$

The flowchart of the algorithm is shown in Fig. 10.11. The two real input data sets, each of length N, are read into real arrays $a(n)$ and $b(n)$. The complex data array $c(n) = a(n) + jb(n)$ is formed, where j is the imaginary unit, $\sqrt{-1}$. The DFT of $c(n)$,

Fig. 10.11 Flowchart of the
DFT algorithm for
computing the transform of
two real-valued data sets
simultaneously

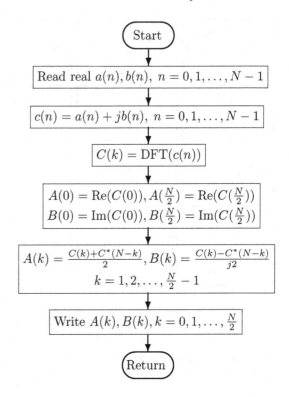

$C(k)$, is computed. The DFTs $A(k)$ and $B(k)$ of $a(n)$ and $b(n)$ are separated from $C(k)$ using Eq. (10.5). The values $A(0)$, $B(0)$, $A(N/2)$, and $B(N/2)$ are readily taken from $C(k)$, since their values are real. Only half of the values of $A(k)$ and $B(k)$ are computed, since the other half is its complex conjugate.

The IDFTs of two DFTs, $A(k)$ and $B(k)$ of real-valued data $a(n)$ and $b(n)$, can be computed simultaneously. The flowchart of the algorithm is shown in Fig. 10.12. As the DFTs are conjugate-symmetric, only one half of the values of $A(k)$ and $B(k)$ are read. The values $A(0)$, $B(0)$, $A(N/2)$, and $B(N/2)$ form the real imaginary parts of $C(0)$ and $C(N/2)$, since their values are real. The second-half of $C(k)$ is found using $C(N - k) = A^*(k) + jB^*(k)$. As the IDFT of $C(k)$ is $c(n) = a(n) + jb(n)$, the real and imaginary parts of $c(n)$ are, respectively, $a(n)$ and $b(n)$.

Example 10.1 Compute the DFT of the sequences $a(n) = \{2, 1, 4, 3\}$ and $b(n) = \{1, 2, 2, 3\}$ using a DFT algorithm for complex data of the same length. Verify that the IDFT of the complex data formed gets back $a(n)$ and $b(n)$.

Solution
Packing the sequences into a complex data, we get

$$c(n) = a(n) + jb(n) = \{2 + j1, 1 + j2, 4 + j2, 3 + j3\}$$

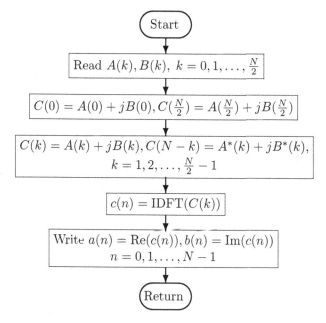

Fig. 10.12 Flowchart of the IDFT algorithm for computing the IDFT of the transform of two real-valued data sets simultaneously

Taking the DFT of $c(n)$, we get

$$C(k) = \{10 + j8, -3 + j1, 2 - j2, -1 - j3\}$$

Using Eq. (10.5), we get

$$A(0) = \operatorname{Re}(C(0)) = 10, \quad A(2) = \operatorname{Re}(C(2)) = 2$$

$$A(1) = \frac{(-3 + j1) + (-1 + j3)}{2} = -2 + j2, \quad A(3) = A^*(1) = -2 - j2$$

$$B(0) = \operatorname{Im}(C(0)) = 8, \quad B(2) = \operatorname{Im}(C(2)) = -2$$

$$B(1) = \frac{(-3 + j1) - (-1 + j3)}{j2} = -1 + j1, \quad B(3) = B^*(1) = -1 - j1$$

Therefore,

$$A(k) = \{10, -2 + j2, 2, -2 - j2\}, \quad \text{and} \quad B(k) = \{8, -1 + j1, -2, -1 - j1\}$$

Given two DFTs, $A(k)$ and $B(k)$, of real-valued data, $a(n)$, $b(n)$, we form $A(k) +$ $jB(k)$ and compute its IDFT yielding the $a(n)$ and $b(n)$ in the real and imaginary parts, respectively. For this example,

$$A(k) + jB(k) = \{10 + j8, -3 + j1, 2 - j2, -1 - j3\}$$

and its IDFT is $a(n) + jb(n) = \{2 + j1, 1 + j2, 4 + j2, 3 + j3\}$.   ∎

Since two DFTs are computed using one DFT for complex-valued data, for computing the DFT of each real-valued data set, the computational complexity is one half of that of the algorithm for complex-valued data. In addition, $N - 2$ addition operations are required in separating each of the two DFTs.

Linear Filtering of Long Data Sequences

In practical applications of linear filtering, the data sequences are real-valued and often very long compared with the impulse response. Typically, convolution is carried out over sections of the input using methods, such as the overlap save method described in Chap. 5. As there are large number of sections, two sections can be packed into a complex data and their transform computed. The DFT of the complex data is multiplied by the DFT of the impulse response, which is computed only once. The IDFT of the product yields the convolution of the two input sequences in the real and imaginary parts. One advantage of this procedure is that no separation and combination of the DFTs are required.

10.4.2   DFT of a Single Real Data Set

The computation of the DFT of a single real data set is referred as RDFT. The inverse DFT of the transform of real data is referred as RIDFT. Since the DFT of real data is conjugate-symmetric, only one half of the number of butterflies in each stage of the DFT algorithm for complex-valued data is necessary. The redundant butterflies can be easily eliminated in the DIT algorithm for complex-valued data in deriving the corresponding RDFT algorithm. Similarly, the RIDFT algorithm can be derived from the corresponding DIF algorithm.

The PM DIT RDFT Butterfly

The equations characterizing the input–output relation of the PM DIT RDFT butterfly, shown in Fig. 10.13, are

Fig. 10.13 Butterfly of the PM DIT RDFT algorithm

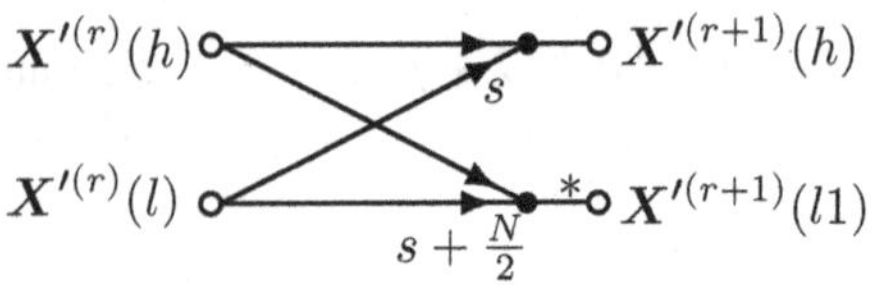

$$X'^{(r+1)}(h) = X'^{(r)}(h) + W_N^s X'^{(r)}(l)$$
$$X'^{(r+1)}(l1) = (X'^{(r)}(h) - W_N^s X'^{(r)}(l))^*$$

where s is an integer whose value depends on the stage of the computation r and the index h. The butterfly for real data is essentially the same as that for complex data with some differences. Only half of the butterflies are computed in each group of butterflies in each stage. Since a butterfly produces one output at the upper-half and another output at the lower-half of a group of butterflies in each stage, the lower-half output has to be conjugated and stored in a memory location in the upper-half.

The flow graph of the PM DIT RDFT algorithm with $N = 16$ is shown in Fig. 10.14. The trace of the RDFT algorithm for computing the DFT with $N = 16$ is shown in Fig. 10.15. The data storage scheme is

$$x'(n) = \left\{ x(n), x\left(n + \frac{N}{2}\right) \right\}, \quad n = 0, 1, \ldots, \frac{N}{2} - 1$$

$$X'(0) = \left\{ X(0), X\left(\frac{N}{2}\right) \right\} \quad \text{and} \quad X'(k) = X(k), \quad k = 1, \ldots, \frac{N}{2} - 1$$

As in the DIT DFT algorithms for complex data, in the last stage, two 8-point DFTs are merged to form a 16-point DFT. The differences are that: (i) only half the number

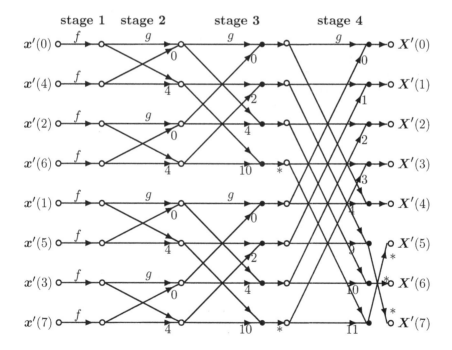

Fig. 10.14 The flow graph of the PM DIT RDFT algorithm with $N = 16$

Input		Stage 1 Output		Stage 2 Output		Stage 3 Output		Stage 4 Output	
$x(0)=1$	$x(8)=0$	1	1	8	-6	16.00	0.00	$X(0)=41$	$X(16)=-9$
$x(1)=3$	$x(9)=2$	7	-1	$1+j1$		$-0.41-j3.24$		$X(1)=1.8162-j7.8620$	
$x(2)=2$	$x(10)=0$	2	2	8	-4	$-6+j4$		$X(2)=-6.71+j4.71$	
$x(3)=5$	$x(11)=2$	6	4	$2-j4$		$2.41-j5.24$		$X(3)=2.2557-j3.3292$	
$x(4)=3$	$x(12)=4$	5	1	11	-1	25.00	-3.00	$X(4)=j3.00$	
$x(5)=4$	$x(13)=2$	6	2	$1-j2$		$3.83-j3.41$		$X(5)=2.5727+j7.1561$	
$x(6)=5$	$x(14)=1$	7	3	14	0	-1		$X(6)=-5.29-j3.29$	
$x(7)=3$	$x(15)=4$	7	-1	$3+j1$		$-1.83+j0.59$		$X(7)=-2.6447-j1.3768$	

Fig. 10.15 Trace of the RDFT algorithm for computing the DFT with $N = 16$

of butterflies are used and (ii) conjugation and swapping operations are required. Due to the lack of symmetry of the data at the input and output end, special butterflies are used. The first-stage butterflies, denoted by f, compute 2-point DFTs. The input and output data are stored as shown in the trace of the algorithm. In addition, the input is also placed in bit-reversed order. Subsequently, there is a g butterfly for each group of butterflies. The g butterflies use the values stored in a pair of nodes. Both the nodes have two real values. The sum and difference of the values stored in the first locations of the top and bottom nodes, respectively, form the output for the first and second location of the top node. For example, in the g butterfly of the first stage, the output values are $1 + 7 = 8$ and $1 - 7 = -6$. The output value at the bottom node is the sum of the second value of the top node and the product of the second value at the bottom node multiplied by $-j$. For the example, $1 + (-j)(-1) = 1 + j1$ is stored in the second node.

The input and output data are stored as shown in the trace of the algorithm. The input and output for the rest of the butterflies are complex and the computation is as shown in Fig. 10.13.

Let the input 16-point real data sequence be

$$\{1, 3, 2, 5, 3, 4, 5, 3, 0, 2, 0, 2, 4, 2, 1, 4\}$$

The first 9 values of the DFT $X(k)$ are

$$\{X(k), k = 0, 1, \ldots, 8\} = \{41, 1.8162 - j7.8620, -6.7071 + j4.7071, 2.2557 - j3.3292,$$

$$j3, 2.5727 + j7.1561, -5.2929 - j3.2929, -2.6447 - j1.3768, -9\}$$

The other values can be obtained using the conjugate symmetry of the DFT of real data. The even-indexed values $xe(n)$ of $x(n)$ are

$$xe(n) = \{1, 2, 3, 5, 0, 0, 4, 1\}$$

The DFT $X(k)$ of $xe(n)$ is

$$\{16, -0.4142 - j3.2426, -6 + j4, 2.4142 - j5.2426, 0,$$
$$2.4142 + j5.2426, -6 - j4, -0.4142 + j3.2426\}$$

The odd-indexed values $xo(n)$ of $x(n)$ are

$$xo(n) = \{3, 5, 4, 3, 2, 2, 2, 4\}$$

The DFT $X(k)$ of $xo(n)$ is

$$\{25, 3.8284 - j3.4142, -1, -1.8284 + j0.5858, -3,$$
$$- 1.8284 - j0.5858, -1, 3.8284 + j3.4142\}$$

The last stage output is the first $(N/2) + 1 = 9$ DFT coefficients. These coefficients are obtained by merging the 5 DFT coefficients of the two 8-point DFTs. The whole computation is similar to complex-valued algorithms with few differences.

The PM DIF RIDFT Butterfly

The RIDFT algorithm is derived from the corresponding DIF DFT algorithm for complex-valued data. Only half of the butterflies of the complex-valued algorithm are necessary. The RIDFT butterfly is shown in Fig. 10.16. It is similar to that of the algorithm for complex-valued data, except that the input data at the bottom node is read from the first-half of the DFT coefficients and conjugated. In addition, the twiddle factors are conjugated compared with those of the RDFT algorithm. The equations governing the butterfly are

$$X''^{(r+1)}(h) = X''^{(r)}(h) + (X''^{(r)}(l1))^*$$
$$X''^{(r+1)}(l) = W_N^{-s}(X''^{(r)}(h) - (X''^{(r)}(l1))^*)$$

The flow graph of the RIDFT algorithm with $N = 16$ is shown in Figure 10.17. The f butterflies compute a 2-point IDFT, without the division operation by 2, of the two real values stored in each node. The g butterflies use the values stored in a pair of nodes. The top node has two real values. The sum and difference of these values are stored in the first location of the top and bottom nodes, respectively. For example, in the g butterfly of the first stage, the output values are $41 - 9 = 32$ and $41 + 9 = 50$. Let the complex value stored in the bottom node is $a + jb$. Then, $2a$ and $-2b$ are

Fig. 10.16 RIDFT butterfly

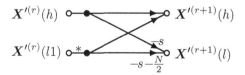

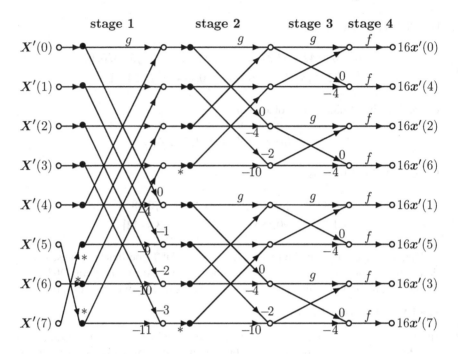

Fig. 10.17 Flow graph of the RIDFT algorithm with $N = 16$

Input		Stage 1 Output		Stage 2 Output		Stage 3 Output		Stage 4 Output	
$X(0)= 41$	$X(16)=-9$	32	0.00	32	-24	8	8	$x(0)=16$	$x(8)=0$
$X(1)=1.8162-j7.8620$		$-0.83-j6.49$		$4+j4$		56	-8	$x(4)=48$	$x(12)=64$
$X(2)=-6.71+j4.71$		$-12+j8$		32	-16	16	16	$x(2)=32$	$x(10)=0$
$X(3)=2.2557-j3.3292$		$4.83-j10.49$		$8-j16$		48	32	$x(6)=80$	$x(14)=16$
$X(4)=j3.00$		50	-6.00	44	-4	40	8	$x(1)=48$	$x(9)=32$
$X(5)=2.5727+j7.1561$		$7.66-j6.83$		$4-j8$		48	16	$x(5)=64$	$x(13)=32$
$X(6)=-5.29-j3.29$			-2	56	0	56	24	$x(3)=80$	$x(11)=32$
$X(7)=-2.6447-j1.3768$		$-3.66+j1.17$		$12+j4$		56	-8	$x(7)=48$	$x(15)=64$

Fig. 10.18 Trace of the RIDFT algorithm

stored in the second location of the top and bottom nodes, respectively. For the first g butterfly, $a + jb = 0 + j3$. Therefore, 0 and -6 are stored.

The trace of the RIDFT algorithm is shown in Fig. 10.18. The input DFT coefficients are the output in Fig. 10.15. The output values have to be divided by 16.

10.5 Summary

- Fourier analysis is dominating in the analysis of linear systems and signals in practice as well, due to the availability of fast algorithms to approximate the Fourier spectrum.
- The computational complexity of computing the 1-D N-point DFT, using fast algorithms, is $O(N \log_2 N)$ for N a power of 2, whereas that of the implementation from the definition is $O(N^2)$.
- The fast algorithms reduce the inherent redundancy of operations in the definition of the DFT. While there are different approaches in designing the DFT algorithms, the power-of-2 algorithms based on the classical divide-and-conquer strategy of developing fast algorithms yield the practically most useful DFT algorithms.
- While there are alternate ways of deriving the power-of-2 algorithms (such as matrix decomposition), the best physical appreciation of the algorithms is obtained using the half-wave symmetry of periodic waveforms.
- The sum and difference of the samples of the first half and the second half of the period of a periodic waveform decompose it into two components. One component is composed of the even-indexed frequency components and the other is composed of the odd-indexed frequency components. This decomposition decomposes a N-point DFT into two $N/2$-point DFTs. Recursive decomposition yields the final algorithm.
- There are $\log_2 N$ stages, each stage with a computational complexity of $O(N)$. Therefore, the computational complexity becomes $O(N \log_2 N)$ for a N-point DFT.
- The decomposition can be carried out by decomposing the frequency components into two sets. One set consists of the even-indexed frequency components and the other consists of the odd-indexed frequency components. The DFTs of the two sets are computed and combined to obtain the whole DFT. This type of algorithms is called the decimation-in-frequency (DIF) algorithms.
- The decomposition can also be carried out by decomposing the time-domain samples into two sets. One set consists of the even-indexed samples and the other consists of the odd-indexed samples. The DFTs of the two sets are computed and merged to obtain the whole DFT. This type of algorithms is called the decimation-in-time (DIT) algorithms.
- As usual, Fourier analysis, in theoretical analysis, is almost always carried out using the complex exponential as the basis waveform. This results in a factor-of-2 redundancy in computing the DFT of real-valued data.
- This redundancy can be reduced by computing the DFT of two sets of real-valued data at the same time.
- An alternate method is to reduce the redundancy in the flow graph of the algorithms for complex data.

Exercises

10.1 Given the samples of a waveform $x(n)$, find the samples of its even half-wave symmetric and odd half-wave symmetric components. Verify that the sum of the samples of the two components add up to the samples of $x(n)$. Compute the DFT of $x(n)$ and its components. Verify that the DFT of the even half-wave symmetric component consists of zero-valued odd-indexed spectral values. Verify that the DFT of the odd half-wave symmetric component consists of zero-valued even-indexed spectral values.

10.1.1 $x(n) = \{\check{0}, 1, 2, 3\}$.

* 10.1.2 $x(n) = \{\check{0}, 1, 0, 1\}$.

10.1.3 $x(n) = \{\check{1}, 3, -1, -3\}$.

10.1.4 $x(n) = \{\check{2}, 1, 3, 4\}$.

10.1.5 $x(n) = \{\check{3}, 1, 2, 4\}$.

10.2 Given a waveform $x(n)$ with period 8, find the samples of the waveform. (a) Give the trace of the PM DIF DFT algorithm in computing its DFT $X(k)$. (b) Give the trace of the PM DIT DFT algorithm in computing its DFT $X(k)$. Verify that both are the same. In both cases, find the IDFT of $X(k)$ using the same DFT algorithms, give the trace and verify that the samples of the input $x(n)$ are obtained.

10.2.1 $x(n) = -2e^{-j\left(\frac{2\pi}{8}n+\frac{\pi}{6}\right)}$

10.2.2 $x(n) = -e^{-j\left(\frac{2\pi}{8}6n-\frac{\pi}{3}\right)}$

* 10.2.3 $x(n) = e^{j\left(\frac{2\pi}{8}0n-\frac{\pi}{3}\right)} + e^{j\left(\frac{2\pi}{8}4n-\frac{\pi}{6}\right)}$

10.2.4 $x(n) = -e^{j\left(\frac{2\pi}{8}n+\frac{\pi}{4}\right)} + e^{j\left(\frac{2\pi}{8}2n+\frac{\pi}{6}\right)}$

10.2.5 $x(n) = 3e^{j\left(\frac{2\pi}{8}6n-\frac{\pi}{4}\right)} + e^{j\left(\frac{2\pi}{8}7n+\frac{\pi}{3}\right)}$

10.3 Given two waveforms $x(n)$ and $y(n)$ with period 8, find the samples of the waveforms. Use the PM DIT DFT algorithm for complex data to find their DFTs using the algorithm for computing the DFTs of two real data sets simultaneously. Compute the IDFTs of the DFTs using a DFT algorithm for complex data sets simultaneously. Verify the DFT $X(k)$ and $Y(k)$ by expressing $x(n)$ and $y(n)$ into its complex exponential components.

10.3.1 $x(n) = \cos\left(\frac{2\pi}{8}1n + \frac{\pi}{6}\right)$ and $y(n) = \cos\left(\frac{2\pi}{8}3n - \frac{\pi}{6}\right)$.

10.3.2 $x(n) = \cos\left(\frac{2\pi}{8}2n - \frac{\pi}{6}\right)$ and $y(n) = \cos\left(\frac{2\pi}{8}7n - \frac{\pi}{3}\right)$.

10.3.3 $x(n) = \cos\left(\frac{2\pi}{8}3n + \frac{\pi}{4}\right)$ and $y(n) = \cos\left(\frac{2\pi}{8}5n - \frac{\pi}{3}\right)$.

10.3.4 $x(n) = \cos\left(\frac{2\pi}{8}0n + \frac{\pi}{6}\right)$ and $y(n) = \cos\left(\frac{2\pi}{8}6n + \frac{\pi}{3}\right)$.

* 10.3.5 $x(n) = \cos\left(\frac{2\pi}{8}6n + \frac{\pi}{4}\right)$ and $y(n) = \cos\left(\frac{2\pi}{8}5n - \frac{\pi}{3}\right)$.

10.4 Given a waveform $x(n)$ with period 8, find the samples of the waveform. Find its DFT $X(k)$ using the PM RDFT algorithm. Find the IDFT of $X(k)$ using the PM RIDFT algorithm to get back the samples of $x(n)$. Verify the DFT $X(k)$ by expressing $x(n)$ into its complex exponential components.

10.4.1 $\cos\left(\frac{2\pi}{8}0n + \frac{\pi}{6}\right)$.

10.4.2 $\cos\left(\frac{2\pi}{8}1n - \frac{\pi}{6}\right)$.

10.4.3 $\cos\left(\frac{2\pi}{8}2n + \frac{\pi}{4}\right)$.

* **10.4.4** $\cos\left(\frac{2\pi}{8}3n - \frac{\pi}{4}\right)$.

10.4.5 $\cos\left(\frac{2\pi}{8}4n + \frac{\pi}{3}\right)$.

Appendix A
Transform Pairs and Properties

See Tables A.1, A.2, A.3, A.4, A.5, A.6, A.7, and A.8.

Table A.1 DFT pairs

$x(n)$, Period $= N$	$X(k)$, Period $= N$
$\delta(n)$	1
1	$N\delta(k)$
$e^{j\left(\frac{2\pi}{N}k_0 n\right)}$	$N\delta(k - k_0)$
$\cos\left(\frac{2\pi}{N}k_0 n + \theta\right)$	$\frac{N}{2}(e^{j\theta}\delta(k - k_0) + e^{-j\theta}\delta(k - (N - k_0)))$
$\cos\left(\frac{2\pi}{N}k_0 n\right)$	$\frac{N}{2}(\delta(k - k_0) + \delta(k - (N - k_0)))$
$\sin\left(\frac{2\pi}{N}k_0 n\right)$	$\frac{N}{2}(-j\delta(k - k_0) + j\delta(k - (N - k_0)))$
$x(n) = \begin{cases} 1 & \text{for } n = 0, 1, \ldots, L - 1 \\ 0 & \text{for } n = L, L + 1, \ldots, N - 1 \end{cases}$	$e^{\left(-j\frac{\pi}{N}(L-1)k\right)}\frac{\sin(\frac{\pi}{N}kL)}{\sin(\frac{\pi}{N}k)}$

© Springer Nature Singapore Pte Ltd. 2018
D. Sundararajan, *Fourier Analysis—A Signal Processing Approach*,
https://doi.org/10.1007/978-981-13-1693-7

Table A.2 DFT properties

Property	$x(n), h(n)$, Period $= N$	$X(k), H(k)$, Period $= N$
Linearity	$ax(n) + bh(n)$	$aX(k) + bH(k)$
Duality	$\frac{1}{N}X(N \mp n)$	$x(N \pm k)$
Time-shifting	$x(n \pm n_0)$	$e^{\pm j\frac{2\pi}{N}n_0 k}X(k)$
Frequency-shifting	$e^{\mp j\frac{2\pi}{N}k_0 n}x(n)$	$X(k \pm k_0)$
Time-convolution	$\sum_{k=0}^{N-1} x(k)h(n-k)$	$X(k)H(k)$
Frequency-convolution	$x(n)h(n)$	$\frac{1}{N}\sum_{v=0}^{N-1} X(v)H(k-v)$
Time-expansion	$h(an) =$ $\begin{cases} x(n) \text{ for } n = 0, 1, \ldots, N-1 \\ 0 \quad \text{otherwise} \end{cases}$ where a is any positive integer	$H(k) = X(k \bmod N)$, $k = 0, 1, \ldots, aN - 1$
Time-reversal	$x(N - n)$	$X(N - k)$
Conjugation	$x^*(N \pm n)$	$X^*(N \mp k)$
Parseval's theorem	$\sum_{n=0}^{N-1} |x(n)|^2$	$\frac{1}{N}\sum_{k=0}^{N-1} |X(k)|^2$

Table A.3 FS pairs

$x(t)$, Period $= T$	$X_{fs}(k)$, $\omega_0 = \frac{2\pi}{T}$
$\begin{cases} 1 \text{ for } |t| < a \\ 0 \text{ for } a < |t| \le \frac{T}{2} \end{cases}$	$\frac{\sin(k\omega_0 a)}{k\pi}$
$\sum_{n=-\infty}^{\infty} \delta(t - nT)$	$\frac{1}{T}$
$e^{jk_0\omega_0 t}$	$\delta(k - k_0)$
$\cos(k_0\omega_0 t)$	$0.5(\delta(k + k_0) + \delta(k - k_0))$
$\sin(k_0\omega_0 t)$	$0.5j(\delta(k + k_0) - \delta(k - k_0))$

Table A.4 FS properties

Property	$x(t), h(t)$, Period $= T$	$X_{fs}(k), H_{fs}(k), \omega_0 = \frac{2\pi}{T}$
Linearity	$ax(t) + bh(t)$	$aX_{fs}(k) + bH_{fs}(k)$
Time-shifting	$x(t \pm t_0)$	$e^{\pm jk\omega_0 t_0} X_{fs}(k)$
Frequency-shifting	$x(t)e^{\pm jk_0\omega_0 t}$	$X_{fs}(k \mp k_0)$
Time-convolution	$\int_0^T x(\tau)h(t - \tau)d\tau$	$T X_{fs}(k)H_{fs}(k)$
Frequency-convolution	$x(t)h(t)$	$\sum_{l=-\infty}^{\infty} X_{fs}(l)H_{fs}(k - l)$
Time-scaling	$x(at), a > 0$, Period $= \frac{T}{a}$	$X_{fs}(k), \omega_0 = a\frac{2\pi}{T}$
Time-reversal	$x(-t)$	$X_{fs}(-k)$
Time-differentiation	$\frac{d^n x(t)}{dt^n}$	$(jk\omega_0)^n X_{fs}(k)$
Time-integration	$\int_{-\infty}^t x(\tau)d\tau$	$\frac{X_{fs}(k)}{jk\omega_0}$, if $(X_{fs}(0) = 0)$
Parseval's theorem	$\frac{1}{T} \int_0^T |x(t)|^2 dt$	$\sum_{k=-\infty}^{\infty} |X_{fs}(k)|^2$
Conjuagte symmetry	$x(t)$ real	$X_{fs}(k) = X_{fs}^*(-k)$
Even symmetry	$x(t)$ real and even	$X_{fs}(k)$ real and even
Odd symmetry	$x(t)$ real and odd	$X_{fs}(k)$ imaginary and odd

Table A.5 DTFT pairs

$x(n)$	$X(e^{j\omega})$, Period $= 2\pi$
$\begin{cases} 1 \text{ for } -N \le n \le N \\ 0 \text{ otherwise} \end{cases}$	$\dfrac{\sin(\omega \frac{(2N+1)}{2})}{\sin(\frac{\omega}{2})}$
$\dfrac{\sin(an)}{\pi n},\ 0 < a \le \pi$	$\begin{cases} 1 \text{ for } \|\omega\| < a \\ 0 \text{ for } a < \|\omega\| \le \pi \end{cases}$
$a^n u(n),\ \|a\| < 1$	$\dfrac{1}{1-ae^{-j\omega}}$
$(n+1)a^n u(n),\ \|a\| < 1$	$\dfrac{1}{(1-ae^{-j\omega})^2}$
$a^{\|n\|},\ \|a\| < 1$	$\dfrac{1-a^2}{1-2a\cos(\omega)+a^2}$
$a^n \sin(\omega_0 n)u(n),\ \|a\| < 1$	$\dfrac{(a)e^{-j\omega}\sin(\omega_0)}{1-2(a)e^{-j\omega}\cos(\omega_0)+(a)^2 e^{-j2\omega}}$
$a^n \cos(\omega_0 n)u(n),\ \|a\| < 1$	$\dfrac{1-(a)e^{-j\omega}\cos(\omega_0)}{1-2(a)e^{-j\omega}\cos(\omega_0)+(a)^2 e^{-j2\omega}}$
$\delta(n)$	1
$\displaystyle\sum_{k=-\infty}^{\infty} \delta(n-kN)$	$\dfrac{2\pi}{N}\displaystyle\sum_{k=-\infty}^{\infty} \delta(\omega - \frac{2\pi}{N}k)$
$u(n)$	$\pi\delta(\omega) + \dfrac{1}{1-e^{-j\omega}}$
1	$2\pi\delta(\omega)$
$sgn(n)$	$\dfrac{2}{1-e^{-j\omega}}$
$e^{j\omega_0 n}$	$2\pi\delta(\omega - \omega_0)$
$\cos(\omega_0 n)$	$\pi(\delta(\omega + \omega_0) + \delta(\omega - \omega_0))$
$\sin(\omega_0 n)$	$j\pi(\delta(\omega + \omega_0) - \delta(\omega - \omega_0))$

Table A.6 DTFT properties

Property	$x(n), h(n)$	$X(e^{j\omega}), H(e^{j\omega})$
Linearity	$ax(n) + bh(n)$	$aX(e^{j\omega}) + bH(e^{j\omega})$
Time-shifting	$x(n \pm n_0)$	$e^{\pm j\omega n_0} X(e^{j\omega})$
Frequency-shifting	$x(n)e^{\pm j\omega_0 n}$	$X(e^{j(\omega \mp \omega_0)})$
Time-convolution	$\sum_{k=-\infty}^{\infty} x(k)h(n-k)$	$X(e^{j\omega})H(e^{j\omega})$
Frequency-convolution	$x(n)h(n)$	$\frac{1}{2\pi} \int_0^{2\pi} X(e^{jv})H(e^{j(\omega-v)})dv$
Time-expansion	$h(n)$ $h(an) = x(n),\ a >$ 0 is an integer and $h(n) = 0$ zero otherwise	$H(e^{j\omega}) = X(e^{ja\omega})$
Time-reversal	$x(-n)$	$X(e^{-j\omega})$
Conjugation	$x^*(\pm n)$	$X^*(e^{\mp j\omega})$
Difference	$x(n) - x(n-1)$	$(1 - e^{-j\omega})X(e^{j\omega})$
Summation	$\sum_{l=-\infty}^{n} x(l)$	$\frac{X(e^{j\omega})}{(1-e^{-j\omega})} + \pi X(e^{j0})\delta(\omega)$
Frequency-differentiation	$(n)^k x(n)$	$(j)^k \frac{d^k X(e^{j\omega})}{d\omega^k}$
Parseval's theorem	$\sum_{n=-\infty}^{\infty} |x(n)|^2$	$\frac{1}{2\pi} \int_0^{2\pi} |X(e^{j\omega})|^2 d\omega$
Conjuagte symmetry	$x(n)$ real	$X(e^{j\omega}) = X^*(e^{-j\omega})$
Even symmetry	$x(n)$ real and even	$X(e^{j\omega})$ real and even
Odd symmetry	$x(n)$ real and odd	$X(e^{j\omega})$ imaginary and odd

Table A.7 FT pairs

$x(t)$	$X(j\omega)$
$u(t+a) - u(t-a)$	$2\frac{\sin(\omega a)}{\omega}$
$\frac{\sin(\omega_0 t)}{\pi t}$	$u(\omega + \omega_0) - u(\omega - \omega_0)$
$e^{-at}u(t), \ \text{Re}(a) > 0$	$\frac{1}{a+j\omega}$
$te^{-at}u(t), \ \text{Re}(a) > 0$	$\frac{1}{(a+j\omega)^2}$
$e^{-a|t|}, \ \text{Re}(a) > 0$	$\frac{2a}{a^2+\omega^2}$
$\frac{1}{a}((t+a)u(t+a) - 2tu(t) + (t-a)u(t-a))$	$a\left(\frac{\sin(\omega\frac{a}{2})}{\omega\frac{a}{2}}\right)^2$
$e^{-at}\sin(\omega_0 t)u(t), \ \text{Re}(a) > 0$	$\frac{\omega_0}{(a+j\omega)^2+\omega_0^2}$
$e^{-at}\cos(\omega_0 t)u(t), \ \text{Re}(a) > 0$	$\frac{a+j\omega}{(a+j\omega)^2+\omega_0^2}$
$\delta(t)$	1
$\sum_{n=-\infty}^{\infty}\delta(t-nT)$	$\frac{2\pi}{T}\sum_{k=-\infty}^{\infty}\delta(\omega - k\frac{2\pi}{T})$
$u(t)$	$\pi\delta(\omega) + \frac{1}{j\omega}$
1	$2\pi\delta(\omega)$
$sgn(t)$	$\frac{2}{j\omega}$
1	$2\pi\delta(\omega)$
$e^{j\omega_0 t}$	$2\pi\delta(\omega - \omega_0)$
$\cos(\omega_0 t)$	$\pi(\delta(\omega + \omega_0) + \delta(\omega - \omega_0))$
$\sin(\omega_0 t)$	$j\pi(\delta(\omega + \omega_0) - \delta(\omega - \omega_0))$

Table A.8 FT properties

Property	$x(t), h(t)$	$X(j\omega), H(j\omega)$
Linearity	$ax(t) + bh(t)$	$aX(j\omega) + bH(j\omega)$
Duality	$X(\pm t)$	$2\pi x(\mp j\omega)$
Time-shifting	$x(t \pm t_0)$	$X(j\omega)e^{\pm j\omega t_0}$
Frequency-shifting	$x(t)e^{\pm j\omega_0 t}$	$X(j(\omega \mp \omega_0))$
Time-convolution	$x(t) * h(t)$	$X(j\omega)H(j\omega)$
Frequency-convolution	$x(t)h(t)$	$\frac{1}{2\pi}(X(j\omega) * H(j\omega))$
Time-scaling	$x(at), a \neq 0$ and real	$\frac{1}{|a|}X(j\frac{\omega}{a})$
Time-reversal	$x(-t)$	$X(-j\omega)$
Conjugation	$x^*(\pm t)$	$X^*(\mp j\omega)$
Time-differentiation	$\frac{d^n x(t)}{dt^n}$	$(j\omega)^n X(j\omega)$
Time-integration	$\int_{-\infty}^{t} x(\tau)d\tau$	$\frac{X(j\omega)}{j\omega} + \pi X(j0)\delta(\omega)$
Frequency-differentiation	$t^n x(t)$	$(j)^n \frac{d^n X(j\omega)}{d\omega^n}$
Parseval's theorem	$\int_{-\infty}^{\infty} |x(t)|^2 dt$	$\frac{1}{2\pi} \int_{-\infty}^{\infty} |X(j\omega)|^2 d\omega$
Autocorrelation	$x(t) * x(-t) =$ $\int_{-\infty}^{\infty} x(\tau)x(\tau - t)d\tau$	$|X(j\omega)|^2$
Conjuagte symmetry	$x(t)$ real	$X(j\omega) = X^*(-j\omega)$
Even symmetry	$x(t)$ real and even	$X(j\omega)$ real and even
Odd symmetry	$x(t)$ real and odd	$X(j\omega)$ imaginary and odd

Appendix B
Useful Mathematical Formulas

Trigonometric Identities

Pythagorean Identity
$$\sin^2 x + \cos^2 x = 1$$

Addition and Subtraction Formulas

$$\sin(x \pm y) = \sin x \cos y \pm \cos x \sin y$$

$$\cos(x \pm y) = \cos x \cos y \mp \sin x \sin y$$

Double-angle Formulas

$$\cos 2x = \cos^2 x - \sin^2 x = 2\cos^2 x - 1 = 1 - 2\sin^2 x$$

$$\sin 2x = 2 \sin x \cos x$$

Product Formulas
$$2 \sin x \cos y = \sin(x - y) + \sin(x + y)$$

$$2 \cos x \sin y = -\sin(x - y) + \sin(x + y)$$

$$2 \sin x \sin y = \cos(x - y) - \cos(x + y)$$

$$2 \cos x \cos y = \cos(x - y) + \cos(x + y)$$

Sum and Difference Formulas

$$\sin x \pm \sin y = 2 \sin \frac{x \pm y}{2} \cos \frac{x \mp y}{2}$$

© Springer Nature Singapore Pte Ltd. 2018
D. Sundararajan, *Fourier Analysis—A Signal Processing Approach*,
https://doi.org/10.1007/978-981-13-1693-7

$$\cos x + \cos y = 2\cos\frac{x+y}{2}\cos\frac{x-y}{2}$$

$$\cos x - \cos y = -2\sin\frac{x+y}{2}\sin\frac{x-y}{2}$$

Other Formulas

$$\sin(-x) = \sin(2\pi - x) = -\sin x$$

$$\cos(-x) = \cos(2\pi - x) = \cos x$$

$$\sin(\pi \pm x) = \mp \sin x$$

$$\cos(\pi \pm x) = -\cos x$$

$$\cos\left(\frac{\pi}{2} \pm x\right) = \mp \sin x$$

$$\sin\left(\frac{\pi}{2} \pm x\right) = \cos x$$

$$\cos\left(\frac{3\pi}{2} \pm x\right) = \pm \sin x$$

$$\sin\left(\frac{3\pi}{2} \pm x\right) = -\cos x$$

$$e^{\pm jx} = \cos x \pm j\sin x$$

$$\cos x = \frac{e^{jx} + e^{-jx}}{2}$$

$$\sin x = \frac{e^{jx} - e^{-jx}}{j2}$$

Series Expansions

$$e^{jx} = 1 + (jx) + \frac{(jx)^2}{2!} + \frac{(jx)^3}{3!} + \frac{(jx)^4}{4!} + \cdots + \frac{(jx)^r}{(r)!} + \cdots$$

$$\cos(x) = 1 - \frac{x^2}{2!} + \frac{x^4}{4!} - \cdots + (-1)^r \frac{x^{2r}}{(2r)!} - \cdots$$

$$\sin(x) = x - \frac{x^3}{3!} + \frac{x^5}{5!} - \cdots + (-1)^r \frac{x^{2r+1}}{(2r+1)!} - \cdots$$

$$\sin^{-1} x = x + \frac{1}{2}\frac{x^3}{3} + \frac{(1)(3)}{(2)(4)}\frac{x^5}{5} + \frac{(1)(3)(5)}{(2)(4)(6)}\frac{x^7}{7} + \cdots, \quad |x| < 1$$

$$\cos^{-1} x = \frac{\pi}{2} - \sin^{-1} x, \quad |x| < 1$$

Summation Formulas

$$\sum_{k=0}^{N} k = \frac{N(N+1)}{2}$$

$$\sum_{k=0}^{N-1}(a + kd) = \frac{N(2a + (N-1)d)}{2}$$

$$\sum_{k=0}^{N-1} ar^k = \frac{a(1 - r^N)}{1 - r}, \quad r \neq 1$$

$$\sum_{k=0}^{\infty} r^k = \frac{1}{1 - r}, \quad |r| < 1$$

$$\sum_{k=0}^{\infty} kr^k = \frac{r}{(1 - r)^2}, \quad |r| < 1$$

$$1 + \cos(t) + \cos(2t) + \cdots + \cos(Nt) = \frac{1}{2} + \frac{\sin(0.5(2N+1)t)}{2\sin(0.5t)}$$

Indefinite Integrals

$$\int u\,dv = uv - \int v\,du$$

$$\int e^{at}\,dt = \frac{e^{at}}{a}$$

$$\int te^{at}\,dt = \frac{e^{at}}{a^2}(at - 1)$$

$$\int e^{bt} \sin(at)\,dt = \frac{e^{bt}}{a^2 + b^2}(b\sin(at) - a\cos(at))$$

$$\int e^{bt} \cos(at)\,dt = \frac{e^{bt}}{a^2 + b^2}(b\cos(at) + a\sin(at))$$

$$\int \sin(at)dt = -\frac{1}{a}\cos(at)$$

$$\int \cos(at)dt = \frac{1}{a}\sin(at)$$

$$\int t\sin(at)dt = \frac{1}{a^2}(\sin(at) - at\cos(at))$$

$$\int t\cos(at)dt = \frac{1}{a^2}(\cos(at) + at\sin(at))$$

$$\int \sin^2(at)dt = \frac{t}{2} - \frac{1}{4a}\sin(2at)$$

$$\int \cos^2(at)dt = \frac{t}{2} + \frac{1}{4a}\sin(2at)$$

$$\int \frac{1}{a^2 + t^2}dt = \frac{1}{a}\tan^{-1}(\frac{t}{a})$$

$$\int \sin(at)\sin(bt)dt = \frac{\sin((a-b)t)}{2(a-b)} - \frac{\sin((a+b)t)}{2(a+b)}, \quad a^2 \neq b^2$$

$$\int \sin(at)\cos(bt)dt = -\left(\frac{\cos((a-b)t)}{2(a-b)} + \frac{\cos((a+b)t)}{2(a+b)}, \quad a^2 \neq b^2\right)$$

$$\int \cos(at)\cos(bt)dt = \frac{\sin((a-b)t)}{2(a-b)} + \frac{\sin((a+b)t)}{2(a+b)}, \quad a^2 \neq b^2$$

Differentiation Formulas

$$\frac{d(uv)}{dt} = u\frac{dv}{dt} + v\frac{du}{dt}$$

$$\frac{d(f(u))}{dt} = \frac{d(f(u))}{du}\frac{du}{dt}$$

$$\frac{d(\frac{u}{v})}{dt} = \frac{v\frac{du}{dt} - u\frac{dv}{dt}}{v^2}$$

$$\frac{d(x^n)}{dt} = nx^{n-1}$$

$$\frac{d(e^{at})}{dt} = ae^{at}$$

$$\frac{d(\sin(at))}{dt} = a\cos(at)$$

$$\frac{d(\cos(at))}{dt} = -a\sin(at)$$

L'Hôpital's Rule

If $\lim_{x \to a} f(x) = 0$ and $\lim_{x \to a} g(x) = 0$, or

If $\lim_{x \to a} f(x) = \infty$ and $\lim_{x \to a} g(x) = \infty$, then

$$\lim_{x \to a} \frac{f(x)}{g(x)} = \lim_{x \to a} \frac{\frac{df(x)}{dx}}{\frac{dg(x)}{dx}}$$

The rule can be applied as many times as necessary.

Bibliography

1. E.A. Guillemin, *Theory of Linear Physical Systems* (Wiley, New York, 1963)
2. E.A. Guillemin, *The Mathematics of Circuit Analysis* (Wiley, New York, 1959)
3. B.P. Lathi, *Linear Systems and Signals* (Oxford University Press, New York, 2005)
4. E.O. Brigham, *The Fast Fourier Transform and Its Applications* (Prentice-Hall, Englewood Cliffs, 1988)
5. R.N. Bracewell, *The Fourier Transform and Its Applications* (McGraw-Hill, New York, 2000)
6. D. Sundararajan, *Signals and Systems – A Practical Approach* (Wiley, Singapore, 2008)
7. D. Sundararajan, *Discrete Fourier Transform, Theory, Algorithms, and Applications* (World Scientific, Singapore, 2001)
8. D. Sundararajan, *Discrete Wavelet Transform, A Signal Processing Approach* (Wiley, Singapore, 2015)
9. D. Sundararajan, *Digital Image Processing - A Signal Processing Approach* (Springer, Singapore, 2017)
10. The Mathworks, *Matlab Image Processing Tool Box User's Guide* (The Mathworks, Inc., U.S.A., 2018)
11. The Mathworks, *Matlab Signal Processing Tool Box User's Guide* (The Mathworks, Inc., U.S.A., 2018)

© Springer Nature Singapore Pte Ltd. 2018
D. Sundararajan, *Fourier Analysis—A Signal Processing Approach*,
https://doi.org/10.1007/978-981-13-1693-7

Answers to Selected Exercises

Chapter 1

1.1.2 0.

1.2.3

$$x(n) = -2\delta(n) - \delta(n-2) + 3\delta(n+2) + 3\delta(n+3)$$

1.3.3

$$\{-1.4575, -1.7292, -1.7320, -1.7321, -1.7321\}$$

1.4.4

$$371.0329$$

1.5.1

$$x(n) = \begin{cases} -2 \text{ for } n = 1, 0, -1 \\ 0 \quad \text{otherwise} \end{cases}$$

1.6.5

$$\{2.5981, 0.7765, -1.5000, -2.8978, -2.5981, -0.7765, 1.5000, 2.8978\}$$

$$x(n) = 2.5981 \cos\left(\frac{\pi}{4}n\right) - 1.5 \sin\left(\frac{\pi}{4}n\right)$$

1.7.3

$$c(n) = 3.8639 \cos\left(\frac{\pi}{4}n + 0.7945\right)$$

© Springer Nature Singapore Pte Ltd. 2018
D. Sundararajan, *Fourier Analysis—A Signal Processing Approach*,
https://doi.org/10.1007/978-981-13-1693-7

The samples of the sinusoid $a(n)$ are

$$\{2.0000, -1.0353, -3.4641 - 3.8637, -2.0000, 1.0353, 3.4641, 3.8637\}$$

The samples of the sinusoid $b(n)$ are

$$\{0.7071, 1.0000, 0.7071, 0.0000, -0.7071, -1.0000, -0.7071, -0.0000\}$$

The samples of the sinusoid $c(n) = a(n) + b(n)$ are

$$\{2.7071, -0.0353, -2.7570, -3.8637, -2.7071, 0.0353, 2.7570, 3.8637\}$$

1.8.4 The waveform is periodic with period 6.
1.9.4
$$x(n) = (-j)(e^{j\left(\frac{2\pi}{8}n - \frac{\pi}{6}\right)} - e^{-j\left(\frac{2\pi}{8}n - \frac{\pi}{6}\right)})$$

The samples of $x(n)$ are

$$\{-1.0000, 0.5176, 1.7321, 1.9319, 1.0000, -0.5176, -1.7321, -1.9319\}$$

1.10.3 The samples of $x(n)$ are

$$\{0.8660, 0.2588, -0.5000, -0.9659, -0.8660, -0.2588, 0.5000, 0.9659\}$$

$$x(-4) = x(4) = -0.8660$$

1.11.4

$$x_e(-3) = 0.5, x_e(-2) = 0.5, x_e(-1) = 0, x_e(0) = 0,$$
$$x_e(1) = 0, x_e(2) = 0.5, x_e(3) = 0.5$$

$$x_o(-3) = -0.5, x_o(-2) = -0.5, x_o(-1) = 0,$$
$$x_o(0) = 0, x_o(1) = 0, x_o(2) = 0.5, x_o(3) = 0.5$$

1.12.3 Neither.
1.13.3 Average power is 2.
1.14.3 $x(an + k) = \sin(\frac{\pi}{8}n - \frac{\pi}{3})$.

$$\{-0.8660, 0, -0.2588, 0, 0.5000, 0, 0.9659, 0, 0.8660,$$
$$0, 0.2588, 0, -0.5000, 0, -0.9659, 0\}$$

1.15.1 $x(an + k) = \{x(1) = 1, x(2) = 2\}$, and 0 otherwise.
1.16.2 $(-2 + j4)$.

1.17.3 $0.3231 + j0.5846$.
1.18.1 $17.0294e^{(-j1.6296)}$.
1.19.3 $5 - j2, 5 + j2$.
1.20.3

$$\{-0.8090 + j0.5878, -0.8090 - j0.5878, 0.3090 + j0.9511, 0.3090 - j0.9511, 1\}$$

Chapter 2

2.1.2 $2^6 2^9 = 2^{15} = 32768$.
2.2.2 $x(n) = e^{j(\frac{\pi}{4}n + \frac{\pi}{6})} + e^{-j(\frac{\pi}{4}n + \frac{\pi}{6})} = (0.5\sqrt{3} + j0.5)e^{j\frac{\pi}{4}n} + (0.5\sqrt{3} - j0.5)e^{-j\frac{\pi}{4}n}$
The samples are

$$\{1.7321, 0.5176, -1.0000, -1.9319, -1.7321, -0.5176, 1.0000, 1.9319\}$$

2.3.5 $\{10 + j2, 4 + j4, -2 + j6, -8\}$.
2.4.2

$$x(n) = \{-0.2679, 3.0000, -3.7321, 5.0000\}$$

$$X(k) = \{4, 3.4641 + j2, -12, 3.4641 - j2\}$$

Chapter 3

3.1.2

$$X(k) = \{10, -2 + j2, -2, -2 - j2\}, Y(k) = \{0, 1 + j3, -10, 1 - j3\},$$
$$Z(k) = \{30, -8, 14, -8\}$$

3.2.3
$$X(k) = \{11, 1, 3, 1\}, \quad X(-7) = 1, \quad x(23) = 2$$

3.3.1
$$\{11, -2 - j1, -3, -2 + j1\}$$

3.4.2
$$\{-3 - j1, 4, -3 + j1, -6\}$$

3.5.2
$$\{2, -4 - j2, -6, -4 + j2\}$$

3.6.3
$$\{16, 4, 8, -8\}$$

3.7.3

$$X(k) = \{-5 + j1, 2 + j4, 3 - j1, 4 - j4\},$$
$$X^*(4 - k) = \{-5 - j1, 4 + j4, 3 + j1, 2 - j4\}$$

3.8.1

$$y(n) = \{16, 15, 6, 13\}$$

3.9.2

$$Y(k) = \{5, -1 + j10, 9, -1 - j10\}$$

3.10.2

$$r_{xh}(n) = \{14, 6, 5, 3\}$$

3.11.3

$$r_{xx}(n) = \{18, 5, 8, 5\}$$

3.12.1

$$X(k) = \{7, -1 - j2, -1, -1 + j2\}$$

3.13.3

$$\{1, 7, 1, 7\}$$

3.14.3

$$\{1, 0, -5, 0\}$$

3.15.1

$$X(k) = \{3, -1\}, \quad XZ(k) = \{3, 1 - j2, -1, 1 + j2\}$$

3.16.5 Real and odd

$$X(k) = \{0, j6, 0, -j6\}$$

3.17.2

$$X(k) = \{14, 3 + j1, 0, 3 - j1\}$$

The power over one period is 54.

Chapter 4

4.1.1

$$x(m, n) = 2 - 1.5e^{j\left(\frac{2\pi}{4}(m+n)+\frac{\pi}{6}\right)} - 1.5e^{-j\left(\frac{2\pi}{4}(m+n)+\frac{\pi}{6}\right)}$$
$$- 0.5je^{j\left(\frac{2\pi}{4}(2m+n)+\frac{\pi}{3}\right)} + 0.5je^{-j\left(\frac{2\pi}{4}(2m+n)+\frac{\pi}{3}\right)}$$

The samples of $x(m, n)$ are

$$x(m, n) = \begin{bmatrix} 0.2679 & 4.0000 & 3.7321 & -0.0000 \\ 2.6340 & 4.0981 & 1.3660 & -0.0981 \\ 5.4641 & 1.0000 & -1.4641 & 3.0000 \\ -0.3660 & -1.0981 & 4.3660 & 5.0981 \end{bmatrix}$$

The DFT is

$$X(k, l) = \begin{bmatrix} 32 & 0 & 0 & 0 \\ 0 & -20.7846 - j12 & 0 & 0 \\ 0 & 6.9282 - j4 & 0 & 6.9282 + j4 \\ 0 & 0 & 0 & -20.7846 + 12 \end{bmatrix}$$

4.2.1

$$X(k, l) = \begin{bmatrix} 40 & -3 + j3 & -2 & -3 - j3 \\ 1 + j1 & -2 + j2 & -7 + j3 & j2 \\ 2 & 5 - j7 & 4 & 5 + j7 \\ 1 - j1 & -j2 & -7 - j3 & -2 - j2 \end{bmatrix}$$

The average power is 122.

4.3.3

$$X(k, l) = \begin{bmatrix} 1 & -j1 & -1 & j1 \\ j1 & 1 & -j1 & -1 \\ -1 & j1 & 1 & -j1 \\ -j1 & -1 & j1 & 1 \end{bmatrix}$$

4.4.2

$$X(k, l) = \begin{bmatrix} 26.00 + j0.00 & 0.00 + j8.00 & -6.00 + j0.00 & 0.00 - j8.00 \\ 6.00 + j0.00 & -4.00 - j4.00 & 2.00 - j4.00 & -4.00 - j4.00 \\ 2.00 + j0.00 & 0.00 + j0.00 & -6.00 + j0.00 & 0.00 + j0.00 \\ 6.00 + j0.00 & -4.00 + j4.00 & 2.00 + j4.00 & -4.00 + j4.00 \end{bmatrix}$$

$$
H(k,l) = \begin{bmatrix}
18.00 + j0.00 & 3.00 - j3.00 & 0.00 + j0.00 & 3.00 + j3.00 \\
3.00 + j5.00 & -6.00 + j2.00 & -5.00 - j1.00 & 0.00 + j2.00 \\
0.00 + j0.00 & -1.00 - j1.00 & -6.00 + j0.00 & -1.00 + j1.00 \\
3.00 - j5.00 & 0.00 - j2.00 & -5.00 + j1.00 & -6.00 - j2.00
\end{bmatrix}
$$

$$
X(k,l)H(k,l) = \begin{bmatrix}
468.00 + j0.00 & 24.00 + j24.00 & -0.00 + j0.00 & 24.00 - j24.00 \\
18.00 + j30.00 & 32.00 + j16.00 & -14.00 + j18.00 & 8.00 - j8.00 \\
0.00 + j0.00 & 0.00 + j0.00 & 36.00 + j0.00 & 0.00 + j0.00 \\
18.00 - j30.00 & 8.00 + j8.00 & -14.00 - j18.00 & 32.00 - j16.00
\end{bmatrix}
$$

$$
y(m,n) = \begin{bmatrix}
40 & 25 & 24 & 37 \\
23 & 24 & 19 & 36 \\
29 & 23 & 33 & 23 \\
37 & 33 & 29 & 33
\end{bmatrix}
$$

$$
X(k,l)H^*(k,l) = \begin{bmatrix}
468.00 + j0.00 & -24.00 + j24.00 & -0.00 + j0.00 & -24.00 - j24.00 \\
18.00 - j30.00 & 16.00 + j32.00 & -6.00 + j22.00 & -8.00 + j8.00 \\
0.00 + j0.00 & -0.00 + j0.00 & 36.00 + j0.00 & -0.00 + j0.00 \\
18.00 + j30.00 & -8.00 - j8.00 & -6.00 - j22.00 & 16.00 - j32.00
\end{bmatrix}
$$

$$
r_{xh}(m,n) = \begin{bmatrix}
31 & 24 & 35 & 36 \\
20 & 32 & 36 & 44 \\
26 & 24 & 34 & 24 \\
28 & 25 & 24 & 25
\end{bmatrix}
$$

$$
H(k,l)X^*(k,l) = \begin{bmatrix}
468.00 + j0.00 & -24.00 - j24.00 & -0.00 + j0.00 & -24.00 + j24.00 \\
18.00 + j30.00 & 16.00 - j32.00 & -6.00 - j22.00 & -8.00 - j8.00 \\
0.00 + j0.00 & 0.00 + j0.00 & 36.00 + j0.00 & -0.00 + j0.00 \\
18.00 - j30.00 & -8.00 + j8.00 & -6.00 + j22.00 & 16.00 + j32.00
\end{bmatrix}
$$

$$r_{hx}(m, n) = \begin{bmatrix} 31 & 36 & 35 & 24 \\ 28 & 25 & 24 & 25 \\ 26 & 24 & 34 & 24 \\ 20 & 44 & 36 & 32 \end{bmatrix}$$

$$X(k, l)X^*(k, l) = \begin{bmatrix} 676 & 64 & 36 & 64 \\ 36 & 32 & 20 & 32 \\ 4 & 0 & 36 & 0 \\ 36 & 32 & 20 & 32 \end{bmatrix} \qquad r_{xx}(m, n) = \begin{bmatrix} 70 & 40 & 38 & 40 \\ 50 & 42 & 34 & 42 \\ 40 & 36 & 40 & 36 \\ 50 & 42 & 34 & 42 \end{bmatrix}$$

Chapter 5

5.1.3
The input sequences are zero padded.

$$X(k) = \{10, -0.1213 - j6.5355, -1 + j3, 4.1213 - j0.5355, 0, 4.1213 \\ + j0.5355, -1 - j3, -0.1213 + j6.5355\}$$

$$H(k) = \{-2, 2.4142 + j1.2426, -2 - j2, -0.4142 + j7.2426, 10, \\ -0.4142 - j7.2426, -2 + j2, 2.4142 - j1.2426\}$$

$$Y(k) = X(k)H(k) = \{-20, 7.8284 - j15.9289, 8 - j4, 2.1716 \\ + j30.0711, \quad 0, 2.1716 - j30.0711, 8 + j4, 7.8284 + j15.9289\}$$

$$y(n) = \text{IDFT}(Y(k)) = \{2, -3, 7, -7, -3, 0, -16, 0\}$$

5.2.1 The input sequences are zero padded.

$$X(k, l) = \begin{bmatrix} 1 & 2 + j1 & 3 & 2 - j1 \\ -2 - j3 & -1 + j2 & 4 + j & 3 - j4 \\ -5 & j5 & 5 & -j5 \\ -2 + j3 & 3 + j4 & 4 - j1 & -1 - j2 \end{bmatrix}$$

$$H(k, l) = \begin{bmatrix} 4 & -j4 & -4 & j4 \\ 5+j1 & 1-j1 & -1+j3 & 3+j5 \\ 6 & 4-j2 & 2 & 4+j2 \\ 5-j1 & 3-j5 & -1-j3 & 1+j1 \end{bmatrix}$$

$$Y(k, l) = X(k, l)H(k, l) = \begin{bmatrix} 4 & 4-j8 & -12 & 4+j8 \\ -7-j17 & 1+j3 & -7+11 & 29+j3 \\ -30 & 10+j20 & 10 & 10-j20 \\ -7+j17 & 29-j3 & -7-j11 & 1-j3 \end{bmatrix}$$

$$y(m, n) = \mathrm{IDFT}(Y(k, l)) = \begin{bmatrix} 2 & -3 & -9 & 0 \\ 0 & 14 & 3 & 0 \\ -2 & -3 & 2 & 0 \\ 0 & 0 & 0 & 0 \end{bmatrix}$$

5.3.2 The input sequences are zero padded.

$$X(k) = \{5, -0.4142 - j2.2426, 3 + j2, 2.4142 - j6.2426,$$
$$-7, 2.4142 + j6.2426, 3 - j2, -0.4142 + j2.2426\}$$

$$H^*(k) = \{6, -0.4142 + j3.2426, 2 - j2, 2.4142 + j5.2426,$$
$$-6, 2.4142 - j5.2426, 2 + j2, -0.4142 - j3.2426\}$$

$$Y(k) = X(k)H^*(k) = \{30, 7.4437 - j0.4142, 10 - j2, 38.5563 - j2.4142,$$
$$42, 38.5563 + j2.4142, 10 + j2, 7.4437 + j0.4142\}$$

$$r_{xh}(n) = \mathrm{IDFT}(Y(k)) = \{23, -6, 6, 4, 0, 4, 7, -8, \}$$

5.4.1 The input sequences are zero padded.

$$X(k,l) = \begin{bmatrix} 1 & 2+j1 & 3 & 2-j1 \\ -4-j5 & -3 & 2-j1 & 1-j6 \\ -9 & -4+j5 & 1 & -4-j5 \\ -4+j5 & 1+j6 & 2+j1 & -3 \end{bmatrix}$$

$$H^*(k,l) = \begin{bmatrix} 2 & -2+j4 & -6 & -2-j4 \\ 4-j2 & 0 & -2-j4 & 2-j6 \\ 6 & 4+j2 & 2 & 4-j2 \\ 4+j2 & 2+j6 & -2+j4 & 0 \end{bmatrix}$$

$$Y(k,l) = X(k,l)H^*(k,l) = \begin{bmatrix} 2 & -8+j6 & -18 & -8-j6 \\ -26-j12 & 0 & -8-j6 & -34-j18 \\ -54 & -26+j12 & 2 & -26-j12 \\ -26+j12 & -34+j18 & -8+j6 & 0 \end{bmatrix}$$

$$r_{xh}(m,n) = \text{IDFT}(Y(k,l)) = \begin{bmatrix} -17 & -9 & 0 & 0 \\ 9 & 2 & 0 & 9 \\ 0 & 0 & 0 & 0 \\ 0 & 9 & 0 & -1 \end{bmatrix}$$

5.5.2

$$X(k) = \{8, 1+j3, -2, 1-j3\}, \quad H(k) = \{0, -j, 0, j\}$$

$$x_h(n) = \{1.5, 0.5, -1.5, -0.5\} \quad \text{DFT}(x(n) + jx_h(n))\{8, 2+j6, -2, 0\}$$

Chapter 6

6.1.1

$$x_0(n) = x_4(n) = x_8(n) = \{1+j1.7321, 1+j1.7321, 1+j1.7321, 1+j1.7321\}$$

$$x_1(n) = x_5(n) = \{1+j1.7321, -1.7321+j1, -1-j1.7321, 1.7321-j1\}$$

$$x_2(n) = x_6(n) = \{1+j1.7321, -1-j1.7321, 1+j1.7321, -1-j1.7321\}$$

$$x_3(n) = x_7(n) = \{1+j1.7321, 1.7321-j1, -1-j1.7321, -1.7321+j1\}$$

6.2.2

$$x_0(n) = x_8(n) = \{-1.7321, -1.7321, -1.7321, -1.7321,$$
$$- 1.7321, -1.7321, -1.7321, -1.7321\}$$

$$-2\cos\left(\frac{2\pi}{8}1n + \frac{\pi}{6}\right) = -2\cos\left(\frac{2\pi}{8}7n - \frac{\pi}{6}\right) = -2\cos\left(\frac{2\pi}{8}9n + \frac{\pi}{6}\right)$$

$$= \{-1.7321, -0.5176, 1.0000, 1.9319, 1.7321, 0.5176, -1.0000, -1.9319\}$$

$$-2\cos\left(\frac{2\pi}{8}2n + \frac{\pi}{6}\right) = -2\cos\left(\frac{2\pi}{8}6n - \frac{\pi}{6}\right) = -2\cos\left(\frac{2\pi}{8}10n + \frac{\pi}{6}\right)$$

$$= \{-1.7321, 1, 1.7321, -1, -1.7321, 1, 1.7321, -1\}$$

$$-2\cos\left(\frac{2\pi}{8}3n + \frac{\pi}{6}\right) = -2\cos\left(\frac{2\pi}{8}5n - \frac{\pi}{6}\right) = -2\cos\left(\frac{2\pi}{8}11n + \frac{\pi}{6}\right)$$

$$= \{-1.7321, 1.9319, -1, -0.5176, 1.7321, -1.9319, 1, 0.5176\}$$

$$x_4(n) = x_{12}(n) = \{-1.7321, 1.7321, -1.7321, 1.7321, -1.7321, 1.7321, -1.7321, 1.7321\}$$

6.3.3

$$w_r(n) = \{1, 1, 1, 1, 1, 0, 0, 0\}$$

$$x(n) = \{1, -0.7071, -0, 0.7071, -1, 0.7071, 0, -0.7071\}$$

$$X(k) = \{0, 0, 0, 4, 0, 4, 0, 0\}$$

$$x_t(n) = \{1, -0.7071, -0, 0.7071, -1, 0, 0, 0\}$$

$$X_t(k) = \{0, 1, j1.4142, 3, 0, 3, -j1.4142, 1\}$$

$$|X_t(k)| = \{0, 1, 1.4142, 3, 0, 3, 1.4142, 1\}$$

6.4.4

$$w_{han}(n) = \{0, 0.3455, 0.9045, 0.9045, 0.3455, 0, 0, 0\}$$

$$x(n) = \{0.5, 0.9659, 0.8660, 0.2588, -0.5, -0.9659, -0.8660, -0.2588\}$$

$$X(k) = \{0, 2 - j3.4641, 0, 0, 0, 0, 2 + j3.464\}$$

$$x_t(n) = \{0, 0.3337, 0.7833, 0.2341, -0.1727, 0, 0, 0\}$$

$$X_t(k) = \{1.1784, 0.2432 - j1.1848, -0.9561 - j0.0996, 0.1023 + j0.3818,$$

$$0.0428, 0.1023 - j0.3818, -0.9561 + j0.0996, 0.2432 + j1.1848\}$$

$$|X_t(k)| = \{1.1784, 1.2095, 0.9612, 0.3953, 0.0428, 0.3953, 0.9612, 1.2095\}$$

Chapter 7

7.1.3

$$x(t) = e^{j(\pi 0 t)} + j2e^{j\left(2t + \frac{\pi}{6}\right)} - j2e^{-j\left(2t + \frac{\pi}{6}\right)} + 4e^{j\left(3t - \frac{\pi}{3}\right)} + 4e^{-j\left(3t - \frac{\pi}{3}\right)}$$

The coefficients are

$$X_{fs}(0) = 1, \, X_{fs}(\pm 1) = 0, \, X_{fs}(\pm 2) = -1 \pm j\sqrt{3}, \, X_{fs}(\pm 3) = 2 \mp j2\sqrt{3}$$

The samples are

$$\{3.0000, -2.5300, 1.3822, 8.3258, -10.5999, 9.2096, -1.7876\}$$

$\omega_0 = 1$.

7.2.2 After canceling common factors, the frequency of the waveforms is $\frac{1}{2}$ and $\frac{2}{9}$. The LCM of the denominators (2,9) is 18. The GCD of the numerators (1,2) is 1. Therefore, the fundamental frequency is $\omega_0 = \frac{1}{18}$ rad/s. The fundamental period is $T = \frac{2\pi}{\omega_0} = \frac{2\pi 18}{1} = 36\pi$. The first sinusoid is the 9th harmonic. The second sinusoid is the 4th harmonic.

$$X_{fs}(0) = 1, \, X_{fs}(\pm 4) = \mp 0.5j, \, X_{fs}(\pm 9) = \pm j$$

7.3.1

$$X_{fs}(k) = \frac{1}{T} \int_{-a}^{a} e^{-jk\omega_0 t} dt$$

$$= \frac{1}{T} \frac{-1}{jk\omega_0} e^{-jk\omega_0 t} \Big|_{-a}^{a} = \frac{1}{T} \frac{-1}{jk\omega_0} (e^{-jk\omega_0 a} - e^{jk\omega_0 a})$$

$$= \frac{2}{T} \frac{1}{j2k\omega_0} (e^{jk\omega_0 a} - e^{-jk\omega_0 a}) = \frac{2}{T} \frac{\sin(k\omega_0 a)}{k\omega_0}$$

With $\omega_0 = \pi/5, \, a = 2, \, T = 10$.

$$X_{fs}(k) = \frac{\sin(0.4k\pi)}{k\pi}$$

$$x(t) = \sum_{k=-\infty}^{\infty} \frac{\sin(0.4k\pi)}{k\pi} e^{jk0.2\pi t}$$

7.4.1

$$x(t) = \frac{2}{\pi} + \frac{4}{\pi} \sum_{k=1}^{\infty} \left(\frac{1}{1-4k^2}\right) \cos\left(2k\left(t - \frac{\pi}{2}\right)\right) =$$

$$\frac{2}{\pi} - \frac{4}{3\pi} \cos\left(2\left(t - \frac{\pi}{2}\right)\right) - \frac{4}{15\pi} \cos\left(4\left(t - \frac{\pi}{2}\right)\right) - \frac{4}{35\pi} \cos\left(6\left(t - \frac{\pi}{2}\right)\right) + \cdots$$

$$= \frac{2}{\pi} + \frac{4}{3\pi} \cos(2t) - \frac{4}{15\pi} \cos(4t) + \frac{4}{35\pi} \cos(6t) + \cdots$$

7.5.2 With $T_s = 1$, the samples are

$$\{3.3660, 3, 4.3660, 1.2679\}$$

With 4 samples, the DFT coefficients are

$$\{12, -1 - j1.7321, 3.4641, -1 + j1.7321\}$$

With $T_s = 0.8$, the samples are

$$\{3.3660, 3.2624, 3.7056, 3.6386, 1.0273\}$$

With 5 samples, the DFT coefficients are

$$\{15, -1.25 - j2.1651, 2.1651 - j1.25, 2.1651 + j1.25, -1.25 + j2.1651\}$$

Chapter 8

8.2

$$j2(\sin(2\omega)) = (e^{j2\omega} - e^{-j2\omega}) \leftrightarrow \delta(n+2) - \delta(n-2)$$

8.4

$$X(e^{j\omega}) = 3 + e^{-j\omega} + 2e^{-j2\omega} \quad \text{and} \quad H(e^{j\omega}) = 1 + 2e^{-j\omega} + 4e^{-j2\omega}$$

$$X(e^{j\omega})H(e^{j\omega}) = 3 + 7e^{-j\omega} + 16e^{-j2\omega} + 8e^{-j3\omega} + 8e^{-j4\omega}$$

$$\{3, 7, 16, 8, 8\}$$

8.6

$$H(e^{j\frac{2\pi}{8}}) = \frac{e^{j\frac{2\pi}{8}}}{e^{j\frac{2\pi}{8}} + 0.7} = 0.6350\angle(0.3197)$$

$$y(n) = 0.6350\sin\left(\frac{2\pi}{8}n - \frac{\pi}{6} + 0.3197\right)$$

8.8

$$y(n) = \left(-3(-0.6)^n + 4(-0.8)^n\right)u(n)$$

8.10.1

$$X(e^{j\omega}) = 2e^{j2\omega} + e^{j\omega} + 3 + 4e^{-j\omega}$$

$$xp(n) = \{x(0) = 3, x(1) - 4, x(2) = 2, x(3) = 1\}$$

$$xp\_t(n) = \{x(0) = 3, x(1) = 4, x(2) = 2, x(3) = 0\}$$

The DFT of $xp\_t(n)$ is

$$\{9, 1 - j4, 1, 1 + j4\}$$

The DFT of $xp(n)$ is

$$\{10, 1 - j3, 0, 1 + j3\}$$

The DFT of the window $w(n) = \{1, 1, 1, 0\}$ is

$$\{3, -j1, 1, j1\}$$

The linear convolution of the two spectra, divided by 4, is

$$\{7.5, 0.75 - j4.75, 1.75 - j0.25, 1 + j4, 1.5, 0.25 + j0.75, -0.75 + j0.25\}$$

The circular convolution output can be obtained by adding the last three terms with the first three terms.

Chapter 9

9.2

$$\frac{2 + j\omega}{(2 + j\omega)^2 + 9}$$

$$X(j0) = 0.1538, \quad X(j\pi) = 0.2727 - j0.0912$$

9.5.1

$$X_{fs}(0) = 0.5 \quad \text{and} \quad X_{fs}(k) = \frac{j}{2k\pi}, \quad k \neq 0$$

9.6.2

$$\frac{1}{2\pi}\left(\pi\delta(t) + \frac{1}{jt}\right) \leftrightarrow u(-\omega)$$

9.7.3

$$X(j\omega) = -j\left(\frac{\sin((\omega - 3)2)}{(\omega - 3)} - \frac{\sin((\omega + 3)2)}{(\omega + 3)}\right)$$

9.8.2

$$\frac{1}{5 + j\omega}\frac{1}{4 + j\omega}$$

9.9.1

$$\frac{\pi}{2}(\delta(\omega - 3) + \delta(\omega + 3)) - \frac{j\omega}{(\omega^2 - 3^2)}$$

9.10.3

$$e^{-j2\omega}\left(\pi\delta(\omega) + \frac{1}{j\omega}\right)$$

9.11.3

$$X(j\omega) = -2\frac{e^{-j2\omega}}{j\omega} - \frac{1}{\omega^2} + \frac{e^{-j2\omega}}{\omega^2}$$

The derivative of $x(t)$ is

$$j\omega X(j\omega) = -2e^{-j2\omega} + \frac{1}{j\omega} - \frac{e^{-j2\omega}}{j\omega}$$

9.12.3

$$0.5\pi\delta(\omega) + \frac{0.5}{j\omega} - \frac{0.5}{2 + j\omega}$$

9.14.1

$$\sin\left(\frac{2\pi}{8}t + \frac{\pi}{6}\right) = \cos\left(\frac{2\pi}{8}t - \frac{\pi}{3}\right) \leftrightarrow \pi\left(e^{-j\frac{\pi}{3}}\delta\left(\omega - \frac{2\pi}{8}\right) + e^{j\frac{\pi}{3}}\delta\left(\omega + \frac{2\pi}{8}\right)\right)$$

9.15.1

$$u(t + 5) - u(t - 5) \leftrightarrow 2\frac{\sin(5\omega)}{\omega}$$

$T_s = 1.$

$$x_s(t) = \sum_{n=-\infty}^{\infty} (u(n+5) - u(n-5))\delta(t-n) \leftrightarrow$$

$$X_s(j\omega) = \sum_{k=-\infty}^{\infty} 2\frac{\sin(5(\omega + 2k\pi))}{\omega + 2k\pi}$$

$T_s = 0.5$.

$$x_s(t) = \sum_{n=-\infty}^{\infty} (u(0.5n+5) - u(0.5n-5))\delta(t - 0.5n) \leftrightarrow$$

$$X_s(j\omega) = 2\sum_{k=-\infty}^{\infty} 2\frac{\sin(5(\omega + 4k\pi))}{\omega + 4k\pi}$$

The other form for $X_s(j\omega)$, with $T_s = 1$, is

$$0.5e^{j5\omega} + e^{j4\omega} + e^{j3\omega} + e^{j2\omega} + e^{j\omega} + 1 + e^{-j\omega} + e^{-j2\omega} + e^{-j3\omega} + e^{-j4\omega} + 0.5e^{-j5\omega}$$

For $T_s = 0.5$, 21 terms are required. With $T_s = 1$, the samples of $X(j\omega)$ and those of the 2 sampled versions are, respectively,

$$\{10, -1.8006, 1.2732, -0.6002, 0\}$$

$$\{10, -1.7125, 1.0151, -0.3090, 0\}$$

$$\{10, -1.7071, 1, -0.2929, 0\}$$

With $T_s = 0.5$, the samples of the 2 sampled versions are, respectively,

$$\{19.9999, -3.5575, 2.4218, -1.0663, 0\}$$

$$\{19.9999, -3.5575, 2.4218, -1.0663, 0\}$$

9.16.3

$$x_s(t) = \sum_{n=-\infty}^{\infty} 2\cos\left(3\frac{2\pi}{8}(0.5n) + \frac{\pi}{3}\right)\delta(t - 0.5n) \leftrightarrow$$

$$X_s(j\omega) = \frac{2\pi}{(0.5)(16)} \sum_{m=-\infty}^{\infty} \left(16e^{j\frac{\pi}{3}}\delta\left(\omega - 3\frac{2\pi}{8} - 4m\pi\right) + 16e^{-j\frac{\pi}{3}}\delta\left(\omega + 3\frac{2\pi}{8} - 4m\pi\right)\right)$$

9.19

$$y(t) = (0.25e^{-t} + 0.75e^{-3t} - 3.5te^{-3t})u(t)$$

9.20.1

$$X(j\omega) = \frac{1}{1 + j\omega}$$

The sample values of the signal are

$$\{0.5000, 0.2865, 0.0821, 0.0235\}$$

The DFT of these values, after scaling by $T_s = 1.25\,\text{s}$, is

$$\{1.1151, 0.5224 - j0.3287, 0.3401, 0.5224 + j0.3287\}$$

The corresponding 4 samples of the FT are

$$\{1, 0.3877 - j0.4872, 0.1367 - j0.3435, 0.0657 - j0.2478\}$$

9.21.1

$$X(j\omega) = \frac{1}{1.5 + j\omega}$$

The sample values of the spectrum are

$$\{0.6667, 0.0235 - j0.1228, 0.0060 - j0.0631, 0.0027 - j0.0423, 0.0027$$
$$+ j0.0423, 0.0060 + j0.0631, 0.0235 + j0.1228\}$$

The IDFT of these values, after scaling by $1/T_s = 1/0.2\,\text{s}$, is

$$\{0.5222, 0.7430, 0.5482, 0.5151, 0.3860, 0.3786, 0.2404\}$$

The first 7 samples of $x(t)$ are

$$\{1, 0.7408, 0.5488, 0.4066, 0.3012, 0.2231, 0.1653\}$$

Chapter 10

10.1.2

$$xe(n) = \{0, 1, 0, 1\}, xo(n) = \{0, 0, 0, 0\}$$

$$X(k) = \{2, 0, -2, 0\}, Xe(k) = \{2, 0, -2, 0\}, Xo(k) = \{0, 0, 0, 0\}$$

10.2.3

$$x(n) = \{1.3660 - j1.3660, -0.3660 - j0.3660, 1.3660 - j1.3660, -0.3660 - j0.3660,$$

$$1.3660 - j1.3660, -0.3660 - j0.3660, 1.3660 - j1.3660, -0.3660 - j0.3660\}$$

$$X(k) = \{4 - j6.9282, 0, 0, 0, 6.9282 - j4, 0, 0, 0\}$$

10.3.5

$$x(n) = \{0.7071, 0.7071, -0.7071, -0.7071, 0.7071, 0.7071, -0.7071, -0.7071\}$$

$$y(n) = \{0.5000, -0.9659, 0.8660, -0.2588, -0.5000, 0.9659, -0.8660, 0.2588\}$$

$$x(n) + jy(n) =$$

$$\{0.7071 + j0.5000, 0.7071 - j0.9659, -0.7071 + j0.8660, -0.7071 \quad j0.2588,$$

$$0.7071 - j0.5000, 0.7071 + j0.9659, -0.7071 - j0.8660, -0.7071 + j0.2588\}$$

$$X(k) + jY(k) = \{0, 0, 2.8284 - j2.8284, -3.4641 + j2, 0, 3.4641 + j2, 2.8284 + j2.8284, 0\}$$

$$X(k) = \{0, 0, 2.8284 - j2.8284, 0, 0, 0, 2.8284 + j2.8284, 0\}$$

$$Y(k) = \{0, 0, 0, 2 + j3.4641, 0, 2 - j3.4641, 0, 0\}$$

10.4.4

$$x(n) = \{0.7071, 0.0000, -0.7071, 1.0000, -0.7071, -0.0000, 0.7071, -1.0000\}$$

$$X(k) = \{0, 0, 0, 2.8284 - j2.8284, 0, 2.8284 + j2.8284, 0, 0\}$$

Index

© Springer Nature Singapore Pte Ltd. 2018
D. Sundararajan, *Fourier Analysis—A Signal Processing Approach*,
https://doi.org/10.1007/978-981-13-1693-7

Printed in the United States
By Bookmasters